Tributes
Volume 9

Acts of Knowledge
History, Philosophy and Logic
Essays dedicated to Göran Sundholm

Volume 1
We Will Show Them! Essays in Honour of Dov Gabbay, Volume 1
S. Artemov, H. Barringer, A. d'Avila Garcez, L. Lamb and J. Woods, eds.

Volume 2
We Will Show Them! Essays in Honour of Dov Gabbay, Volume 2
S. Artemov, H. Barringer, A. d'Avila Garcez, L. Lamb and J. Woods, eds.

Volume 3
Probability and Inference: Essays in Honour of Henry E. Kyburg
Bill Harper and Greg Wheeler, eds.

Volume 4
The Way Through Science and Philosophy:
Essays in Honour of Stig Andur Pedersen
H. B. Andersen, F. V. Christiansen, K. F. Jørgensen, and V. F. Hendricks, eds.

Volume 5
Approaching Truth: Essays in Honour of Ilkka Niiniluoto
Sami Pihlström, Panu Raatikainen and Matti Sintonen, eds.

Volume 6
Linguistics, Computer Science and Language Processing.
Festschrift for Franz Guenthner on the Occasion of his 60th Birthday
Gaston Gross and Klaus U. Schulz, eds.

Volume 7
Dialogues, Logics and Other Strange Things.
Essays in Honour of Shahid Rahman.
Cédric Dégremont, Laurent Keiff and Helge Rückert, eds.

Volume 8
Logos and Language.
Essays in Honour of Julius Moravcsik.
Dagfinn Follesdal and John Woods, eds.

Volume 9
Acts of Knowledge: History, Philosophy and Logic.
Essays dedicated to Göran Sundholm
Giuseppe Primiero and Shahid Rahman, eds.

Tributes Series Editor
Dov Gabbay   dov.gabbay@kcl.ac.uk

Acts of Knowledge
History, Philosophy and Logic
Essays Dedicated to Göran Sundholm

edited by

Giuseppe Primiero

and

Shahid Rahman

© Individual author and College Publications 2009. All rights reserved.

ISBN 978-1-904987-92-5

College Publications
Scientific Director: Dov Gabbay
Managing Director: Jane Spurr
Department of Computer Science
King's College London, Strand, London WC2R 2LS, UK

http://www.collegepublications.co.uk

Original cover design by orchid creative   www.orchidcreative.co.uk
Printed by Lightning Source, Milton Keynes, UK

All rights reserved. No part of this publication may be reproduced, stored in a retrieval system or transmitted in any form, or by any means, electronic, mechanical, photocopying, recording or otherwise without prior permission, in writing, from the publisher.

CONTENTS

GIUSEPPE PRIMIERO, SHAHID RAHMAN
Preface   vii

PART I   FROM HISTORY TO LOGIC   1

MARK VAN ATTEN
Monads and Sets. On Gödel, Leibniz, and the Reflection Principle   3

BENOÎT CASTELNÉRAC, MATHIEU MARION
Arguing for Inconsistency:
Dialectical Games in the Academy   35

JACQUES P. DUBUCS, WIOLETTA MISKIEWICZ
Logic, Act and Product   73

WOLFGANG KÜNNE
Bolzano and (Early) Husserl
on Intentionality   95

KEVIN MULLIGAN
Tractarian Beginnings and Endings
Worlds, Values, Facts and Subjects   141

JAN VON PLATO
Gentzen's Original Proof of the
Consistency of Arithmetic Revisited   159

PART II   FROM PHILOSOPHY TO LOGIC   181

CATARINA DUTILH NOVAES
Judgements, Contents
and their Representations   183

GIUSEPPE PRIMIERO
Epistemic Modalities   207

SARA NEGRI
Kripke Completeness Revisited   233

MARIA VAN DER SCHAAR
Judgement, Belief and Acceptance   267

PART III   FROM LOGIC TO PHILOSOPHY   **287**

HELGE RÜCKERT
Sundholm's Paradox of Knowability:
A Novel Paradox?   289

SHAHID RAHMAN, NICOLAS CLERBOUT, LAURENT KEIFF
On Dialogues and Natural Deduction   301

BJØRN JESPERSEN, MARIE DUŽÍ, PAVEL MATERNA
'π' in the Sky   337

KAI F. WEHMEIER
On Ramsey's 'Silly Delusion' Regarding Tractatus 5.53   353

INDEX   369

CONTACT INFORMATION
Authors   377

Preface
GIUSEPPE PRIMIERO, SHAHID RAHMAN

We, the editors, see this volume as a non-standard tribute, for various reasons. The editors and authors wanted to acknowledge Göran Sundholm and his work, but not on the occasion of his 60th birthday or retirement, as it is usually the case: we actually hope he will be around doing philosophy for a very long time to come! Rather, the editors thought — in a certain moment of Göran's life, maybe one characterized by some mental occupations lying outside the struggle of philosophical thinking — that now was the moment to bring together at least few of the people, his colleagues and his students, who had, in their own academic research, examined the variety of topics and interests typical of Göran's scientific activity. Hence, this volume also needs to be a non-standard selection of contents, sought from the academics who have worked, studied, collaborated and disagreed with Sundholm; engaging in debated issues and exploring untouched areas maybe only suggested or hinted at in his own work.

With this premise, it was natural to think of both the title of this volume, and its structure. In the first place, *Acts of Knowledge* wants to characterize the papers contained in this volume as bringing something scientifically valuable in their respective fields: all the papers present cutting-edge research in their own style, contributing to very lively debates occurring in the literature in logic, philosophical logic and history of logic. But this title is at the same time reminiscent of Göran's philosophical attitude towards logic and mathematics, a tribute to his constructivist background, which has been an influence or a challenge for any of the contributors. Hence, the subtitle "History, Philosophy and Logic": it refers directly to Göran's broad interests into the various aspects of the Philosophy of Logic, Mathematics, and Language, their origins and development, especially with the focus on the Modern History of Logic and the philosophical implications thereof. The readers will find scattered all along this volume pieces of — and reflections on — all these themes.

The first section of the volume explores the borders between philosophy and history of logic. Mark van Atten reflects on the almost theological implications of Leibnizian monads and Gödelian sets; Benoît Castelnérac & Mathieu Marion argue on the formulation of Socratic dialogues by means of rules inspired by dialogical logics; Jacques Dubucs & Violetta Miskiewicz suggest an analysis of knowledge acts inspired by Twardowski's logical reflections; in Wolfgang Künne's contribution, the dialogue between logic and philosophy takes the words of Bolzano and Husserl, for a suggestive meeting on the debated issue of intentionality; Kevin Mulligan gives us an exploration into subjectivity and objective values, inspired by Tractarian passages (but not only); finally, Jan von Plato slides towards the more formal reconstruction of a piece of history of logic, namely Gentzen's Consistency Proof of Arithmetic.

The second section collects more conceptual analyses, in particular on some philosophical implications of formal representation of knowledge processes. Catarina Dutilh Novaes explores the notion of knowledge act and the problems related to both its classical and constructivist representation; Giuseppe Primiero extends the analysis of modalities done by Sundholm in his typical constructivist vein, suggesting the formulation of an epistemic theory to compete with the mainstream semantic interpretation; Sara Negri follows the very same area of research, by analysing a result of Kripke concerning the completeness of modal logics; Maria van der Schaar returns on the more philosophical side, with a clarification of some epistemic acts tipically accompanying the all-invasive notion of judgment, namely belief and acceptance.

In the third section, we close the circle and focus on more formal matters. Helge Rückert comments on a Paradox of Knowability recently formulated by Sundholm himself; Shahid Rahman, Nicolas Clerbout and Laurent Keiff explore aspects of the dialogical semantics that lies just beyond the realism/anti-realism divide; Bjørn Jespersen, Marie Duží and Pavel Materna challenge the constructivist approach with a realist-inspired procedural semantics for mathematical objects; finally, Kai Wehmeier designs a semantics for a first-order logic with an exclusive interpretation of the variables.

As one realizes, many of the topics that characterize today's research in (philosophical) logic are here coherently collected. The related variety of perspectives and approaches shall not surprise those who know Göran and his work: these only represent some of the themes related to his research, our hope is to have moved a few steps forward. Starting with his Oxford doctorate in proof theory in 1983, Sundholm's career proceeded from being Fellow by Examination at Magdalen College in Oxford, then Lecturer at Radboud University in Nijmegen and Reader in Theoretical Philosophy in Stockholm. He was appointed Professor in Philosophy at Leiden University in 1987, he has accepted visiting professorships at Siena, Rio de Janeiro/Campinas, Stockholm, and more recently he has been Fellow at the Netherlands Institute for Advanced Studies (NIAS) and at the Swedish Collegium for Advanced Study. In all these years and throughout these prestigious positions he has built his characteristic systematic and historical research within the philosophy of logic, what he likes to describe with the *"conceptual architecture versus conceptual archaeology"* motto. We have tried to reconstruct the very

~~same mix of interests and ideas, and we hope he will like this result.~~

In this collection, we hope to have assembled a collection of contributions from whose variety and quality scholars and researchers can profit. We believe that historians and philosophers of logic, logicians and even computer scientists can find something to their taste in this collection.

This volume is also incomplete. One of the expected contributions is missing, and it will also never appear anywhere else. On April, 12, 2009 Paolo Casalegno, full professor of Philosophy of Language at the University of Milan died after a long illness, unknown to most of us. He had enthusiastically reacted to our offer to participate in this project, having met Göran at the various scientific meetings of the League of Research Universities. Let us use these few lines also to remember him.

Thanks go in the first place to the authors who have reacted to our offer to participate in this project, and who have made this volume possible. It has been exciting to work with all of them, and it has definitely been worth the time spent on it. A great thanks goes also to all the readers and referees who have helped us in the editorial process, commenting and providing suggestions on previous versions of each paper. We are grateful to them for their silent contribution to this volume.

Enough for introductory words, now, we hope you will enjoy the reading.

PART I

FROM HISTORY TO LOGIC

Monads and Sets. On Gödel, Leibniz, and the Reflection Principle

MARK VAN ATTEN

ABSTRACT. Gödel once offered an argument for the general reflection principle in set theory that took the form of an analogy with Leibniz' Monadology. I discuss the mathematical and philosophical background to Gödel's argument, reconstruct the proposed analogy in detail, and argue that it has no justificatory force.

Voor Göran, in dank en vriendschap

1 Introduction

Gödel described his general philosophical theory to Hao Wang as "a monadology with a central monad [...] like the monadology of Leibniz in its general structure."[1] At the same time, he believed that Cantorian set theory is a true theory, which describes some "well-determined reality."[2] I will first discuss the embedding of Cantorian set theory in a Leibnizian metaphysics that the combination of these two beliefs of Gödel's requires.[3] Then I turn to an attempt by Gödel to justify (a particular form of) the reflection principle in set theory by drawing an analogy to the monadology. Of this attempt I will argue that, although its success might not depend on whether the monadology is true or not, it fails. More generally, I defend the claim that while a Leibnizian metaphysics is compatible with Cantorian set theory, by itself it provides no clues that can be used in justifying set-theoretical principles, be it by analogy or directly.[4]

---

[1] Wang (1996), 0.2.1.
[2] Gödel (1990), p.181.
[3] Paul Benacerraf kindly allowed me to relate the following. At a dinner in 1974 or 1975, Gödel had conversations with Gerald Sacks on large cardinals and with Benacerraf on the mind-body problem. In the latter, he made reference to 'monads.' Gödel carried on these two conversations *simultaneously*, turning from left to right and back. (One argument advanced by Gödel was this: (1) the monads that our minds are have unambiguous access to the full set-theoretic hierarchy; (2) the full set-theoretic hierarchy cannot be adequately represented physically; therefore, (3) the mind cannot be reduced to a physical structure.)
[4] A monograph on the monadology in relation to Cantorian set theory is Osterheld-Koepke (1984). However, the reflection principle is not discussed there. On another note, it is argued there (p.128)

2 Fitting Cantor's sets into Leibniz' metaphysics

One immediate obstacle to the project of relating Cantorian set theory to Leibniz' metaphysics in any positive way would seem to be this. Cantor defines a set as a "many, which can be thought of as a one"[5] and as "each gathering-together ['Zusammenfassung'] M into a whole of determined and well-distinguished objects m of our intuition or of our thought (which are called the 'elements' of M)."[6] Cantorian set theory being largely about infinite sets, it is a theory of certain infinite wholes. But Leibniz denies the existence of infinite wholes of any kind.[7] For example, he says that one has to acknowledge that there are infinitely many numbers,[8] but he denies that they can be thought of as forming a unity:

> "I concede [the existence of] an infinite multitude, but this multitude forms neither a number nor one whole. It only means that there are more elements than can be designated by a number, just as there is a multitude or complex of all numbers; but this multitude is neither a number nor one whole."[9]

The distinction Leibniz draws between aggregates that are unities and aggregates that are mere multitudes is somewhat similar to the one Cantor would later draw between sets and proper classes, but their reasons are very different. Leibniz arrives at this distinction by a general argument that would rule out any infinite set altogether. He argues that there can be no infinite wholes or unities of any kind. It is not the notion of infinity as such that poses the problem for him, as is clear from

---

that on monadological grounds we can never decide the Continuum Hypothesis; one may well doubt that Gödel's understanding of the monadology and its relation to set theory would have had such a consequence. Gödel paired his belief in the monadology to a conviction that in principle a rational mind could decide every mathematical proposition. (He believed that "Leibniz did not in his writings about the Characteristica universalis speak of a utopian project" and that this would provide a means "to solve mathematical problems systematically," Gödel (1990), p.140. He realized that, because of his own incompleteness theorem, such a Characteristica could not assume the form of an entirely formal system.) In particular, he worked hard (but unsuccessfully) at deciding the Continuum Hypothesis. For further discussion of Gödel's belief in the solvability of all mathematical problems, see Kennedy & Van Atten (2004).

[5] Cantor (1932) p.204n.1.
[6] Cantor (1932), p.282; trl. modified from Grattan-Guinness (2000), p.112.
[7] Friedman (1975), p.338, suggests that even so, Leibniz might have been willing to accept the for him inconsistent notion of infinite whole as a fiction that may prove useful in calculations, on a par with his acceptance of imaginary roots in algebra. To illustrate this point, Friedman refers to Leibniz (1705), II, ch.17, §3.
[8] Leibniz to Des Bosses, March 11/17, 1706, Leibniz (1875–1890), II, p.305: "One cannot deny that the natures of all possible numbers are indeed given, at least in God's mind, and that as a consequence the multitude of numbers is infinite." ("Neque enim negari potest, omnium numerorum possibilium naturas revera dari, saltem in divina mente, adeoque numerorum multitudinem esse infinitam.") Where translations are my own, I give the original as well.
[9] Leibniz to Joh. Bernoulli, February 21, 1699, Leibniz (1849–1863), III/2, p.575: "Concedo multitudinem infinitam, sed haec multitudo non facit numerum seu unum totum; nec aliud significat, quam plures esse terminos, quam numero designari possint, prorsus quemadmodum datur multitudino seu complexus omnium numerorum; sed haec multitudo non est numerus, nec unum totum."

this exchange between Philalethe and Theophile (who represents Leibniz) in the *New Essays*:

> "PH: We have no idea of an infinite space, and nothing is clearer than the absurdity of an actual idea of an infinite number.
>
> TH: I agree. But the reason for this is not that one could have no idea of the infinite, but that an infinity cannot be a true whole."[10]

Specifically, Leibniz holds that the notion of an infinite whole contradicts the axiom that the whole is greater than the part. It is well known that Leibniz' argument is not sound and rests on an equivocation on 'greater than,' once defined in terms of the notion of proper superset and once defined in terms of the notion of non-surjective injection.[11] It can be shown, although for limitations of space I will not do so here, that Leibniz himself had all the means to see that his argument is not sound. The importance of that fact is that it shows that Leibniz' denial of infinite wholes does not reflect a limitation intrinsic to his philosophical system.

In Gödel's notebooks, I have so far not found a specific comment on Leibniz' argument that there can be no infinite wholes. But in the Russell paper from 1944 he wrote:

> "Nor is it self-contradictory that a proper part should be identical (not merely equal) to the whole, as is seen in the case of structures in the abstract sense. The structure of the series of integers, e.g., contains itself as a proper part."[12]

Among other things, Gödel says here that it is consistent that an equality relation holds between proper part and the whole. This entails a rejection of Leibniz' argument. And in a very similar note from 1944, again without mentioning Leibniz, Gödel adds: "the same can be contained as a part in 2 different ways."[13] That same consideration can be used to show that Leibniz' argument is not valid. Of course, the incorrectness of Leibniz' argument against infinite wholes implies nothing as to whether its conclusion is true or false. But clearly it will not be this argument that poses an obstacle to combining, as Gödel did, a belief in monadology with a belief in Cantorian set theory.

I now turn to the status of pure sets in Leibniz' metaphysics itself. Leibniz calls collections 'aggregates' or 'multitudes.' In his philosophical remarks on them, he usually discusses aggregates of objects in the world; but from these remarks

---

[10] Leibniz (1875–1890), V, p.146: "PH : Nous n'avons pas l'idée d'un espace infini, et rien n'est plus sensible que l'absurdité d'une idée actuelle d'un nombre infini. TH : Je suis du même avis. Mais ce n'est pas parcequ'on ne sauroit avoir l'idée de l'infini, mais parcequ'un infini ne sauroit estre un vrai tout."

[11] See, for example, the refutation in Benardete (1964), pp.47–48.

[12] Gödel (1990), p.130.

[13] "dasselbe [kann] auf 2 verschiedene Weisen als Teil enthalten sein," Gödel's Notebook *Max XI* (1944), p.18.

together with what he says about pure numbers, one can derive what his philosophical views on pure sets would have been.[14]

In a letter to De Volder of 1704, Leibniz writes that

> "Whatever aggregates out of pluralities there are, they are *unities* only in thought. They have no other reality than a borrowed one or that of the things out of which they are composed."[15]

Note the similarity with Cantor's definitions of a set that were quoted in the previous section; there with an emphasis on sets being a 'one' or a 'whole,' here on the fact that for Leibniz as for Cantor, the unity of an aggregate consists its elements being thought or considered together. Therefore, Leibniz says, an aggregate has the character of a relation:

> "Being and one are reciprocal notions, but where a being is given by aggregation, we also have one being, even though that entity and that unity are semi-mental.
>
> Numbers, units, and fractions have the nature of relations. And to that extent, they may in a sense be called beings."[16]

Leibniz here qualifies a unified aggregate as a semi-mental entity because he is thinking of aggregates of objects in the world. But an aggregate of mental objects would be entirely mental. The pure sets as we know them from Cantor's set theory, then, for Leibniz would fundamentally be pure relations that are entirely in the mind. Not in the human mind, but in God's mind, for, as Leibniz writes in the *New Essays*:

> "The relations have a reality that is dependent on the mind, as do truths; but not on the human mind, as there is a supreme intelligence that determines all of them at all times."[17]

Correspondingly, the truths about these pure relations have their existence in God's mind:

---

[14]Gödel makes some remarks on monads and sets on Wang (1996), p.296, but not so much on the relation between them.

[15]January 21, 1704, Leibniz (1875–1890), II, p.261: "quaecunque ex pluribus aggregata sunt, ea non sunt unum nisi mente, nec habent realitatem aliam quam mutuatam seu rerum ex quibus aggregantur." Also Leibniz (1705), p.133: "Cette unité de l'idée des Aggregés est tres veritable, mais dans le fonds il faut avouer que cette unité des collections n'est qu'un rapport ou une relation dont le fondement est dans ce qui se trouve en chacune des substances singulieres à part. Ainsi ces Estres par Aggregation n'ont point d'autre unité achevée que la mentale ; et par consequent leur Entité aussi est en quelque façon mentale ou de phenomene, comme celle de l'arc en ciel."

[16]Leibniz to Des Bosses, March 11, 1706, Leibniz (1875–1890), II, p.304: "Ens et unum convertuntur, sed ut datur Ens per aggregationem, ita et unum, etsi haec Entitas Unitasque sit semimentalis. Numeri, Unitates, fractiones naturam habent Relationum. Et eatenus aliquo modo Entia appellari possunt."

[17]Leibniz (1705), II, ch.30, §4: "Les rélations ont une réalité dépendante de l'esprit commes les Verités ; mais non pas de l'esprit de l'homme, puisqu'il y a une suprême intelligence, qui les détermine toutes en tout temps."

"One must not say, with some Scotists, that the eternal verities would exist even though there were no understanding, not even that of God.

For it is, in my judgement, the divine understanding which gives reality to the eternal verities, albeit God's will have no part therein. All reality must be founded on something existent. It is true that an atheist may be a geometrician: but if there were no God, geometry would have no object. And without God, not only would there be nothing existent, but there would be nothing possible. That, however, does not hinder those who do not see the connexion of all things with one another and with God from being able to understand certain sciences, without knowing their first source, which is in God."[18]

And, in "On the radical origination of things" from 1697:

"Neither these essences nor the so-called eternal truths about them are fictitious but exist in a certain region of ideas, if I may so call it, namely, in God himself, who is the source of all essence and of the existence of the rest [...] and since, furthermore, existing things come into being only from existing things, as I have also explained, it is necessary for eternal truths to have their existence in an absolutely or metaphysically necessary subject, that is, in God, through whom those possibilities which would otherwise be imaginary are (to use an outlandish but expressive word) realized."[19]

Leibniz even explicitly draws the conclusion that the eternal truths are invariant with respect to possible worlds:

"And these [propositions] are of eternal truth, they will not only obtain as long as the world will remain, but they would even have obtained, if God had created the world in another way."[20]

As Robert Adams has pointed out, Leibniz' thesis that mathematical objects have their existence in God's mind might well be acceptable to a mathematical Platonist, given the necessary existence of God, given the independence of God's thought from, in particular, human thought, and given the independence of eternal truths of God's will.[21] It is therefore not surprising to see the Platonist Gödel remark in a notebook from 1944, at the time, that is, when he was studying Leibniz intensely (1943–1946), that "the ideas and eternal truths are somehow parts of God's substance," that "one cannot say that they are created by God," and that they

---

[18]Leibniz (1710), §184; trl. Leibniz (1991), p.158. See also Leibniz (1705), II, 25, §1 and Leibniz (1875–1890), VII, p.111.

[19]Leibniz (1875–1890), VII, pp.302–8; trl. Leibniz (1969), p.488.

[20]Leibniz (1903), p.18: "Et hae sunt aeternae veritatis, nec tantum obtinebunt, dum stabit Mundus, sed etiam obtinuissent, si Deus alia ratione Mundum creasset."

[21]Adams (1983), p.751. For Descartes, in contrast, mathematical truth is a matter of God's will, and hence on a Cartesian conception God could *choose* to make reflection true, perhaps for similar reasons as why according to Leibniz (1991), §46, Leibniz (1710), §380, God favours reflection in the physical world. See also footnote 55 below. A particularly interesting comment by Leibniz on the relation between God's will, mathematics, and creation is found in Leibniz (1695), p.57. He there says that, although irrational numbers are to some extent imperfect because they cannot be expressed as fractions, this imperfection "comes from their own essence and cannot be blamed on God;" and that, although God could have avoided creating objects (in the world) with irrational measures, if He has nevertheless done so, it is because it results in a universe with a greater variety of forms.

rather "make up God's essence."[22] Gödel also writes that, of the mappings from propositions to states of affairs "the correct one" is "the one which is realized in God's mind."[23]

This aspect of Leibniz' views on mathematical objects therefore will have provided an additional interest for Gödel in a Leibnizian proof of God's existence: a corollary of such a proof for him would be that a single, fixed universe of all sets V indeed exists, and hence that there is a privileged model for the axioms of set theory. Gödel describes his belief in such a privileged model in, for example, his Cantor paper from 1947.[24]

3 The reflection principle

There is an attempt of Gödel's to justify, by drawing an analogy to Leibniz' monadology, the reflection principle in set theory. Gödel never published the argument but he did present it to Wang;[25] here it will be quoted in section 4.1 below.

The basic idea behind the reflection principle is that the universe V of all sets is in some sense too large to be adequately conceivable or definable in set-theoretic terms. From this observation, one concludes to

(1) If a clearly conceived, set-theoretical property holds of V, this property cannot be unique to V and will also characterize a set contained in it.

With respect to that property, that set is then said to 'reflect' the universe.[26] (Again by reflection one then also sees that that set is not the only one to reflect the universe in that way, and that there are many more.)

Well-known applications of this informal principle are the following. The universe contains (set-theoretic encodings of) the natural numbers, hence there is also a set that contains the natural numbers (and so, by separation, there exists a set that contains nothing but the natural numbers). This use of reflection is already found in Cantor.[27] Or: of any given set, the universe contains all its subsets, hence there

---

[22]"Die Ideen und ewigen Wahrheiten sind irgendwie Teile der göttlichen Subst[anz]. Daher kann man nicht sagen, daß sie von Gott geschöpft wurden (denn Gott wurde nicht von Gott geschöpft), sondern sie machen das Wesen Gottes aus." Gödel's Notebook *Max XI* (1944), p.31]. Compare Leibniz (1875–1890), VII, p.305, lines 1–4, which Gödel copied in a note (item 050130 in his archive), Leibniz (1710), §§335,380, and the passage in Leibniz' letter to Wedderkopf, quoted on p.23 below.

[23]"Daß eine gewisse Kombination von Begriffen oder Symbolen 'wahr' ist, bedeutet, daß sie ein adäquates Bild von etwas Existierendem ist, hängt also von der Abbildungsrelation ab. Manche Abbildungsrelationen können wir selbst konstruieren, manche (und insbesondere 'die richtigen', nämlich die im Verstand Gottes realisierten) finden wir vor," Gödel's Notebook *Phil XIV*, p.7, July 1946 or later.

[24]Gödel (1990), p.181.

[25]Wang (1996), 8.7.14.

[26]E.g., Lévy (1960a), p.228, and Lévy (1960b), p.1. For two recent monographs on the reflection principle, diametrically opposed to one another in their philosophical approach, see Roth (2002) and Arrigoni (2007). The former corresponds more closely to Gödel's view as described here.

[27]In note 2 to his paper from 1883, "On infinite, linear point manifolds 5": "Whereas, hitherto, the infinity of the first number class [...] has served as [a symbol of the Absolute], for me, precisely

is also a set that contains all subsets of the given set (and so, by separation, there exists a set that contains nothing but the subsets of the given set). Or: the universe is inaccessible, hence there is an inaccessible cardinal.[28]

The first two of these applications yield justifications of two axioms of Zermelo-Fraenkel set theory, the axiom of infinity and the axiom of the powerset. Regarding the latter, note that it is not particularly clear (although for Gödel himself it apparently was) that, as the standard iterative concept of set has it, the collection of all subsets of an infinite set is a set as opposed to a proper class.[29] The informal reflection principle is a means to provide the justification needed. It is of course not excluded that alternative ways to convince ourselves of the truth of these (and other) axioms exist. Regarding the justification of the existence of inaccessibles, Gödel stated his preference for reflection over other methods in a letter to Paul Cohen of August 13, 1965:

"As far as the axiom of the existence of inaccessibles is concerned I think I slightly overstated my view.[30] I would not say that its evidence is due *solely* to the analogy with the integers. But I do believe that a clear analogy argument[31] is much more convincing than the quasi-constructivistic argument in which we imagine ourselves to be able somehow to reach the inaccessible number. On the other hand, Levy's principle[32] might be considered more convincing than analogy."[33]

Indeed, as Wang reports, Gödel said that the justification of axioms by an appeal to reflection is the fundamental one:

"All the principles for setting up the axioms of set theory should be reducible to Ackermann's principle: The Absolute is unknowable. The strength of this principle increases as we get stronger and stronger systems of set theory. The other principles are only heuristic principles. Hence, the central principle is the reflection principle, which presumably will be understood better as our experience increases. Meanwhile, it helps to separate out more specific principles which either

---

because I regarded that infinity as a tangible or comprehensible idea, it appeared as an utterly vanishing nothing in comparison with the absolutely infinite sequence of numbers.", Cantor (1932), p.205n.2; trl. Hallett (1984), p.42. See also Hallett (1984), pp.116–117.

[28] A cardinal κ is inaccessible if it is regular (i.e., not the supremum of k ordinals all smaller than k) and a limit (i.e., not the next cardinal greater than some cardinal λ.)

[29] For Gödel's justification of the power set axiom on the iterative conception of set (not by reflection), see Wang (1974), p.174, and Wang (1996), p.220. For criticism, see e.g., Parsons (1977), p.277 and Hallett (1984), pp.236-238. Gödel's comment on an early version of Parsons (1977) seems to me to be instructive but also indicative of a weakness of Gödel's own use of idealization: "he does not understand 'idealization' broadly enough" Gödel (2003b), p.390. On a different occasion, Gödel acknowledged that there are cases where idealization is understood too broadly to be very convincing; see the quotation from his letter to Cohen that follows in the main text.

[30] Given the beginning of the preceding paragraph in the letter, "When we spoke about the power set axiom..." (p.385), presumably Gödel here refers to that same conversation.

[31] [Gödel's footnote] such as, e.g., the one obtained if an inaccessible α is defined by the fact that sums and products of fewer than α cardinals $< \alpha$ are $< \alpha$.

[32] A formulation of the idea of the unknowability of V that one also finds in Cantor and Ackermann (quoted elsewhere in this paper); in Levy's words, "the idea of the impossibility of distinguishing, by specified means, the universe from partial universes," Lévy (1960b), p.1. Levy in that paper studies four specific versions of that principle.

[33] Gödel (2003a), p.386.

give some additional information or are not yet seen clearly to be derivable from the reflection principle as we understand it now."[34]

Ackermann had stated that the notion of set is open-ended and that therefore the universe of all sets does not admit of a sharp definition (and is in that sense unknowable) Ackermann (1956), p.337. (This is a reflection principle because it means that if we do find a set-theoretic property of V, this cannot be a definition of it, and hence there is a set that shares the property.) He also took this to be in accord with Cantor's intentions; and, although Ackermann does not point this out, this is indeed the principle that Cantor had used to justify the existence of the *set* of all natural numbers (see footnote 27 above).

In some sense one could say that, if Gödel's belief in this reducibility of the principles for setting up the axioms to reflection is correct, then the informal reflection principle captures the concept of set. Note that the reflection principle that Gödel has in view here is not to be confused with reflection principles that are provable in a particular formal system, such as the Montague-Levy reflection theorem in *ZF*.[35] By Gödel's incompleteness theorem, no single formal system for set theory can be complete, and the reflection principle Gödel is speaking about is precisely meant as the fundamental way to arrive at further axioms to extend any given system. His principle therefore has to be, and to remain, informal. Its strength increases with every application because the resulting stronger system in turn gives rise to the formulation of stronger properties to reflect.

In its fully general form (1), the principle of course cannot be upheld. For example, the property of containing every set in the universe is not reflected by any set contained in it, as such a set would have to contain itself. Reflection principles will therefore have to be precise or restrictive about the properties for which they are supposed to hold. Gödel suggested that reflection holds for *structural* properties.[36] The property of containing all sets is not structural, because it does not specify a property of all sets that might define a structure that they instantiate or exemplify. A sufficiently rich positive characterization of the notion of structural property is still wanting, but the present consideration illustrates why Gödel included it in the reflection principle that I will discuss here (the label is mine):

(2) A *structural* property, possibly involving V, which applies only to elements of V, determines a set; or, a subclass of V thus definable is a set.[37]

Gödel's realist conception of V permits him to look for properties of V directly; this marks a deep difference with the kind of thinking about reflection that had

---

[34] Wang (1996), 8.7.9. See also Wang (1996), 8.7.16.

[35] In the context of a particular formal system, the properties of V that can be reflected are of course limited by what can be expressed and defined in that system. That should contribute much to the principle's being provable, in case it is.

[36] See Wang (1977), section 3, Wang (1996), pp.283–285 and Reinhardt (1974), p.189n.1.

[37] Wang (1996), 8.7.10.

been introduced by Zermelo.[38] Zermelo saw set theory as describing rather an open-ended, always extendable series of ever larger universes. Like Gödel, he accepted a version of the reflection principle, but, because of his different idea of what set theory is about, his principle is justified and used in a somewhat different way.[39] According to Zermelo, V does not really exist and hence there are no literal truths to be found about it. Talk of properties of V must really be talk about the limited set of principles used in the construction of some initial segment of the open-ended series of universes.[40] This limited set of principles remains available in the construction of any longer segment of the series, and this is why the property in question will persist. In other words, we have a justification of Zermelo's reflection principle by a continuity argument.[41] Gödel, on the other hand, is not forced to construe talk of properties of V as talk about something limited; hence, reflection as exemplified by Gödel's principle has been characterized as 'top-down,' Zermelo's as 'from below.'[42] Potentially, top-down reflection is the more powerful of the two. But in its use the principle is correspondingly more difficult, as it requires one to sort out those properties of V that are not reflectable from those that are; hence Gödel's quest for 'structural properties.' Moreover, one might think that Zermelo's conception is to be preferred on philosophical grounds, as by accepting it, one is freed from the demands for an argument for the existence of V and for an account of how can we come to know truths about it.[43]

But it is precisely here that Gödel will have seen an advantage for his view. As Hellman, who supports and develops Zermelo's conception, has noted, that conception requires that one accepts a notion of possible objects that does not imply the existence of possibilia.[44] But as we saw in section 2, from a Leibnizian point of view such a notion of possibility cannot be accepted, and talk of possibilities that are not grounded in something existent is ultimately unintelligible. The same criticism would be applied to any other interpretation of set theory in which

---

[38] Zermelo (1930).

[39] On their differences, see also the extensive discussion (from a somewhat different perspective) by Tait (1998).

[40] 'Construction' in the sense that the existence of this segment is derived from specific axioms by specific principles. In classical set theory, such axioms and principles will themselves generally not be 'constructive' in the sense in which that term is used to characterize varieties of mathematics such as intuitionism.

[41] The logic of open-ended series is intuitionistic rather than classical. This type of reasoning we will see again later on in this paper, see footnote 84. For more on this type of argument and its justifications, see Van Atten & Van Dalen (2002). Georg Kreisel wrote to me in a letter of March 7, 2006, that in the period that he knew Gödel, the latter was "sympathetic to a justification by intuitionistic logic (in terms of not necessarily constructive knowledge)" of set-theoretic reflection principles.

[42] E.g., Hellman (1989), p.90.

[43] From the point of view of constructive mathematics in the sense explained in footnote 40, what remains to be accounted for in Zermelo's conception would of course still be far too much.

[44] Hellman (1989), pp.57,58.

commitment to the existence of V is avoided by resorting to modal notions.[45] According to Gödel, the open-endedness of the notion of set that motivates resorting to notions of possibility is not the correlate of an ontological fact: "To say that the universe of all sets is an unfinishable totality does not mean objective undeterminedness, but merely a subjective inability to finish it."[46] (Here, 'subjective' seems to refer to the act, however idealized, of obtaining a collection by putting it together from elements which are considered to be given prior to that act. Cantor's notion of set (quoted above) contains a subjective element in just this particular sense. The universe V can never be obtained in such an act, as V cannot be a set.) A closely related Leibnizian observation is made by Mugnai:

> "In man's limited intellect there is a distinction between the 'capacity to think' and the 'actual exercise' of that capacity. This distinction is not met within God. If the ideas *in Mente Dei* are conceived as 'dispositional properties' then we must also postulate a 'state' of the divine intellect in which it carries out a limited activity, during which all the totality of ideas are never present all at once. This is surely unacceptable from the theological point of view, however, since it limits the divine powers and assimilates the psychological and reasoning activity of God to the example of human activity."[47]

4   Gödel's analogy argument for the reflection principle

4.1   Presentation of the argument

Gödel's argument for principle (2) that I should like to analyze (not his only one) consists in drawing an analogy to Leibniz' monadology. Here I will present that argument, try to fill in the details, consider the question whether it is a good argument, and conclude that it is not. In doing so, I will not be arguing that the alternative arguments that Gödel had for the validity of reflection principles are incompatible with a Leibnizian metaphysics. What I am going to argue is that the one argument we know of in which Gödel explicitly tries to argue from a Leibnizian metaphysics to a form of the reflection principle in set theory does not work.

A note on the sources that will be used here: as yet, Gödel's philosophical notebooks have been transcribed only partially. For all I know there may be material in those untranscribed parts that is relevant to the matter at hand. As a principle of interpretation, I will assume that the argument that Gödel presented to Wang in the 1970s, when he had perfect access to his notebooks from the 1940s (except for the one from 1945–1946 that he reported lost), is the version that he considered best. As for Leibniz, I have tried to use, whenever possible, writings from 1686 and later, as that is the phase in the development of Leibniz' philosophy that in

---

[45] Yourgrau's criticism of Parsons' position is of the same type. See Parsons (1977), pp.268–297 and Yourgrau (1999), p.177–185.

[46] Wang (1996), 8.3.4. Tait (1998), p.478, wishes to leave open the same possibility of objective undeterminedness that Gödel denies.

[47] Mugnai (1992), p.24. Similarly, Jolley (1990), p.138, notes: "Now Leibniz might be more reluctant than Mates to allow that divine ideas are dispositions, for this may be difficult to reconcile with the traditional view that God is pure act."

1714 culminated in the *Monadology*. But in particular some earlier texts may be relevant as well.

What might motivate one to draw an analogy between monadology and set theory is that in both cases we have a universe of objects, the objects resemble in some sense the whole, and the actual universe is in some sense the best out of a collection of possible universes. In the monadology, God chooses a universe or world to actualize from out of the collection of possible worlds, according to some criteria for which one is best; in set theory, models for ZFC are known which are generally not believed to correspond to set-theoretical reality (e.g., the so-called 'minimal model' is considered not to be the 'best' model because it is too small). The themes of reflection and mirroring occur often in Leibniz' writings. A typical example is Leibniz' formulation of his Principle of Harmony in section 56 of the *Monadology*:[48]

"Each simple substance has relations that express all the others, and is in consequence a perpetual living mirror of the universe."[49]

One could use the monadology as a means to generate structural principles for monads and their relations, substitute in such a principle the notion of set for that of monad, and then seek independent reasons why the set-theoretical principle thus obtained should be true. The justification one might then come up with will not depend on an analogy between the universes of monads and sets. This merely heuristic approach was followed by Joel Friedman in his paper "On some relations between Leibniz' monadology and transfinite set theory"[50] where he obtained maximizing principles in set theory on the basis of maximizing principles of harmony in the monadology. A similar somewhat loose (but not necessarily less fruitful) approach was taken by Wim Mielants in his paper "Believing in strongly compact cardinals", where "Leibniz's philosophy is only a source of inspiration for the maximization properties we use here."[51] One conclusion that may be drawn from the present paper is that such a heuristic approach will probably be more fruitful than an analogy of the type Gödel wished to draw.

Gödel's analogy is one that he takes to be by itself a justification of a form of the reflection principle, without the need to adduce independent reasons. As will be discussed later, a convincing analogy argument does not always require that the situation with which an analogy is drawn is, in its full extent, actual or real. But, to look ahead a bit, Gödel's use of his analogy as a sufficient justification is based on the idea that the reflection principle is true in set theory for exactly the same reason why a certain monadological proposition is true. As long as it is not clear that such a general reason, should it exist at all, would involve no

---

[48] The actual name is given to it in section 78.
[49] Leibniz (1991), section 56.
[50] Friedman (1975).
[51] Mielants (2000), p.290.

specifically monadological notions, it is not clear whether here, too, justification can be treated independently of a justification of the monadology. For the moment I will leave it an open question whether one has to accept the monadology as the true metaphysics in order to be convinced by Gödel's argument, and concentrate rather on the prior task of filling in the details of the analogy that he indicates.

Hao Wang recorded Gödel's argument in item 8.7.14 of his *Logical Journey*. For clarity, I quote the preceding item as well:

> "8.7.13 [...] Consider a property $P(V,x)$, which involves V. If, as we believe, V is extremely large, then x must appear in an early segment of V and cannot have any relation to much later segments of V. Hence, within $P(V,x)$, V can be replaced by some set in every context. In short, if P does not involve V, there is no problem; if it does, then closeness to each x helps to eliminate V, provided chaos does not prevail."

> "8.7.14 There is also a theological approach, according to which V corresponds to the whole physical world, and the closeness aspect to what lies within the monad and in between the monads. According to the principles of rationality,[52] sufficient reason, and preestablished harmony, the property $P(V,x)$ of a monad x is equivalent to some *intrinsic* property of x, in which the world does not occur. In other words, when we move from monads to sets, there is some set y to which x bears intrinsically the same relation as it does to V. Hence, there is a property $Q(x)$, not involving V, which is equivalent to $P(V,x)$. According to medieval ideas, properties containing V or the world would not be in the essence of any set or monad."[53]

So in the case for sets, the claim is that $P(V,x) \equiv Q(x)$, where $Q(x) = \exists y P(y,x)$ and x and y are sets. (Certainly, the fact that $Q(x)$ is a one-place predicate does not suffice to make it express a non-relational property.[54])

The approach is 'theological' because in the monadological setting, it is a central monad or God who creates a universe of objects.[55] To make Gödel's analogy more explicit, I propose to put it in a slightly different form, the rationale of which will be explained as we go along. As Gödel adds the explanation that "according to medieval ideas, properties containing V or the world would not be in the essence of any set or monad." it is clear that he in this analogy argument considers only essential properties. He first presents, in effect, the following monadological proposition:

---

[52] By this, I take it, Gödel means the principle of contradiction.

[53] Wang (1996), 8.7.14.

[54] See Ishiguro (1990), ch.6.

[55] A curious example of a theological approach by Gödel to a mathematical question is found in his notebook *Max X* (1943–1944), p.18: "Does the commandment that one shall make neither likeness nor image perhaps also mean, that type theory must be accepted and that any formalisation of the all leads to a contradiction?" ("Bedeutet vielleicht das Gebot, du sollst dir kein Gleichnis noch Bildnis machen, auch, daß die Typentheorie anzunehmen ist und jede Formalisierung des Alls zu einem Widerspruch führt?"). The inference from a commandment to a mathematical truth would seem to fit a Cartesian view of the relation between God and mathematics better than a Leibnizian one. For Descartes, mathematical truth was determined by God's will; Leibniz contested this. For an analysis of this difference between Descartes and Leibniz, see Devillairs (1998). More positive statements by Gödel on type-free logic occur in, for example, his correspondence with Gotthard Günter, see Gödel (2003a), pp.527,535.

"Essential, relational properties of (created) monads are intrinsic properties in which the universe as a whole does not occur but part of it does."

'Part' here is meant in the proper sense according to which no part of the universe expresses the whole universe perfectly; this is in fact implied by the condition that in the properties in question "the universe as whole does not occur." The notion of expression Leibniz describes as follows:

> "That is said to express a thing in which there are relations which correspond to the relations of the thing expressed."[56]

> "It is sufficient for the expression of one thing in another that there should be a certain constant relational law, by which particulars in the one can be referred to corresponding particulars in the other."[57]

> "One thing expresses another (in my terminology) when there exists a constant and fixed relationship between what can be said of one and of the other."[58]

Clearly, a perfect expression of x by y requires a 1-1 correspondence between all properties of x and (some) properties of y.

Let us call the above monadological proposition the 'reflection principle for (created) monads.' Gödel then proposes that we move from monads to sets and obtain from this, by analogy, the reflection principle for sets:

> "Essential, relational properties of sets are intrinsic properties in which V does not occur but a set does."

In the move from monads to sets, the immediate analogue of a part (in the strong sense) of the universe of monads (a collection of monads) is a part of the universe of sets, hence a collection of sets and not an individual set. But this actually suffices, because of the following principle that Gödel accepted: any collection that is properly contained in V and that cannot be mapped 1-1 to it (and in that sense cannot perfectly 'express' V), is not a proper class but a set. This is known as 'Von Neumann's axiom'.[59] So although the immediate analogue of a collection of monads that does not perfectly express the universe of monads is a collection of sets that does not perfectly express V, by Von Neumann's axiom the latter collection is itself a set. It is this set the existence of which Gödel's analogy argument concludes to.

---

[56] Leibniz (1969), p.207.
[57] Leibniz (1903), p.15; trl. Rutherford (1995), p.38.
[58] Leibniz (1875–1890), II, p.112; trl. Mates (1986), p.38n.11.
[59] The idea had already been formulated by Cantor in a letter to Dedekind of July 28, 1899, first published in Cantor (1932), seven years after Von Neumann's paper (1925). For a clear and detailed discussion of this axiom, see Hallett (1984), section 8.3.

Gödel commented on Von Neumann's axiom:

> "As has been shown by Von Neumann, a multitude is a set if and only if it is smaller than the universe of all sets.[60] The great interest which this axiom has lies in the fact that it is a maximum principle, somewhat similar to Hilbert's axiom of completeness in geometry. For, roughly speaking, it says that any set which does not, in a certain well defined way, imply an inconsistency exists."[61]

This fits well into Leibniz's picture according to which mathematical existence is equivalent to mathematical possibility, and the latter is wholly determined by a (global) principle of non-contradiction; we will come back to this later.

4.2 The analogy is ineffective

The conception of analogy arguments I will use here is Kant's, who in section 58 of the *Prolegomena* writes: "Such a cognition is one by analogy, which does not signify for example, as the word is commonly understood, an imperfect similarity of two things, but a perfect similarity of two relations between entirely dissimilar things."[62] If the similarity in question is perfect, it will be embodied in a general principle that governs both of the domains involved in the analogy. Only the existence of such an underlying general principle can give an analogy argument genuine force. Of course, once such a general principle has been identified, it can be used to construct a direct argument for the desired conclusion, and the analogy is no longer necessary. The function of the analogy will then have been to have pointed to the relevant general principle.[63]

So in order to show that the similarity claimed by Gödel is not arbitrary or superficial, but does indeed carry argumentative weight, it would have to be shown that the reflection principle holds for monads because they instantiate a more general principle that implies reflection for universes of objects satisfying certain conditions. Applying that same more general principle to the universe of sets should then yield the reflection principle for sets.[64]

---

[60] Wang (1996), 8.3.7.

[61] Wang (1996), 8.3.8. The inconsistency Gödel refers to here is the inconsistency arising from conceiving of a particular kind multitude as set. As we saw above, for Gödel V genuinely exists, but as a mere multitude and not as a set.

[62] "Eine solche Erkenntnis ist die nach der Analogie, welche nicht etwa, wie man das Wort gemeiniglich nimmt, eine unvollkommene Ähnlichkeit zweier Dinge, sondern eine vollkommene Ähnlichkeit zweier Verhältnisse zwischen ganz unähnlichen Dingen bedeutet." Gödel will surely have known this passage; but in his copy of the Reclam 1888 edition of the *Prolegomena*, there are no reading marks to it. (I am grateful to Marcia Tucker at the Historical Studies-Social Science Library of the IAS for having verified this.)

[63] To emphasize that this is the function of an analogy, St. Augustine classified it with the signs, cf. Maurer (1973).

[64] In his formulation of reflection principle (2) on p.10 above, Gödel mentions a restriction on the properties that can be reflected, saying that they should be 'structural.' I will come back to the possible role of this restriction in the analogy later.

But such a principle, I claim, cannot exist. In a first step, I argue that it is consistent with the purely metaphysical principles of the monadology to assume that the reflection principle for monads holds but the reflection principle for sets fails. In the second step, I explain why this entails that Gödel's analogy is ineffective, whether the monadology is true or not.

That in the monadology the reflection principle for monads is consistent follows from the fact that, as I will now argue, in the monadology that principle is true.

As a preliminary, the meaning of the term 'essence' has to be clarified. Leibniz uses it in different ways. Sometimes he defines the essence of a monad as simply the collection of all its properties, considered in abstraction from the existence of that monad. As he holds that each monad expresses the whole universe or world, by this definition it is trivially false that the essence of a monad does not involve the world.[65] But Leibniz also has another notion of essence, which is the one that will be relevant here. This notion is defined as the collection of all the necessary properties of that substance. For example, in 1676 Leibniz first defines an 'attribute' as "a necessary predicate conceived through itself, or that cannot be analysed into several others" and then "an *essence* is [...] the aggregate of all the attributes (of a thing)."[66] In 1678 he defines the "essence of a thing" as "the specific reason of its possibility" and specifies that what is true in the region of essences is "unconditionally, absolutely and purely true."[67] This definition he repeats two decades later, in 1701, "the essence of the thing being nothing but that which makes its possibility in particular."[68] Of particular interest for its idealistic content is Leibniz' remark in the *New Essays* (1705) that possibility is the same as being distinctly intelligible (which intelligibility is ruled out for contingent properties).[69] Finally,

---

[65] While reading Leibniz (1903), Gödel noted: "The proposition that every thing involves all others, can be understood purely logically. Namely: It involves all accidents, among these however also the relations to all other things; these however involve the other things. But that is only an accidental, no necessary involvement. But to the extent that to the essence belongs the reaction in arbitrary situations, it also involves essentially—also through knowledge (mirror)—accidentental involvement." ("Die Aussage, daß jedes Ding alle andere involviert, kann rein logisch verstanden werden. Nämlich: Es involviert alle Acc[identia], unter diesen aber auch die Beziehungen zu allen anderen Dingen; diese involvieren aber die anderen Dinge. Das ist aber nur ein accident[elles], kein notwendiges Involvieren. Aber insofern zum Wesen die Reaktion in beliebigen Lagen gehört, involviert [es?] sie auch essentiell—auch durch Erkenntnis (Spiegel)— acci[dentelles] Involvieren.") Gödel's Notebook *Max X* (1943–1944), pp.70–71. Here Gödel must be referring to Leibniz' statement on p.521 of that edition, "Every singular substance involves in its perfect notion the whole universe" ("Omnis substantia singularis in perfecta notione sua involvit totum universum.")

[66] Leibniz (1923–), VI, iii, p.574, as quoted in Adams (1994), p.127.

[67] Leibniz (1923–), II, i, pp.390 and 392, as quoted in Adams (1994), pp.136,138.

[68] Leibniz (1875–1890), IV, p.406: "l'essence de la chose n'étant que ce qui fait sa possibilité en particulier."

[69] Leibniz (1875–1890), V, p.246: "But whether they depend on the mind or not, it suffices for the reality of their ideas, that these modes are *possible* or, which is the same thing, distinctly intelligible." ("Mais soit qu'ils dependent ou ne dependent point de l'esprit, il suffit pour la realité de leur idées, que ces Modes soyent possibles ou, ce qui est la même chose, intelligibles distinctement.") Gödel noted

in 1714, he writes that

> "I consider possible everything that is perfectly conceivable, and which therefore has an essence, an idea; without taking into consideration whether the other things allow for it to come into being."[70]

With this notion of essence in place, the argument for Reflection for created monads proceeds as follows:

1. All properties of monads consist in their own perceptions; this does not rule out relational properties as these are intrinsic too. (Premise)

2. Essential properties correspond to distinct perceptions. (Premise)

3. No created monad can distinctly perceive the whole universe. (Premise)

4. Essential, relational properties of (created) monads are intrinsic properties in which the as a whole does not occur but part of it does. (From 1, 2 and 3)

In the opening sections of the *Monadology*, Leibniz says that monads are the ultimate constituents of reality. They are simple in the sense that they are not composed out of parts (section 1). Elsewhere, Leibniz also says that the monads are not in space and time, but that space and time are rather phenomena that depend on the way monads represent reality to themselves. Although monads are simple, they do have inner states, and these can change. This does not contradict the fact that they have no parts, if this is understood to mean (in terms of Husserl's third *logische Untersuchung*) that they have no independent parts but only dependent ones, like a continuum.[71] The changes arise within the monad itself and do not come from outside, for monads have no parts that can be acted upon from outside; they "have no windows" (section 7). Only God can be said to act upon the created monads directly. Leibniz identifies the specification and variety of simple substances with the internal complexity of these inner states (section 12), and calls these transitory states 'perceptions' (section 14). The properties of a monad consist in its proper perceptions.[72] Perceptions "enfold and represent a multiplicity in a unity, or in

---

this one, see item 050131 in his archive.

[70] To Bourguet, December 1714, Leibniz (1875–1890), III, pp.573–574: "J'appele possible tout ce qui est parfaitement concevable, et qui a par consequent une essence, une idée : sans considerer, si le reste des choses luy permet de devenir existant." See also Leibniz (1991), section 43, Leibniz (1710), section 390.

[71] Leibniz used the absence of independent parts as an argument against the conception of the mind as a machine or mechanism: the mind is a unity, whereas a machine has (independent) parts, e.g., in his *New System of the Nature and Communication of Substances* from 1695, Leibniz (1969), p.456. Gödel appealed to the very same argument: "Consciousness is connected with one unity. A machine is composed of parts," Wang (1996), 6.1.21.

[72] The special case of reflexive knowledge or consciousness that some monads sometimes have of their inner states, apperception, plays no role in Gödel's analogy.

the simple substance" (section 14), and in fact each monad perceives or represents the whole universe.[73] Various crucial points for Gödel's analogy are now made in section 60:

> "For in regulating the whole, God has had regard for each part, and in particular for each monad, which, its very nature being representative, is such that nothing can restrict it to representing only part of things. To be sure, this representation is only confused regarding the detail of the whole universe. It can only be distinct in regard to a small part of things, namely those that are nearest or most extensively related to each monad. Otherwise each monad would be a deity. It is not in their object [namely the whole universe], but in the particular mode of knowledge of this object that the monads are restricted. They all reach confusedly to the infinite, to the whole; but they are limited and differentiated by the degrees of their distinct perceptions."[74]

If monads did not differ this way, they would all be one and and the same, by identity of indiscernables (which is a consequence of Sufficient Reason). For the only properties monads have are perceptual, and perceptions differ only in degree of distinctness.[75] Only the monad which is God perceives the whole universe perfectly; the perception of the universe by created monads necessarily is (partly) confused, because their receptivity is necessarily limited (secton 47).[76] It follows that the perceptions of no created monad can exhaust the universe. This precludes that the perceptions of a created monad stand in 1-1 relation to the elements of the universe, and therefore no created monad expresses the universe perfectly.

Note in passing how the fact that monads have no windows and only God acts directly upon them explains, when combined with the idea that sets are objects in God's mind, Gödel's assertion to Paul Benacerraf that the monads have unambiguous access to the full set-theoretic hierarchy.[77] As Leibniz wrote around 1712:

> "I am convinced that God is the only immediate external object of souls, since there is nothing except him outside of the soul which acts immediately upon it. Our thoughts with all that is in us, in so far as it includes some perfection, are produced without interruption by his continuous operation. So, inasmuch as we receive our finite perfections from his which are infinite, we are immediately affected by them. And it is thus that our mind is affected immediately by the eternal ideas which are in God, since our mind has thoughts which are in correspondence with them and participate in them. It is in this sense that we can say that our mind sees all things in God."[78]

---

[73] Compare also the earlier *On Nature's Secrets* from around 1690: "Indeed, the multiple finite substances are nothing other than diverse expressions of the same universe according to diverse respects and each with its own limitations." Leibniz (1875–1890), VII, p.311n., trl. Leibniz (1991), p.217.

[74] Leibniz (1991), section 60.

[75] Gödel writes in his Notebook *Max X* (1943–1944), p.20: "Almost any property can be had to different degrees" ("Man kann fast alle Eigenschaften in verschiedenen Graden haben.")

[76] Necessarily, for by identity of indiscernables God is unique; section 39 cites, alternatively, the principle of sufficient reason.

[77] See footnote 3 above; this part of the anecdote is also reported in Maddy (1990), p.79.

[78] "Conversation of Philarète and Ariste" (one of the direct forerunners of the *Monadology*), Leibniz (1875–1890), VI, p.593; trl. Leibniz (1969), p.627. Gödel seems to have had this or a similar passage, e.g., Leibniz (1686), section 28, in mind when he remarked in his letter to Gotthard Günther of April 4, 1957: "That abstract conceptual thought enters individual monads only through the central monad is a truly Leibnizian thought.", Gödel (2003a), p.527.

The fact that all of a monad's properties are internal to it might seem to rule out relational properties, in which case Gödel's analogy argument would not work, for if there are no relations between monads then there is no basis for an analogy concluding to the existence of relations between sets. In fact, on Leibniz' understanding of relations, relational properties are not at all ruled out: a monad x will have a relational property P if x expresses the relata in the way characteristic for P. But to express other monads this way is an entirely internal property; it does by itself not guarantee that these other monads indeed exist. This is indeed what Leibniz meant, as he makes clear in his reply to an objection made by his correspondent Des Bosses. Des Bosses had written to Leibniz:

> "If the monads of the universe get their perceptions out of their own store, so to speak, and without any physical influence of one upon the other; if, furthermore, the perceptions of each monad correspond exactly to the rest of the monads which God has already created, and to the perceptions of these monads, and are harmonized so as to represent them; it follows that God could not have created any one of these monads which thus exist without constructing all the others which equally exist now, for God can by no means bring it about that the natural perception and representation of the monads should be in error; their perception would be in error, however, if it were applied to nonexistent monads as if they existed."[79]

And Leibniz replied:

> " He can do it absolutely [i.e., as far as logic is concerned]; he cannot do it hypothetically [i.e., when also God's will is taken into account], because he has decreed that all things should function most wisely and harmoniously. There would be no deception of rational creatures, however, even if everything outside of them did not correspond exactly to their experiences, or indeed if nothing did, just as if there were only one mind; because everything would happen just as if all other things existed, and this mind, acting with reason, would not charge itself with any fault. For this is not to err. [...] Not from necessity, therefore, but by the wisdom of God does it happen that judgements formed upon the best appearances, and after full discussion, are true."[80]

So in what Leibniz calls an 'absolute' sense, a monad can have a relational property without that relation obtaining in the world. But in the actually created world this is excluded, for in choosing that world God sees to it that the perceptions of its monads are in harmony with one another.[81] This depends on God's will instead of logic and that is why Leibniz says that it is not 'absolutely' but 'hypothetically' necessary that relational properties express relations that indeed obtain. In the presence of this principle of harmony, the circumstance that a monad x in a world truly has relational property P not only implies, but is equivalent to, the circumstance that it has an appropriate intrinsic property. This explains why Gödel mentions the principle harmony in his analogy argument: as he wishes to reason

---

[79]Des Bosses to Leibniz, April 6, 1715, Leibniz (1875–1890), II, p.493; trl. Leibniz (1969), p.611.

[80]Leibniz to Des Bosses, April 29, 1715, Leibniz (1875–1890), II, p.496; trl. Leibniz (1969), p.611. See also Leibniz (1686), section 14, Leibniz (1710), section 37, and Leibniz (1875–1890), IV, p.530.

[81]Also: "It can be said that God arranges a real connection by virtue of that general concept of substances which implies perfect interrelated expressions between all of them, though this connection is not immediate, being based on what God has wrought in creating them." Leibniz (1875–1890), II, pp.95–96; trl. Rutherford (1995), p.146.

by analogy that there exists a set y that is related to the set x by $P(y,x)$, he needs, in the domain to which the analogy is drawn, the existence of a monad (or collection of monads; see below) y for the monad x to relate to. Without a principle of harmony, that existence would not be guaranteed.

The following step is to see that, more specifically, properties that are essential correspond to perceptions that are distinct. Leibniz understands by necessary properties those that admit of finite analysis into primitive ones (section 33). They cannot involve confused perceptions, as those combine many perceptions into one in such a way that there is no complete, finite analysis into distinct perceptions. In the *Monadology*'s twin, the paper *Principles of Nature and Grace* from the same year, 1714, Leibniz states in section 13 that "Our confused perceptions are the result of the impressions which the whole universe makes upon us."[82] They therefore correspond to, or express, contingent truths (*Monadology* section 36). God knows contingent truths a priori, but not by demonstration. An infinite demonstration is impossible according Leibniz, as such an object would form an infinite whole, which he believed could not exist; rather, God knows contingent truths by a (direct) "infallible vision."[83] There is a continuum of qualities of perception, of which complete distinctness is one extreme. The more distinct a perception is, the more it contributes to the individuality of a monad, to the point where complete distinctness corresponds to essential properties.

In particular, a relational property of a monad that is part of its essence demands that its expression of all relata is clear and distinct. It follows that, as Gödel says, it cannot be an essential property of any monad x to stand in a relation P to the universe. A monad may well stand in a relation P to the universe but this will then not be an essential property of the monad. Suppose that one finds a necessarily true proposition A that says of a created monad x that it stands in a relation P to the universe. For Leibniz, that A is a necessary truth means that A expresses an essential property of x. For the reason just given, what specifically makes A true cannot involve the whole universe but only a proper part of it. Hence, A is equivalent to a proposition B that says that x stands in a necessary relation P to just part of the universe. By the principle of harmony, a part of the universe such as perceived by x indeed exists. Thus we have arrived at what we have called a 'reflection principle for (created) monads.'[84] As noted above (p.15), this argument

---

[82]Leibniz (1875–1890), VI, p.604.

[83]"On Freedom" (1689), Leibniz (1973), p.111.

[84]This principle is of course closely related to the ancient and medieval idea that things are known according to the capacities of the knower, and that hence a lower being's knowledge of a higher being is necessarily incomplete. A difference between that idea and reflection is that only the latter explicitly concludes to the existence of a third object (with a certain property). But in a formulation of Odo Reginaldus from around 1243–1245, that conclusion is more or less present: "How can a finite being reach the infinite? About this, some others have said that God will present himself to us moderated, and that he will show himself not in his essence, but in a creature," Côté (2002), p.78, trl. mine. ("Quomodo

does not yield the conclusion that there is a monad to which x is related, but that there is a part of the universe (in the sense of a collection of monads that does not express the universe perfectly) to which it is related; we also saw why, in the presence of Von Neumann's axiom, this suffices for Gödel's analogy. If an individual monad z such that x stands in the same relation to z as it does to the collection of monads y is possible, then it could be argued that God would go on actually to create that monad z, on the ground of a principle of maximality or plenitude (which is a form of the principle of harmony).[85]

At this point, the following might seem to be a quick argument against Gödel's analogy. Reflection for monads depends on God's will (namely, on His choice to create a universe that is harmonious), and is in that sense contingent; reflection for sets, on the other hand, is supposed to be a necessary principle. But then these two forms of reflection cannot be true on the ground that both instantiate one and the same general principle. However, this argument does not succeed, because the general principle might be (or could be made) conditional on harmony: 'For all harmonious universes, ...' In the case at hand, all that harmony amounts to is the requirement that, if an object in a universe has a relational property, the relata also exist in that universe. For the universe of monads this needs, as we saw, some argument, while for the universe of sets it seems trivial. But for the applicability of the general principle the reason why a universe is harmonious would not matter, only that it is.

Instead, the argument against Gödel's analogy proceeds from the fact that, in contrast to reflection for monads, it is consistent with the monadology that no reflection principle for sets holds. This is because the monadology poses no metaphysical constraints on the essential mathematical properties that a set (or any object of pure mathematics) can have. The explanation for this is as follows.

As we saw in section 2, the objects of pure mathematics are, for Leibniz, entirely mental objects, and have their primary and original existence in God's mind. As a consequence, the existence of relations between pure sets or collections (in particular, V) will have no foundation in a created monad. Relations between pure sets or collections are, ontologically, relations between God and himself. Relations have their ultimate reality in God's being able to think them. But, contrary to the case of created substances, for Leibniz there are no intrinsic limitations to

---

potest finitum attingere ad infinitum? Propter hoc dixerunt alii quod deus contemperatum se exhibebit nobis, et quod ostendet se nobis non in sua essentia, sed in creatura." – Odo then comments that this opinion has fallen from favour ("Sed hec opinio recessit ab aula"), which, theologically, is not surprising.). From here it is only a small step to: "Suppose creature A has a perception of God. Then God is capable of making a creature B such that A's perception cannot distinguish between God and B." The argumentation here is reminiscent of continuity arguments. Côté's monograph (2002) is an invaluable analysis of the medieval discussion of finite beings' knowledge of an infinite God.

[85]Leibniz (1923–), VI, iii, 472: "After due consideration I take as a principle the Harmony of things, that is, that the greatest amount of essence that can exist does exist." (1671), trl. Mercer (2001), pp.413–414.

God's thinking other than non-contradiction. "Possible things are those which do not imply a contradiction," he says,[86] and God thinks all possibilities:

> "The infinity of possibles, however large it may be, is not larger than that of the wisdom of God, who knows all possibles."[87]

What is true in mathematics, and in particular what relations can obtain between mathematical objects, depends only on the Principle of Contradiction. The Principle of Sufficient Reason and its consequences have no influence on what is or is not the case in mathematics. Leibniz explains this in his second letter to Clarke, from 1715:

> "The great foundation of mathematics is the *principle of contradiction or identity*, that is, that a proposition cannot be true and false at the same time and that therefore A is A and cannot be non-A. This single principle is sufficient to demonstrate every part of arithmetic and geometry, that is, all mathematical principles. But in order to proceed from mathematics to natural philosophy, another principle is requisite, as I have observed in my *Theodicy*; I mean the *principle of a sufficient reason* [...] By that principle, viz., that there ought to be a sufficient reason why things should be so and not otherwise, one may demonstrate the being of a God and all the other parts of metaphysics or natural theology."[88]

Leibniz says this to support his contention that the mathematical principles of the materialist philosophers are the same as those of Christian mathematicians, the difference between them rather being the metaphysical one that the Christians admit immaterial substances. As Leibniz sees it, the truths of metaphysics (i.e., the principles specifically about monads and their relations to one another) all follow from the principle of sufficient reason (together with the principle of contradiction), but the principle of contradiction is prior to the principle of sufficient reason. In particular, then, sufficient reason and its consequences are compatible with any relation that obtains between pure possibilities. As a special case, no metaphysical principle constrains what is true in pure mathematics. This idea one finds in both early and late Leibniz. For example, the young Leibniz wrote to Magnus Wedderkopf (May 1671),

> "No reason can be given for the ratio of 2 and 4 being the same as that of 4 and 8, not even in the divine will. This depends on the essence itself, or the idea of things. For the essences of things are numbers, as it were, and contain the possibility of beings which God does not make as he does existence, since these possibilities or ideas of things coincide rather with God himself."[89]

---

[86] Leibniz to Joh. Bernoulli, February 21, 1699, Leibniz (1849–1863), III, p.574: "Possibilia sunt quae non implicant contradictionem."

[87] Leibniz (1710), section 225: "L'infinité des possibles, quelque grande qu'elle soit, ne l'est pas plus que celle de la sagesse de Dieu, qui connaît tous les possibles."

[88] Leibniz (1969), pp.677–678. Also section 9 in the fifth letter to Clarke Leibniz (1969), p.697, and Leibniz (1710), section 351.

[89] Leibniz (1969), p.146.

And, much later, in a letter to Pierre Varignon of June 20, 1702,

"Entre nous je crois que Mons. de Fontenelle, qui a l'esprit galant et beau, en a voulu railler, lorsqu'il a dit qu'il vouloit faire des elemens metaphysiques de nostre calcul."[90]

As Michel Fichant has concluded,

"The idea of a metaphysics of the calculus of the infinite, or of a metaphysical transposition of a consideration on the calculus of the infinite, is entirely alien to Leibniz; whenever someone ventured in that area, he has always objected to it."[91]

This absence of a metaphysical constraint on mathematical truth implies that no description or reasoning in purely metaphysical terms can lead us to the discovery of an underlying general principle that would imply that reflection holds for sets, too, as such a metaphysical description will be equally compatible with the falsehood of reflection for sets. Yet, Gödel's analogy argument in effect precisely attempts to draw attention to a general principle in this way. Gödel first describes a purely metaphysical fact, namely the reflection principle for monads, and then arrives at the desired mathematical conclusion by, as he says, "moving from monads to sets." This analogy argument therefore fails. Of the reason for this failure, i.e., the fact that metaphysical principles do not constrain pure possibilities,[92] two further consequences should be noted. First, adding, in particular, a metaphysical principle that somehow corresponds to the restriction in Gödel's formulation of reflection principle (2) on p.10 above that the set-theoretical properties to be reflected should be *structural* will not help making the analogy work. A second consequence is that the monadology will not suggest a *disanalogy* with the reflection principle for sets either. No truth about monads and their relations

---

[90] Leibniz (1849–1863), IV, p.110.

[91] Fichant (2006), pp.29–30: "L'idée d'une métaphysique du calcul de l'infini ou d'une transposition métaphysique d'une réflexion sur le calcul de l'infini est totalement étrangère à Leibniz ; il l'a toujours récusée chaque fois que quelqu'un s'est aventuré dans ces parages." A few lines further on, he writes: "It is true that he says in a famous letter to Varignon that 'the real never fails to be perfectly governed by the ideal and the abstract', on account of which, in effect, mathematical calculations are applicable to nature, but, basically, to nature inasmuch as the real in question is at the level of the phenomena, not at that of the substances." ("Il est vrai qu'il dit dans une lettre célèbre à Varignon, que 'le réel ne laisse pas de se gouverner parfaitement par l'idéal et l'abstrait', ce qui fait que, effectivement, les calculs mathématiques sont applicables à la nature, mais, au fond, à la nature pour autant que le réel dont il est alors question se situe sur le plan du phénomène, et non sur celui des substances.") A curious exception to Leibniz's advocated practice of keeping metaphysical principles out of purely mathematical arguments occurs in his attempts to show that absolute space is Euclidian, which appeal to the principle of sufficient reason. For a full discussion of this exception, see De Risi (2007), pp.252–264.

[92] One of Leibniz' manuscripts of around the time of the *Monadology* is titled "The metaphysical foundations of mathematics", Leibniz (1969), pp.666–674. But in it, Leibniz actually proceeds by defining and developing pure concepts; it is metaphysical on account of the great generality of them. An example is his argument for the proposition that the whole is always greater than the part. But there is no mention whatsoever of monads and the principles governing them and their relations, and therefore not metaphysical in that sense.

can contradict mathematics, for on Leibniz' conception, God's acts of creation are (voluntary) acts of applying mathematics.⁹³

The argument against Gödel's analogy does not depend on Leibniz' specific construal of mathematical possibility in terms of non-contradiction; what it depends on is the more general condition that the notion of pure possibility that defines mathematical truth is a boundary condition on the possible worlds out of which the metaphysical principles select one. Any notion of mathematical possibility that guarantees invariance of mathematical truth with respect to possible worlds will satisfy this condition.⁹⁴

Generally speaking, a successful analogy from a state of affairs in one domain to a state of affairs in another may or may not presuppose that the first domain actually exists. The only function of the description of a state of affairs in the first domain is to suggest to us the relevant general principle governing it, so that we can apply that to the second domain. Such a principle may well hold in merely possible or fictional domains as well as in actual ones. Gödel's analogy argument may or may not presuppose that the monadology is true. That would seem to depend on whether the general principle required should involve notions specific to the monadology or not. The reason just presented why the analogy is ineffective does not turn on the answer to this question, however, for it was argued that there can be no such principle anyway. This also means that, if one makes the assumption that Leibniz' monadology (or something sufficiently close to it) is the

---

⁹³"God makes the world while calculating and exercising knowledge" ("Cum Deus calculat et cogitationem exercet, fit mundus"), Leibniz (1875–1890), VII, p.191n, and "Necessity in geometry is absolute, but it follows that this is also the case in physics, because the supreme Wisdom, who is the source of things, acts as the most perfect geometer and observes harmony" ("Absolutae est necessitatis in Geometria, sed tamen succedit et in Physica, quoniam suprema Sapientia, quae fons est rerum, perfectissimum Geometram agit et Harmoniam observat"), Leibniz (1849–1863), VI, p.129. (The notion of absoluteness here must be a wider one than the one in Leibniz' letter to Des Bosses, quoted on p.20 above, from which harmony is explicitly excluded.) For general discussion of this idea, see Osterheld-Koepke (1984), pp.138–144. The idea also contributes to an explanation of the following observation by Gödel from 1942, Notebook *Max VI*, p.380: "The principle that every math[ematical? metaphysical?] proposition has a generalisation for arbitrary higher cardinality (but not the other way around) expresses one of the most general properties of the structure of the world. Namely: Everything is mirrored in everything. (The symbol and the reference are structur[ally] the same?) God created man to His likeness. The same thing appears at different levels. Here we have an 'unfolding'." ("Das Prinzip, daß jeder math[ematische? metaphysische?] Satz eine Verallgemeinerung für beliebig höhere Mächtigkeit hat (aber nicht umgekehrt) drückt eine der allgemeinsten Eigenschaften des Aufbaus der Welt aus. Nämlich: Alles spiegelt sich in allem. (Das Symbol und die Bedeutung sind struktur[ell] gleich?) Gott schuf den Menschen sich zum Bild. Dasselbe erscheint auf verschiedene Niveaus. Es handelt sich um eine 'Entfaltung'.") Compare *Monadology* section 83.

⁹⁴Note that for Leibniz, what makes mathematical truths true has nothing to do with possible worlds, only with the principle of contradiction. For an argument that the notion of possibility that defines mathematical truth in Husserl's transcendental idealism satisfies the condition mentioned, see Van Atten (2001). The particular relevance of this fact is that after 1959 Gödel adopted Husserl's transcendental idealism as a means to develop Leibniz' monadology scientifically. See the Concluding remark, below.

true metaphysics, there is no direct argument either: knowing the details of exactly how sets fit into this metaphysics yields no additional means to determine the truth value of the reflection principle. Both the analogy argument and a direct argument will fail for the same reason, namely, that in Leibniz' system the specifically metaphysical principles do not imply constraints on what can be true about pure sets and collections. More generally, as we have seen, Leibniz' specifically metaphysical principles do not imply constraints on what can be true in any part of pure mathematics. The present considerations on Gödel's analogy argument and on the possibility of a direct argument are therefore not really specific to sets and reflection, and can be expected to have wider application.

In the light of the absence of implied metaphysical constraints on mathematics, it is not surprising that when Leibniz attempts to show that there can be no infinite wholes, he proceeds from logical truths and not from metaphysics or properties of minds. Contrast this to, for example, Brouwer, who based his idea that in mathematics there exist only potentially infinite constructions (and hence no constructed infinite wholes) not on a conceptual argument but on an observation about the human mind.

4.3 'Medieval ideas'

After having presented the analogy with the monadology, Gödel adds that "according to medieval ideas, properties containing V or the world would not be in the essence of any set or monad." As the reflection principle for monads follows from the monadology itself, and the analogy should then directly lead to the reflection principle for sets, this remark on medieval ideas does not seem to play a role in the argument. It seems rather an afterthought, a corraboration of the argument and its conclusion from medieval quarters.

A characteristically medieval idea (in the Christian world) is that the world and its creation are radically contingent. If the essence of any object in the world would involve the whole world, that essence might be taken to put limits on that contingency, and hence on God's freedom in creating the world. A related point is that if the essence of an object would involve the world, understood as the totality of all actual objects, it would in particular involve its own existence, but for the medievals this is only the case for God. To the extent however that one is looking for medieval ideas that could be applied to set theory, where truths are necessary and contingency plays no role whatsoever, this seems not the right suggestion for what Gödel may have had in mind.

The only idea I have been able to find that does not depend on contingency would be the idea that 'being' is what medieval philosophers called a transcendental notion. This means that the notion of 'being' (ens), and for example others such 'one' or 'true,' transcend the categories into which reality can be classified because they are too general notions to define a category. The extension of the concent of being coincides with, or (if one assumes God exists but does not fall under the categories) even properly includes, the extensions of the categories combined. Aristotle already recognized the existence of such notions (*Metaphysics* 1003b25, 1061a15). The idea is therefore not medieval in the sense of having been introduced in the Middle Ages; but it is typically medieval in that the development of theories about transcendentals did not begin until then. The first systematic treatment of transcendentals is taken to be *Summa de Bono* by Philip the Chancellor, written between 1228 and 1236; but the best known passages dealing with this notion are those in Aquinas' *De Veritate* (1256–1259) and the *Summa Theologica* (1265–1272).[95]

Aquinas specifies that "the individual essence of an object is what is given by the definition [of that object]."[96] In turn, that definition consists in a specification of the genus of the object and of the specific differences that distinguishes it from other objects of the same genus. The argument that being cannot be a genus is the following: "Every genus has differences distinct from its generic essence. Now no difference can exist distinct from being; for non-being cannot be a difference."[97] The idea is that genera and differences serve to distinguish the objects that exist from one another, and hence correspond to asymmetries between them; however, no two objects that both have being can be related to being asymmetrically. Therefore, on the Aristotelian model of definitions, the concept of being cannot contribute to the definition of any object. If one understands by 'the world' 'all that has being,' this means that the essence of no object involves the world. Indeed, Aquinas calls the multitude that results from dividing being according to all its forms the 'transcendent multitude.' points out that like being itself it is not a genus, and distinguishes this from 'numerical multitudes' (Aquinas (1265-1274), I, 30, a.3).

Leibniz also recognizes that 'being' is a transcendental notion. Usually he refers to the characteristic property of transcendentals that they are all convertible with being: that is, the transcendental terms (e.g., being, one, true) differ from one another intension but not in extension. To Des Bosses, Leibniz wrote on February 14, 1706: "I agree with you that being and one are convertible;"[98] and some twenty years earlier, on April 30, 1687, to Arnauld:

> "I regard as an axiom this proposition of which the two parts differ only by their emphasis, namely, that what is not really *one* being is not really one *being* either. It has always been believed that one and being are reciprocal."[99]

---

[95] From notes in his archive, it is known that Gödel read works of Aquinas.

[96] Aquinas (1265-1274), I, 29, a.2 ad 3: "essentia proprie est id quod significatur per definitionem."

[97] Aquinas (1265-1274), I, 3, a.5: "omne enim genus habet differentias quae sunt extra essentiam generis; nulla autem differentia posset inveniri, quae esset extra ens; quia non ens non potest esse differentia." See also Aristotle's argument in *Metaphysics* 998b21–28.

[98] "Ens et unum converti tecum sentio." Leibniz (1875–1890), II, p.300.

[99] "Je tiens pour un axiome cette proposition identique qui n'est diversifiée que par l'accent, sçavoir

In this last sentence, Leibniz makes an implicit reference to Aristotle and the scholastics. I do not know whether it was this reference that led Gödel to consider medieval philosophy in this context. Be that as it may, Leibniz' reason for considering 'being' a transcendental was different from that of the scholastics.[100] Where the scholastics considered the notion of being as it applies to an object in the actual world, Leibniz considered the notion of being as it applies to a possible object. This notion corresponds to that of being one, as a possible object is determined by one complete concept. Leibniz' conception in terms of purely possible as opposed to actual beings (in the world) comes closer to what Gödel says when he invokes these 'medieval ideas,' as he wants to include sets, which for Leibniz are always possible but, being 'incomplete' (i.e., never concrete), never actual objects.

The conception of being (or the world) as a transcendental, whether construed in the scholastic or in the Leibnizian sense, would indeed have the consequence that Gödel mentions, namely that no essence of a substance involves the world. But the reason why this is so would hardly be suggestive of the reflection principle. The argument from the transcendental nature of being would go through regardless of the exact properties of the universe (or of the realm of possible objects), for it depends only on an intrinsic characteristic of Aristotelian definitions. No aspect of inexhaustibility or inconceivability of the universe plays a role in it. It would seem, then, that the transcendental nature of being is compatible with both the failure and the correctness of the reflection principles for sets and for monads.

5 Concluding remark

As we have seen, Leibniz' monadology is compatible with whatever the truths of pure mathematics may turn out to be. A positive consequence of this fact is that, should a purely conceptual or internal justification for the reflection principle be found[101] this will fit into the monadology immediately. But Gödel was also interested in yet another approach. The idea here is to deepen Leibniz' monadology by considering that concepts and possibilities, though not created by God, are constituted in his mind. To Hao Wang, Gödel once complained that "some of the concepts, such as that of possibility, are not clear in the work of Leibniz",[102] and he stressed that "Leibniz had not worked out the theory."[103] As a means to develop Leibniz' philosophy, Gödel came to embrace and recommend Husserl's transcendental phenomenology from 1959 onward.[104] The suggestion, then, is that a phe-

---

que ce qui n'est pas veritablement *un* estre, n'est pas non plus veritablement un *estre*. On a tousjours crû que l'un et l'estre sont des choses reciproques." Leibniz (1875–1890), II, p.97.

[100] See also Kaehler (1979), p.119n.39.
[101] James Van Aken (1986), p.1001, observes that such an internal argument would be 'a coup.'
[102] Wang (1996), p.310.
[103] Wang (1996), p.87.
[104] For an analysis of Gödel's turn to phenomenology, see Van Atten & Kennedy (2003). He praised Dietrich Mahnke's *Neue Monadologie* (1917), a version of Leibniz' monadology written from a largely

nomenological analysis of the types of acts and powers involved in the constitution of possibilities may lead to sufficient clarification of the notion of mathematical possibility to lead to a (direct) justification of the reflection principle.[105]

Acknowledgments

Earlier versions of this paper were presented at a colloquium on Gödel's philosophy, Boston University, February 27, 2006; in the seminar of the philosophy department at Seattle University, March 2, 2006; in the IHPST seminar on the philosophy of science, ENS, March 20, 2006; at the University of Leuven, May 5, 2006; at the international symposium "Gödel: the texts", Lille, May 18–20, 2006; at the REHSEIS seminar, Université Paris 7, June 19, 2006; and at the VIIIth International Congress of Ontology, San Sebastián, September 29–October 3, 2008. I thank the audiences for their questions, comments and criticisms. I have benefited from exchanges with Paul Benacerraf, Leon Horsten, Hidé Ishiguro, Juliette Kennedy, Georg Kreisel, Nico Krijn, Göran Sundholm, Robert Tragesser, and Jennifer Weed. The Institute for Advanced Study, Princeton, kindly permitted to quote from Gödel's notebooks. For the transcriptions from the Gabelsberger shorthand I am grateful to Robin Rollinger, who in turn benefited from earlier transcriptions generously provided by Cheryl Dawson; microfilms of the relevant pages of the notebooks were kindly made available to Rollinger for this purpose by Gabriella Crocco.

BIBLIOGRAPHY

Ackermann, W. (1956). "Zur Axiomatik der Mengenlehre", *Mathematische Annalen*, 131:336-345.

Adams, R.M. (1983). "Divine necessity", *Journal of Philosophy*, 80:741–751.

Adams, R.M. (1994). *Leibniz. Determinist, Theist, Idealist*, Oxford University Press, Oxford.

van Aken, J. (1986). "Axioms for the set-theoretic hierarchy", *Journal of Symbolic Logic*, 51:992–1004.

Aquinas (1265-1274). *Summa Theologiae*. In: *S. Thomae de Aquino Opera Omnia Iussu Impensaque (Leonis XIII P.M. edita)*, vols.4–12. Roma, 1888-1906.

Arrigoni, T. (2007). *What is Meant by V? Reflections on the Universe of All Sets*, Mentis, Paderborn.

van Atten, M. (2001). "Gödel, mathematics, and possible worlds", *Axiomathes*, 12(3–4):355–363.

van Atten, M. and Kennedy, J. (2003). "On the philosophical development of Kurt Gödel", *The Bulletin of Symbolic Logic*, 9(4):425–476.

---

phenomenological point of view, as 'vernünftig!,' Van Atten & Kennedy (2003), p.457.

[105] As Gödel (1961), pp.383–385, suggests for mathematical axioms in general.

van Atten, M. and van Dalen, D. (2002). "Arguments for the Continuity Principle", *The Bulletin of Symbolic Logic*, 8(3):329–347.

Benardete, J.A. (1964). *Infinity. An Essay in Metaphysics*, Clarendon Press, Oxford.

Cantor, G. (1932). *Gesammelte Abhandlungen mathematischen und philosophischen Inhalts* ed. E. Zermelo. Springer, Berlin.

Castañeda, H.-N. (1976). "Leibniz's syllogistico-propositional calculus", *Notre Dame Journal of Formal Logic*, XVII(4):481–500.

Côté, A. (2002). *L'infinité divine dans la théologie médiévale (1220–1255)*, Vrin, Paris.

Devillairs, L. (1998). *Descartes, Leibniz. Les vérités éternelles*, Presses Universitaires de France, Paris.

Fichant, M. (2006). "La dernière métaphysique de Leibniz et l'idéalisme", *Bulletin de la société française de philosophie*, 100(3), pp.1–37.

Friedman, J.I. (1975). "On some relations between Leibniz' monadology and transfinite set theory", in K. Müller, H. Schepers, and W. Totok, editors, *Akten des II. Internationalen Leibniz-Kongresses. Hannover, 19–22 Juli 1972*, volume XIV (Teil 3) of *Studia Leibnitiana Supplementa*, Wiesbaden.

Gödel, K. (1961). "The modern development of the foundations of mathematics in the light of philosophy", Manuscript (published in Gödel (1995)pp. 374–387).

Gödel, K. (1990). *Collected Works. II: Publications 1938–1974*. (eds. S. Feferman et al.), Oxford University Press, Oxford.

Gödel, K. (1995). *Collected Works. III: Unpublished essays and lectures*. (eds. S. Feferman et al.), Oxford University Press, Oxford.

Gödel, K. (2003). *Collected Works. IV: Correspondence A-G*. (eds. S. Feferman et al.), Oxford University Press, Oxford.

Gödel, K. (2003). *Collected Works. V: Correspondence H-Z*. (eds. S. Feferman et al.), Oxford University Press, Oxford.

Grattan-Guinness, I. (2000). *The Search for Mathematical Roots 1870–1940. Logics, Set Theories and the Foundations of Mathematics from Cantor through Russell to Gödel*, Princeton University Press, Princeton.

Hallett, M. (1984). *Cantorian Set Theory and Limitation of Size*, Clarendon Press, Oxford, 1984.

Hellman, G. (1989). *Mathematics Without Numbers. Towards a Modal-Structural Interpretation*, Clarendon Press, Oxford.

Ishiguro, H. (1990). *Leibniz's Philosophy of Logic and Language (2nd ed.)*, Cambridge University Press, Cambridge.

Jolley, N. (1990). *The Light of the Soul. Theories of Ideas in Leibniz, Malebranche, and Descartes*, Clarendon, Oxford.

Kaehler, K. (1979). *Leibniz. Der methodische Zwiespalt der Metaphysik der Substanz.* Meiner, Hamburg.

Kant, I. (1783). *Prolegomena zu einer jeden künftigen Metaphysik, die als Wissenschaft wird auftreten können*, J.F. Hartknoch, Riga.

Kennedy, J., van Atten, M. (2004). "Gödel's modernism: on set-theoretic incompleteness", *Graduate Faculty Philosophy Journal*, 25(2), pp.289–349. Erratum facing page of contents in 26(1), 2005.

Leibniz, G.W. (1665). *De conditionibus*, 1665. French translation *Des conditions*, ed., trl. P. Boucher. Vrin, Paris, 2002.

Leibniz, G.W. (1686). "Discours de métaphysique", in Leibniz (1875–1890), vol.2, pp.12–14 and vol.4, pp.427–463.

Leibniz, G.W. (1695). "Dialogue effectif sur la liberté de l'homme et sur l'origine du mal", in G.W. Leibniz, *Système nouveau de la nature et de la communication des substances et autres textes 1690–1703* (ed. C. Frémont), Flammarion, Paris, 1994, pp.49–58.

Leibniz, G.W. (1705). "Nouveaux essais sur l'entendement", in Leibniz (1875–1890), vol.5, p.39–509.

Leibniz, G.W. (1710). "Essais de theodicée sur la bonté de Dieu, la liberté de l'homme et l'origine du mal", in Leibniz (1875–1890), vol.6, pp.1–436.

Leibniz, G.W. (1849–1863). *Leibnizens mathematische Schriften* ed. C.I. Gerhardt, 7 vols., Asher (later Schmidt), Berlin (later Halle). Cited by volume and page.

Leibniz, G.W. (1875–1890). *Die philosophischen Schriften von Gottfried Wilhelm Leibniz*, ed. C.I. Gerhardt, 7 vols., Weidmann, Berlin. Cited by volume and page.

Leibniz, G.W. (1903). *Opuscules et fragments inédits* ed. L. Couturat, Presses Universitaires de France, Paris.

Leibniz, G.W. (1923–). *Sämtliche Schriften und Briefe*, Akademie der Wissenschaften, Darmstadt, Leipzig and Berlin. Cited by series, volume, and page.

Leibniz, G.W. (1969). *Philosophical Papers and Letters (2nd ed.)* ed., trl. L.E. Loemker, D. Reidel Publishing Company, Dordrecht.

Leibniz, G.W. (1973). *Philosophical Writings.* ed. G.H.R. Parkinson, trl. M. Morris and G.H.R. Parkinson, J.M. Dent and Sons, London.

Leibniz, G.W. (1991). *G.W. Leibniz's Monadology. An Edition for Students*, ed. N. Rescher, University of Pittsburgh Press, Pittsburgh.

Leibniz, G.W. (2005). *The Labyrinth of the Continuum: Writings on the Continuum Problem, 1672–1686* ed., trl. R.T.W. Arthur, Yale University Press, New Haven.

Lévy, A. (1960a). "Axiom schemata of strong infinity in axiomatic set theory", *Pacific Journal of Mathematics*, 10, pp.223–238.

Lévy, A. (1960b). "Principles of reflection in axiomatic set theory", *Fundamenta Mathematicae*, 49, pp.1–10.

Maddy, P. (1990). *Realism in mathematics*, Clarendon Press, Oxford.

Mahnke, D. (1917). *Eine neue Monadologie*, Reuther & Reichard, Berlin. Kantstudien Ergänzungsheft, 39.

Mates, B. (1986). *The Philosophy of Leibniz. Metaphysics and Language*, Oxford University Press, Oxford.

Mates, B. (2001). *Leibniz's Metaphysics: Its Origins and Development*, Cambridge University Press, Cambridge.

Maurer, A. (1973). "Analogy in patristic and medieval thought", in P. Wiener, editor, *Dictionary of the History of Ideas*, vol.1, pp.64–67. Charles Scribner's Sons, New York.

Mielants, W. (2000). "Believing in strongly compact cardinals", *Logique et Analyse*, 43(171–172), pp 283–300.

Mugnai, M. (1992). *Leibniz' Theory of Relations*, Franz Steiner, Stuttgart. Studia Leibnitiana Supplementa 28.

von Neumann, J. (1925). "Eine Axiomatisierung der Mengenlehre", *Journal für die reine und angewandte Mathematik*, 154, pp.219–240.

Osterheld-Koepke, M. (1984). *Der Ursprung der Mathematik aus der Monadologie*, Haag und Herchen, Frankfurt/Main.

Parsons, C. (1977). "What is the iterative conception of set?", in R.E. Butts and J. Hintikka, editors, *Logic, Foundations of Mathematics and Computability Theory*, pp.335–367. D. Reidel Publishing Company, Dordrecht. Page references to the reprint in Parsons (1983), pp.268–297.

Parsons, C. (1983). *Mathematics in Philosophy. Selected Essays*, Cornell University Press, Ithaca.

Reinhardt, W.N. (1974). "Remarks on reflection principles, large cardinals, and elementary embeddings", in T. Jech, editor, *Axiomatic Set Theory*, volume 13 (part II) of *Proceedings of Symposia in Pure Mathematics*, pp.189–205, American Mathematical Society, Providence RI.

de Risi, V. (2007). *Geometry and Monadology. Leibniz's Analysis Situs and Philosophy of Space*, Birkhäuser, Basel.

Roth, D. (2002). *Cantors unvollendetes Projekt. Reflektionsprinzipien und Reflektionsschemata als Grundlagen der Mengenlehre und großer Kardinalzahlaxiome*, Herbert Utz, München.

Rutherford, D. (1995). *Leibniz and the Rational Order of Nature*, Cambridge University Press, Cambridge.

Tait, W. (1998). "Zermelo's conception of set theory and reflection principles", in M. Schirn, editor, *The Philosophy of Mathematics Today*, pp.469–483. Clarendon Press, Oxford.

Wang, H. (1974). *From Mathematics to Philosophy*, Routledge and Kegan Paul, London.

Wang, H. (1977). "Large sets", in R.E. Butts and J. Hintikka, editors, *Logic, Foundations of Mathematics and Computability Theory*, pp.309–333. D. Reidel Publishing Company, Dordrecht.

Wang, H. (1987). *Reflections on Kurt Gödel*, MIT Press, Cambridge, MA.

Wang, H. (1996). *A Logical Journey From Gödel to Philosophy*, MIT Press, Cambridge, MA.

Yourgrau, P. (1999). *Gödel Meets Einstein*, Open Court, Chicago.

Zermelo, E. (1930). "Über Grenzzahlen und Mengenbereiche. Neue Untersuchungen über die Grundlagen der Mengenlehre", *Fundamentae Mathematicae*, 16, pp.29–47.

Arguing for Inconsistency: Dialectical Games in the Academy

BENOÎT CASTELNÉRAC, MATHIEU MARION

ABSTRACT. In this paper we propose, along the model of modern dialogical games (in the school of Lorenzen), a set of structural and particle rules, for dialectical games as played at the time when both Plato and Aristotle were active in the Academy. This is part of an effort to understand how these games functioned, as a prerequisite for an interpretation of Plato's philosophy, as well as a proper understanding of pre-Socratic philosophies. We argue that dialectical games are consistency-management games, in which, for a given thesis A asserted at the outset by the proponent, the opponent (usually Socrates) argues for the inconsistency of a set of beliefs of the proponent that includes A, and nothing more. We thus criticize Vlastos' influential interpretation of the Socratic *elenchus*, according to which Socrates meant to 'refute' the proponent, i.e., to show that A is false and, therefore, that $\neg A$ is true. Finally, we offer some suggestions concerning the point of playing these games and the birth of logic in the context of dialectical games.

> Socrates: So now please do whichever of these you like:
> either ask questions or answer them.
> Plato, Gorgias 462b

> How did philosophy, from these interactive beginnings,
> develop into what Popper once described as a cult of great
> philosophers preaching Sermons on the Mount?
> Johan van Benthem

1 Introduction: the study of dialectics as games

This paper is part of a larger research programme concerning dialectical games played in Ancient Greece, their origin, their rules, their influence on philosophy, and their role in the origin of logic. By dialectical games, we mean the sort of games exemplified in the dialogues of Plato, with Socrates as the main character trying to 'drive his opponent into an *elenchus*' (ἔλενχος)[1] and for which Aristotle wrote later on his textbook, the *Topics*.[2] These are games where, typically, the proponent of a thesis, say, *A* is asked a series of questions by an opponent - the role usually assumed by Socrates in Plato's dialogues - whose task is to show that, in holding *A*, the proponent will contradict some other beliefs that he also happens to hold, i.e., if successful, the opponent shows that *A* is part of an inconsistent set of premises. In other words, these games can be described as games of 'consistency management,'[3] in accordance with Aristotle's definition at the beginning of his *Topics*, according to which *"we shall ourselves, when standing up to an argument, avoid saying anything contrary to it."*[4] We wish merely to offer here a list of their apparent rules, with a discussion that shows their rationale. We will provide a snapshot of games that were played in the mid-4th century BC, therefore at a time when both Plato and Aristotle where at the Academy. Within the context of our research programme this period should be seen as a *terminus ante quem*, from which one could begin the historical search, going back to Parmenides and Zeno. We will use as primary source the set of rules provided by Aristotle in *Topics* VIII and the numerous meta-discussions about dialectical games in Plato's dialogues, and offer what amounts to a composite picture. We are not, however, denying that there are important differences between Plato and Aristotle that need to be addressed, most of which we will for reason of space gloss over. Although we shall refer often to Aristotle, it is the games as played in Plato's dialogues, from his standpoint, that we wish to understand: we should warn the reader at the outset about distortions introduced by Aristotle - the reasons for these are already an interesting philosophical topic in itself. The development of Plato's thought, between the early 'Socratic' dialogues and later ones such as *Republic* also raises some further difficulties which we do not wish to deny, but that cannot be addressed at this early stage. It suffices to say that we do not think that Plato changed

---

[1] We take this expression from Ryle (1966), p.105. Although we shall use it as a matter of convenience, we should point out that it is far from ideal, as it keeps mum about differences in the active and the passive uses of the verb *elenchein* (ἐλέγχειν); We should perhaps clear these terminological matters at the outset to prevent confusion. '*Elenchus*' refers to the 'method,' here defined through a set of rules by which one 'tests' (*elenchein*) a proponent's ability to defend a given thesis *A* - for the reasons behind our translation of '*elenchus*' by 'test,' see section 2. (Cognate words are also used at times, e.g., *exetazein* (ἐξετάζειν) or 'to examine' at Plato, *Apology* 29e.) The active use of *elenchein* refers to the action of putting a thesis *A* to the test, while the passive use refers to the thesis or its proponent being submitted to the test, e.g., respectively, at Plato, *Charmides* 166c and 166d. But the passive uses of *elenchein* (*elenchesthai*, ἐλέγχεσθαι) *also* include reference to failure to pass the test, e.g., at Plato, *Apology* 17b, *Gorgias* 473b or *Theaetetus* 161a - hence Ryle's translation. Active and passive uses of *exelenchein* (ἐξελέγχειν), 'to go over the test,' 'to bring the test to its end' are also joined in, e.g., Plato, *Euthydemus* 303d; see also *Gorgias* 473e and 522d for uses of the passive form *exelenchesthai* (ἐξελέγχεσθαι).

[2] There are of course other sources, such as Xenophon's *Memorabilia* (IV, 4, §9).

[3] Using here an expression we owe to Dutilh-Novaes (2007).

[4] Aristotle, *Topics*, I, 1, 100a 20-21.

his conception of 'dialectics' from early 'Socratic' dialogues to later ones such as *Sophist* or *Republic*, because it was not his or Socrates' to change at will: They were not the inventor of that practice, they had no patent on it and, to a large extent, they could not modify it as they go along, as if it were their own. We will therefore quote from early and late, in order to show that, indeed, all meta-discussions of the *elenchus* show a remarkable, perfectly explainable, unity. What Plato could do, however, is to reflect on what can be achieved through dialectical games, their limits, and so forth.[5] We merely claim, therefore, that only after one has investigated and obtained a good grasp of the dialectical games played in his days, can the evolution of Plato's thought, e.g., the introduction of the theory of Forms,[6] properly be plotted and understood.

We will study dialectical games from the point of view of dialogical logic, invented and developed by Paul Lorenzen and Kuno Lorenz.[7] In other words, we shall use an approach related to what should be called 'game semantics,'[8] in opposition to the better-known 'proof-theoretical semantics' pioneered by Dag Prawitz and Sir Michael Dummett, or the usual 'truth-conditional semantics' and we wish to explore how far one can look at dialectical games as dialogical games, and we offer the following as proof of the fruitfulness of the 'game semantical' approach: Where other approaches have precious little to say, many crucial dimensions of Plato's dialogues are now thrown into relief which would otherwise pass unnoticed, with deep consequences on our understanding of Plato's views. Our approach will thus raise numerous interesting philosophical issues that cannot be discussed, however, for reasons of space, and we shall merely conclude with some tentative remarks concerning the point of playing these games and the origins of logic.

After some brief comments on the meaning of the word '*elenchus*'[9] and the

---

[5]The cautionary words at *Republic* 539a-540a are good example of this: They do not imply at all that Plato's conception of dialectical games has changed since the early dialogues.

[6]On this point, see Nehamas (1990).

[7]Lorenzen and Lorenz (1978). Some knowledge of dialogical logic is thus presupposed in what follows. For an introduction, see Rückert (2001), Rahman and Keiff (2004), Fontaine and Redmond (2008). Catarina Dutilh Novaes has recently formalized the obligational games of medieval philosophy in Part 3 of Dutilh-Novaes (2007), but our approach will differ from hers, because we believe, without being wedded to the opinion, that, contrary to their medieval counterparts, dialectical games were not regimented enough for a full formalization to be possible; It would induce distortions. So we eschew any use of models and stick only to a description of game rules.

[8]The expression is likely to mislead, as we are not referring to Jaakko Hintikka's game-theoretical semantics, but to the burgeoning game semantics in theoretical computer science, as well as the recent renewal of dialogical logic primarily at the hands of Shahid Rahman and Helge Rückert.

[9]To investigate the role of dialectical games at the origin of logic is not new. To the best of our knowledge, the earliest claim that logic originated in dialectical games is in Ernst Kapp's *Greek Foundations of Traditional Logic* (Kapp (1942)), but it is limited to a discussion of Aristotle, and does not even contain a detailed discussion of the relationship of the *Topics* to the *Analytics*. Furthermore,

principle of non-contradiction, we will simply list what appears to us to be the rules setting up dialectical games, commenting them as we go along. Our main subsidiary scholarly objective will be to refute Gregory Vlastos' influential interpretation of the Socratic *elenchus* according to which Socrates, when driving the proponent of a thesis A to an *elenchus*, was actually arguing for the truth of $\neg A$.[10] This interpretation has been convincingly criticized for its weak textual basis, especially by Hugh Benson,[11] but to our knowledge no one seems to have insisted on the fact that it makes little sense from a logical point of view, i.e., from the point of view of the dialectical games themselves, when the way they actually function is properly understood.

2 *Elenchus* and non-contradiction

The key to a proper understanding of dialectical games is a correct understanding of the meaning of the word '*elenchus*,' which is usually translated by 'refutation.' It may be said that, in driving the proponent of a thesis A into an '*elenchus*,' Socrates has not 'refuted' his interlocutor himself but the thesis A. One can argue, however, that, on philological grounds, this translation of '*elenchus*' by 'refutation' is misleading. This has been shown by James Lesher,[12] who has investigated shifts in meaning of the term from early occurrences in the writings of Homer down to Plato and Aristotle. He has shown that in Homer and other early poets it is connected with the idea of 'shame' and, later, in Pindar and others, it is used as meaning a 'shaming in a contest' or a 'test that reveals the true nature of a thing or person.'[13] Lesher's intention in that paper was, however, merely to argue against translating the occurrence of '*poluderis elenchus*' in Parmenides' Fragment 7 as 'strife-encompassed refutation' or some equivalent, and he claimed that the word '*elenchus*' acquired the meaning of 'refutation' for the first time only in early Platonic dialogues in passages such as *Protagoras* 344b or *Gorgias* 473e.[14] The interpretation of Parmenides is of no immediate concern to us at this stage, but we

---

our approach is clearly distinct, as we wish to make essential use of modern game-theoretic tools in the study of dialectical games. This is not entirely new from the standpoint of dialogical logic, since Lorenzen himself already referred to dialectical games in the very first paper in which he introduced his dialogical logic, "Logik und Agon", Lorenzen (1960), p. 187. There were since numerous other perceptive remarks, e.g., at Kamlah and Lorenzen (1984), p. 142, Lorenz (1973), pp. 355-358, Lorenzen (1987), p. 78, or Krabbe (2006), pp. 666-670, but these were never followed by the sort of investigation of which this paper is the beginning. There are also more recent discussions from the point of view of game semantics. Jaakko Hintikka had some discussion from the standpoint of his own 'interrogative model,' which is derived from is own game-theoretical approach to semantics, see Hintikka (1993) and Hintikka (2007). Our proposal does not rely, however, on anything specific to this approach.

[10] Vlastos (1994), p. 11.
[11] Benson (1995).
[12] Lesher (1984), pp. 3-7 and Lesher (2002), pp. 20-26.
[13] Lesher (1984), p. 7 and Lesher (2002), p. 25.
[14] Lesher (1984), p. 8 and Lesher (2002), p. 26.

disagree about the remark about Plato.

First, there is no reason *not* to believe that earlier elements of meaning - that of a *test* and of the *shame* caused by failing it, as well of the fact that failure *reveals something* about the one who failed - are still present in Plato's dialogues. Indeed, as Lesher himself points out, as late as *The Republic* the meaning of '*elenchus*' appears clearly to owe more to 'test' - failure of military or athletic valour - than to 'refutation,' although it is so translated in the standard translation, which we modify here:

> "Unless someone can distinguish in an account the form of the good from everything else, can survive all [tests] (*elenchon*, ἐλέγχων), as if in a battle, striving to [examine] (*elenchein*, ἐλέγχειν) things not in accordance with opinion but in accordance with being, and one can come through all this with his account still intact, you'll say that he doesn't know the good itself or any other good."[15]

Secondly, the idea of 'shaming in a contest' is of course certainly present in Plato's dialogues: Those who are refuted more often than not get angry, and anger is just the normal reaction when one does not wish to recognize one's own feeling of shame. There are numerous other remarks to that effect in the dialogues, including at *Sophist* 230b to be quoted in a moment. In the *Apology*, Socrates argues that his numerous *elenchi*, and those performed by his followers, made him unpopular (*Apology* 22e), precisely because they caused anger:

> "The result is that those whom [Socrates' followers] question are angry, not with themselves but with me. [...] they would not want to tell the truth, I am sure, that they have been proved to lay claim to knowledge when they know nothing."[16]

Thirdly, the earlier meaning of 'testing that reveals the true nature of a thing or person' is also present in the *elenchi* to which Socrates submitted 'those reputed wise,'[17] in order to try in vain to prove the oracle wrong (*Apology* 21c): These victims are shown through failure to pass the test not to possess the wisdom or knowledge that they are said to possess, therefore not to be wise (*Apology* 21c and 23b); Their true nature is revealed.

This is also consistent with Socrates' concern for the soul, e.g., in the *Apology* (29d, 30a-b and 41e) or the *Sophist* (230d), since an *elenchus* causes a 'cleansing' or *catharsis* of the soul in this that the soul is 'cleansed' as one gets rid of false

---

[15] Plato, *The Republic*, 534b-c. The clause "not in accordance with opinion but in accordance with being" has to do with the introduction of the theory of Forms. Again, this is a topic which we cannot discuss here. To see why the introduction of the theory of Forms does not induce any change in the underlying conception of dialectical games, see Nehamas (1990), pp. 7f.

[16] Plato, *Apology*, 23c-d.

[17] In the *Apology* alone, Socrates lists politicians (21c-22a), poets (22a-c) and craftsmen (22c-e), an incomplete list, as one needs to add rhapsodes in *Ion*, generals in *Laches*, some sophists such as Gorgias and Protagoras in eponymous dialogues, etc.

beliefs about oneself.[18] The *Sophist* is worth quoting at length, as, once more, it shows as a late dialogue a remarkable unity in the underlying conception of the *elenchus* in Plato's dialogues:

> "They [anyone performing an *elenchus*] cross-examine someone when he thinks he's saying something though he's saying nothing. Then, since his opinions will vary inconsistently, these people will easily scrutinize them. *They collect his opinions together during the discussion, put them side by side, and show that they conflict with each other at the same time on the same subjects in relation to the same things and in the same respects.* The people who are being examined see this, *get angry at themselves*, and become calmer towards others. *They lose their inflated and rigid beliefs about themselves that way*, and no loss is pleasanter to hear or has more lasting effect on them. Doctors who work on the body think it can't benefit from any food that's offered to it until what's interfering with it from inside is removed. *The people who cleanse the soul, my young friend, likewise think the soul too, won't get any advantage from any learning that's offered to it until someone [tests it (*elenchon, ἐλέγχων*) and shames it through failure to pass the test (*elenchomenon, ἐλεγχόμενον*)], removes the opinions that interferes with learning, and exhibits it cleansed*, believing that it knows only those things that it does know, and nothing more.
>
> [...] For all these reasons, Theaetetus we have to say that [the test (*elenchon, ἔλεγχον*)] *is the principal and most important kind of cleansing*."[19]

One should note *en passant* that the Visitor, who just spoke here, also describes this very practice of the *elenchus* for the purpose of cleansing the soul as sophistry "of noble lineage" (*Sophist*, 231b). This remark implies that Socrates was not the only practitioner of dialectical games. If so,[20] then this remark at *Sophist* 230b-e informs us about the history of the *elenchus* and it fits perfectly with Lesher's views on the etymology of the word. It thus seems quite normal that Plato would use the expression '*elenchus*' in his dialogues to describe both the dialectical games and their result[21] as *a test that shames those who fail it*, and which *cleanses their soul through that shaming*.

The point of dialectical games is therefore that the proponent of a thesis A has to pass *a test of consistency*: A has to fit into a set of beliefs Γ held by its proponent in such a way that the addition of A to Γ does not render that set inconsistent. If and when the opponent unveils an inconsistency, it is shameful for the proponent. For example, at *Charmides* 169c, Critias becomes silent, because

---

[18] See also Plato, *Meno* 84b, where the point is expressed thus: Anyone who thinks that he does not know is in a better position than those who think falsely that they know.

[19] Plato, *Sophist*, 230b-e. Translation modified, our emphasis.

[20] There is a well-known debate on this point between G. Kerferd (1954), who argues in favour and J. R. Trevaskis (1955), whose contrary opinion is now widespread. This debate cannot be arbitrated here, although it should be clear that we side with Kerferd. In the context of this paper, we should at least point out that Trevaskis actually relies in his argument, at Trevaskis (1955), p.47, on an interpretation of the '*elenchus*' which is conform to that of Vlastos, which we criticize in section 4 below. This is not the only point we would make, but it shows that this debate ought to be revisited. Alexander Nehamas also provides us with many supporting arguments, although he also agrees with Vlastos, see Nehamas (1990), p.7.

[21] See footnote 1, above.

"[...] his consistently high reputation made him feel ashamed in the eyes of the company and he did not wish to admit to me that he was incapable of dealing with the question I had asked him."[22]

In Plato's dialogues, shame is largely linked with the fact that Socrates drives experts into *elenchi* about their own domain of expertise, thus showing them publicly not to know what they pretend to know, i.e., not to be truly competent. So the proper etymology of '*elenchus*' fits Plato's use of that word to describe the conclusion of dialectical games. There appears to be no need, at least as far as Plato's dialogues are concerned - Aristotle being another matter - to introduce any new shade of meaning. (If, as we believe, those games were practiced before Socrates, then this should not come as a surprise.) The translation of '*elenchus*' by '*refutation*' in Plato's dialogues has been agreed upon for quite some time now but, over and above that fact, there is *no* reason to accept it, other than the belief, against which we will argue in section 4, that in winning a dialectical game, Socrates has not just shamed his proponent but that he has somehow 'refuted' the thesis A held by the proponent, i.e., he has shown the A is false and $\neg A$ true. We argue below that this betrays a complete misunderstanding of the very nature of dialectical games; These are only a *test*, one of *consistency*, and the end of which neither the falsity of A, nor the truth of $\neg A$ is proven.

A form of the principle of non-contradiction or $\neg(A \& \neg A)$ is the key to any test of consistency, without it no such test could be devised, so one must show that it was stated explicitly and agreed upon, therefore operative. The origin of this principle is sometimes erroneously attributed to Aristotle, who states it in *Metaphysics* Γ, 3, 1005b 23 and *Prior Analytics* I, 46, 52a 2-3. It is also often attributed to Plato, who also states it in many places, such as *Republic* 436e-437a or *Sophist* 230b, just quoted, where he puts opinions that "conflict with each other at the same time on the same subjects in relation to the same things and in the same respect" at the heart of the *elenchus*. Some scholars would claim that Socrates introduced a new form of *elenchus*, e.g., in his cross-examination of Meletus in the *Apology* 24d-28a.[23] This claim would cohere very well with the idea that the *elenchus* as a 'refutation' also appeared for the first time in Plato's early dialogues, a view that we just rejected, so we have to show further, to uproot this interpretation, that the principle of non-contradiction was already understood and in use prior to Socrates. In order fully to make this case, one would need to study the origins of the principle of non-contradiction (and those of dialectics by the same token) in Parmenides and Zeno, a task that cannot be undertaken here, although one should at least note the obvious, namely that Zeno's method of *reductio ad absurdum* presupposes the principle of non-contradiction. At all events, we need only to point out that, *prior*

---

[22] Plato, *Charmides* 169c.
[23] For example, Dorion (1990).

to Plato's Socrates, Gorgias already made a substantial use of the principle of non-contradiction in a legal context, in his *Defence of Palamedes*:

> "You have accused me in the indictment we have heard of two most contradictory things, wisdom and madness, things which cannot exist in the same man. When you claim that I am artful and clever and resourceful, you are accusing me of wisdom, while when you claim that I betrayed Greece, you accused me of madness. For it is madness to attempt actions which are impossible, disadvantageous and disgraceful, the results of which would be such as to harm one's friends, benefit one's enemies and render one's own life contemptible and precarious. And yet *how can one have confidence in a man who in the course of the same speech to the same audience makes the most contradictory assertions about the same subjects?*"[24]

The point here is that someone who asserts A and $\neg A$ in front of a court, 'in the course of the same speech to the same audience,' loses all credibility. This passage is all the more interesting since the occurrence of the principle of non-contradiction in the *Apology* also is in a legal context, when Socrates replies to his cross-examination:

> "You cannot be believed, Meletus, even, I think, by yourself. The man appears to me, men of Athens, highly insolent and uncontrolled. He seems to have made his deposition out of insolence, violence and youthful zeal. He is like one who composed a riddle and is trying it out: 'Will the wise Socrates realize that I am jesting and contradicting myself, or shall I deceive him and others?' I think he contradicts himself in the affidavit, as if he said: 'Socrates is guilty of not believing in gods but believing in gods,' and surely that is the part of a jester.
>
> Examine with me, gentlemen, how he appears to contradict himself, and you, Meletus, answer us."[25]

Here too, acquiescing to self-contradiction is to lose all credibility in front of the audience. This is the same principle in virtue of which the dialectical games and the resulting *elenchus* draw their strength. In the cases where Socrates is out to 'test' those 'reputed wise,' and when they are driven to an *elenchus*, i.e., to admit both A and $\neg A$ on a topic on which they are reputed for their knowledge, they are shown to all not to know what they are talking about, and they lose their credibility (why they should, we will not enquire into at this stage). With this principle of non-contradiction made explicit, dialectical games may indeed proceed, with the proponent as sole witness, so to speak. Recall here Socrates to Callicles in the *Gorgias*:

> "And yet for my part, my good man, I think it's better to have my lyre or a chorus that I might lead out of tune and dissonant, and have the vast majority of men disagree with me and contradict me, than to be out of harmony with myself, to contradict myself, though I'm only one person."[26]

Or to Polus earlier in the same dialogue:

---

[24] Gorgias (DK) 82 B.11a, 25, our emphasis; translation Dillon & Gergel, pp. 90-91. The *Encomium of Helen* contains a reference to dialectical games, see Gorgias (DK), 82 B.11, 13.

[25] Plato, *Apology* 26e-27b.

[26] Plato, *Gorgias* 482b-c.

For my part, if I don't produce you as a single witness to agree with what I'm saying, then I suppose I've achieved nothing worth mentioning concerning the things we've been discussing. And I suppose you haven't either, if I don't testify on your side, though I'm just one person."[27]

"For I do know how to produce one witness to whatever I'm saying and that's the man I'm having a discussion with."[28]

Still in the same dialogue, Socrates also refuses at 471e to admit that Polus could even begin to 'test' him by listing counter-examples to his claims instead of trying to show that *he*, Socrates, could not consistently maintain them. With these passages, we can see how the principle of non-contradiction is embedded at the very heart of dialectical games, and clearly understood as such.

3 Structural rules for dialectical games

We may now describe a minimal set of rules for these games. We shall propose a 'composite,' taken from the rules one may extract from Aristotle's *Topics* VIII[29] as well as from the numerous meta-discussions in Plato's dialogues, along with what can be gathered as being implicit within the remarkable set of examples that the latter contain. Of course, we need a disclaimer: We do not claim that all examples of dialectical games in Plato follow exactly the following set of rules. The fact that rules are explicitly stated in *Topics* VIII and the existence of numerous other textbooks, now lost, indicates that the Greeks were self-consciously regimenting their discussions. What we do argue for is simply that (for reasons not to be entered into here) the practice of public discussion became regimented in Ancient Greece, and that this regimentation followed in broad outline these rules at the time of Socrates, Plato and Aristotle.

(i) Games always involve two players: a proponent **P** and an opponent **O**.

(ii) A game begins with **O** eliciting from the **P** his commitment to an assertion or thesis *A*.

As already pointed out, Socrates is almost always the opponent **O**. Most dialogues begin with his use of some trick, e.g., flattery or an avowal of ignorance, to elicit from his interlocutor commitment to a thesis such as 'Virtue can be taught,' a commitment which then becomes the topic of the dialogue. However, roles can be reversed during the course of the dialogue and it is suddenly Socrates who has to answer questions. Therefore, as with modern dialogical logic, the players **O** and **P** in rule (i) must be distinguished from their *roles as questioner and answerer* (in

---

[27] Plato, *Gorgias* 472b.
[28] Plato, *Gorgias*, 474a.
[29] Here Le Blond (1939), chap. 1, Moraux (1968) and Bolton (1990) remain good commentaries. See also Slomkowski (1997), chap. 1, which does not cover well the game rules but gives a detailed analysis of the content of admissible initial theses.

the jargon of dialogical logic: as attacker and defender).[30] In some dialogues, e.g., *Gorgias*, numerous role reversals occur, with Socrates starting as the questioner, then being put on the defensive, having to answer, and then becoming the questioner again, etc. One should note, however, that in dialogical logic, role reversal is linked to the rule for negation: if **P** asserts $\neg A$, **O** can only attack it by asserting A, but then **O** must now defend A against attacks from **P**. Such reversals occur in Plato's dialogues, see section 6 below. But role reversal also occurs for other reasons that must be clearly distinguished, e.g., in relation to rule (vi), below. In such cases, the role reversal is temporary, in order to open and close a sub-game in which Socrates answers to objections to his move, so that the game can resume.

It is also often emphasized in meta-discussions in Plato's dialogues that role reversal occurs simply because what is most important in the dialectical games is not just that **P** is shamed and cleansed, but also the discovery of logical relations between theses: As Socrates puts it, if **P** is driven into an *elenchus*, a result is thus obtained of importance to players and the audience alike or for the 'common good,' as Socrates puts it in *Charmides* 166d-e (to be quoted below). For example, towards the beginning of the *Philebus*, Socrates is quite plain about this cooperative endeavour, or 'spirit of enquiry' as Aristotle would later say (*Topics* VIII, 5, 159a, 25):

> "We are not contending here out of love for victory for my suggestion to win or for yours. We ought to act together as allies in support of the truest one."[31]

It does not matter, therefore, if it is Socrates or the interlocutor who takes on the role of defender, and at times Socrates willingly plays the role of the answerer in order to prove that point, e.g., in the *Protagoras*:

> "If Protagoras is not willing to answer questions, let him ask them, and I will answer, and at the same time try to show him how I think the answerer *ought* to answer. When I've answered all the questions he wishes to ask, then it's his turn to be accountable to me in the same way."[32]

In rule (ii), the nature of the thesis A to be defended needs to be clarified. According to Aristotle, at *Topics*, I, 1, 100a 29, such theses are *endoxa* (ἔνδοξα) or 'probable' opinions, which he defines later as that which is "accepted by everyone or by the majority or by the wise - i.e., by all, or by the majority, or by the most notable and reputable of them."[33] Aristotle even says of his *endoxa* that they could

---

[30]For the possibility of switching roles, see Plato, *Protagoras* 338d; for actual cases, see, e.g., *Gorgias*, where numerous changes occur, e.g., at 462b and 470c. Interestingly, at *Euthydemus* 297c Plato portrays Socrates as trying illegitimately to switch roles.

[31]Plato, *Philebus* 14b.

[32]Plato, *Protagoras* 338d.

[33]Aristotle, *Topics* I, 1, 100b 20. Aristotle actually distinguishes between a *protasis* (πρότασις), which resembles this definition (*Topics* I, 10, 104a 8), a *problema* (πρόβλημα) about which there is a lack of unity among, say, the wise (*Topics* I, 10, 104b 2) and a *thesis* (θέσις) or paradoxical opinion

be about any subject presented to us."[34] It looks, therefore as if for both Plato and Aristotle only propositions that can be verified with certainty, on the spot, so to speak, are to be excluded. This is indeed in accordance with passages in Plato's dialogues where it is admitted that some theses, e.g., concerning the number of teeth in Euthydemus' mouth (*Euthydemus* 294c) or the road from Athens to Larissa (*Meno* 97a-b), are such that they can be verified and stand in no need certification by dialectical games. Socrates and his interlocutor almost always debate over theses, often under the form of definitions such as 'Justice is the interest of the stronger,' pertaining to the moral domain, for which there could not be any verification.[35] However, it should be clear that the theses discussed are not limited to the moral domain, since, e.g., ontological and logical theses are discussed in the *Parmenides*, mathematical ones in the *Meno*. (For a further discussion of the point of playing dialectical games for such classes of propositions, see section 7 below.)

The play then proceeds according to these rules:

(iii) The play then proceeds through a series of alternate questions and answers. **O** asks questions such that **P** may give a 'short answer:' ideally, but not necessarily, 'yes' or 'no.'

(iv) Proceeding thus, **O** elicits further commitments from **P**, e.g., commitment to assertions B, C, etc., which can be conceived as added to **P**'s 'commitment store' or 'scoreboard.'[36]

There are numerous passages testifying to rule (iii) in Plato, e.g., *Gorgias* 449b-d. Gilbert Ryle was of the opinion that dialectical games proceed only with what we would call today 'yes/no interrogatives' and that this would exclude 'factual,' 'arithmetical' and 'technical' questions, thus limiting the topics available for discussion to 'conceptual questions.'[37] This remark seems without merit,[38] as '*wh*-interrogatives' also frequently occur; We phrased rule (iii) in order not to exclude them. Aristotle, however, insisted at *Topics* VIII, 2, 158a 16-17 that dialectical games should contain only 'yes/no interrogatives' and showed how transform some '*wh*-interrogatives' into 'yes/no interrogatives:' instead of asking 'What is

---

such as 'contradiction is impossible' (*Topics*, I, 1, 104b 19); Platonic dialogues contain dialectical games concerning all these types. One should note here another difference with obligational games, where the *positum* that **P** must defend may be a proposition recognized at the start as false, i.e., the negation of a *casus* assumed as true, see Dutilh-Novaes (2007), p.155. This shifts the interest of the obligational games to the ability for **P** to defend it. In dialectical games, it is never known if the thesis to which **P** commits himself as the opening move is true or not, and the thesis is thus put to the test, along with **P**.

[34] Aristotle, *Topics*, I, 1, 100a 19.
[35] These could be *protasis*, *problema* or *thesis*, as defined in footnote 33.
[36] These two notions are taken from, respectively, Hamblin (1970) and Lewis (1983), essay 13.
[37] Ryle (1966), p. 105.
[38] See Hintikka (1993), pp. 6-7, 12-13 for a critique of Ryle on this point.

F?', e.g., 'What is courage?', one simply has to ask **P** a disjunction 'Is courage this or that?' and carry on from there.

In Plato's dialogues, Socrates more often than not encounters recalcitrant proponents, who do not wish to submit themselves to rule (iii). This recalcitrance is understandable in light of the fact that there is a fundamental asymmetry between **O** and **P**. Needless to say, there are numerous meta-discussions where Socrates has to convince his partner to carry on answering his questions. It is worth noting that Socrates often insists on the fact that dialectical games as test are a common endeavour, and they are not directed just at **P**: The result is as important for **O**. As Socrates points out to Protagoras, who asks if he should answer according to the majority opinion or to his own beliefs:

> "It makes no difference to me, provided you give the answers, whether it is your own opinion or not. I am primarily interested in testing the argument, although it may happen both that the questioner, myself, and my respondent wind up being tested."[39]

We already noted the same idea in *Philebus*, 14b and *Protogoras*, 338d, we find it again in an exhortation served to Critias in the *Charmides*:

> "How could you possibly think that even if I were to refute everything you say, I would be doing it for any other reasons than the one I would give for a thorough investigation of my own statements - the fear of unconsciously thinking I know something when I do not. And this is what I claim to be doing now, examining the argument for my own sake primarily but perhaps also for the sake of my friends. Or don't you believe it to be for the common good, or for that of most men, that the state of each existing thing should become clear?"
> "Very much so Socrates," he said
> "Pluck up courage then, my friend, and answer the questions as seems best to you, paying no attention to whether it is Critias or Socrates who is being refuted. Instead, give your attention to the argument itself to see [how the one who is tested will fare]."[40]

This is related to an important fact about dialectical games, which is difficult to grasp, namely that **P** must answer not just faithfully in the sense that he must believe the answer,[41] but also in accordance to the logical relation with what has been conceded before. There is an ambiguity here, which is not resolved in Plato, between what has been called the 'say what you believe'[42] or 'doxastic' constraint[43], and what could be called a 'logical' constraint. The former is a well-known aspect of the Socratic method, we certainly do not deny its presence: It is enshrined in passages such as *Gorgias* 472b-c quoted at the end of section 2, and we even give below an explanation of its presence in terms of the very important rule (viii), below, which guarantees, if strictly adhered to, that **O** argues effectively for inconsistency. But the 'logical constraint' is often unjustly ignored. To put it simply, a

---

[39] Plato, *Protagoras*, 333c.
[40] Plato, *Charmides* 166d-e. Translation modified.
[41] Even this is too strong, as we just saw in *Protagoras*, 333c.
[42] Vlastos (1994), pp. 13-17.
[43] Benson (1987), p. 70.

proponent already committed to A should, upon conceding $A \to B$, defend B, etc. As even Vlastos points out: "In an argument your only means of compulsion are logical."[44] That the 'logical constraint' is present in Plato is borne out by Socrates' already-quoted comment in *Protagoras*:

> "If Protagoras is not willing to answer questions, let him ask them, and I will answer, and at the same time try to show him how I think the answerer *ought* to answer."[45]

The two constraints are not incompatible: if, by hint of logic, **O** gets **P** to recognize a new thesis as a consequence of other theses already committed to, then **P** is *by the same token* committing himself to it, and can be said to 'believe' it. The emphasis in Plato is, however, definitely on the 'doxastic constraint', while Aristotle emphasizes instead the 'logical constraint:'

> "The business of the questioner is so to develop the argument as to make the answerer utter the most implausible of the necessary consequences of his thesis; while that of the answerer is to make it appear that it is not he who is responsible for the impossibility or paradox, but only his thesis; for one may, no doubt, distinguish between the mistakes of taking up a wrong thesis to start with, and that of maintaining it properly, when once taken up."[46]

Here, it is open for **P** to put the blame on the thesis that he has undertaken to defend; There is no room for this in Plato's dialogues, where the emphasis on shame might have prevented a full understanding of the 'logical constraint,' and therefore blocked the way to the discovery of logical form.[47]

At all events, the presence of the 'logical constraint' legitimizes this remark from the late Michael Frede:

> "The point of this exercise is clear: over the course of time one learns how propositions are logically related to each other, which follows from which, and which are incompatible with which."[48]

Plato's or Xenophon's Socrates almost exclusively deals with moral issues, but a very good example of the subtle logical mistakes to which Socrates alludes at *Cratylus* 436d and that can be unearthed by repeated plays of dialectical games is provided by Aristotle in *Sophistical Refutations*, 11, 171b 16, with the squaring of the circle by means of the lunules. It has nothing 'sophistical' about it. Aristotle does not say much - he even assumes that his reader knows the proof - but we can

---

[44] Vlastos (1994), p. 20.
[45] Plato, *Protagoras* 338d. Our emphasis.
[46] Aristotle, *Topics*, VIII, 4, 159a 18-24.
[47] See Dorion (1997) for a discussion. One should note that for this very reason, Aristotle's dialectical games are already closer to obligational games, which are even open to a deterministic interpretation, where **P**'s answers would be constrained by the inferential relations. See Dutilh-Novaes (2007), pp. 156f. for more details.
[48] Frede (1992), p. 213.

reconstruct his argument from Simplicius' copy of a lost commentary by Alexander of Aphrodisias, which shows the proof to contain one faulty assumption.[49] One may assume from the context that the faulty assumption was unearthed in a dialectical game where the opponent argued against the squaring of the circle.

We should add here, since Frede did not have in mind our own approach in terms of dialogical games, that it would be better to read in the passage just quoted 'over the course of many plays' for 'over the course of time:' indeed, it is important to understand that there is a distinction between the *game* - e.g., chess or tennis - and the *matches* or plays,[50] a distinction easily missed when talking about dialogical or dialectical games. It is often necessary to play many times over one thesis in order to be able to ascertain anything about it. This is one of the reasons why Socrates always claims his ignorance: The result of some plays, even if pointing in the same direction, might never give strong enough a result for it to be called knowledge. There are many passages corroborating this, such as *Gorgias* 508e-509a, to be discussed below. At the end of *Lysis*, it is suggested that the interlocutors may have gone astray and that they "look over" all that was said, "like the able speakers in the law courts" (*Lysis* 222e). This suggests that they could have another go at it to see if they could obtain a different result, with another move or combination of moves at a given stage. In *Theaetetus*, 172d, Socrates mentions to Theodorus that they are beginning their "third discussion." These passages suggest strongly that Plato understood the distinction between game and matches. Dialogues such as the *Laches* can be seen as repeated plays of the same game, for the definition of courage. If anything, this shows how remarkable the depth of the thinking of the Greeks (at least of Plato and Aristotle) on their dialectical games was.

To come back to our list of rules, and the frequent recalcitrance of **P** to submit to rule (iii). Some room for manoeuvring is in fact provided for **P**, with these two further rules:[51]

(v) **P** has a right to clarify or adjust an earlier concession, whenever **O** twists this concession in an undesirable direction.

(vi) **P** can protest against **O**'s questions by formulating objections.

Aristotle even allows as a formal move for the proponent that he claims 'I do not understand:'

> "The questioner should be met in a like manner also in the case of terms used obscurely and in several ways. For the answerer, if he does not understand, is always permitted to say 'I do not understand,' and he is not compelled to reply 'Yes' or 'No' to a question which may mean different things.[52]

---

[49] See, e.g., Dorion (1995), pp. 283-285 for details.
[50] In French, the distinction is captured by the distinction between *un jeu* and *une partie*.
[51] Taken from Krabbe (2006), p.668. On clarity, see Plato, *Lesser Hippias* 364b-d.
[52] Aristotle, *Topics* VIII, 7, 160a 18-23.

What the whole passage from which this quotation is extracted shows is that Aristotle is granting this formal move to the proponent as a precaution, so that the proponent can avoid being trapped by the opponent's use of the fallacy of equivocation.

The possibility of doubting instead of merely answering 'yes' or 'no' raises yet another issue. Indeed, to doubt may be a strategy for **P**, when he feels that committing himself to a given answer might give the advantage to **O**. Provided there is an agreed limit on the length of the play, not even systematic use, but even simply frequent recourse to 'I doubt' would give an unfair advantage to **P**, who could thus buy time. From this, one assumes that in dialectical games the possibility of doubting is not an option for **P**.[53] Enforcing the following rule, which fits the spirit of Socratic enquiry, would settle this problem:

(vii) If **O**'s questions are admissible in light of (vi), then **P** must give an answer.

The next two rules are probably the most crucial ones, philosophically speaking:

(viii) **O** may not introduce any thesis A, **P** must first commit himself to any thesis A.

(ix) Having elicited from **P** commitment to, say, B and C, **O** can then propose that they 'add up' or 'syllogize,'[54] inferring absurdity (*adunaton*, ἀδύνατον) from A, B and C.

Rule (viii) enforces the asymmetry in favour of **O**. It is a peculiar form of what is known in dialogical logic as the 'formal rule.' In dialogical logic, it is normally stated the reverse way:

---

[53] In this, they differ from obligational games, where doubt is admitted (and where it has interesting epistemic properties). One should note, however, that the possibility of doubting opens the door to a 'lazy' strategy for **P**, namely, to doubt systematically any thesis to which he is asked to commit himself, see Dutilh-Novaes (2007), sec. 3.3. There are ways to circumvent this in obligational games – Dutilh-Novaes (2007), p. 158 –, but, in the case of dialectical games, the possibility of systematic doubting might have been a reason for excluding doubt as a possible move for **P**. This point appears to be linked with Socrates' 'doxastic constraint,' i.e., the request that the proponent answers questions with full honesty, in this limited but interesting way: At least when one honestly believes the answer to a given question to be, say, 'yes,' one should not claim 'I doubt' in order simply to gain strategic advantage.

[54] This is the original meaning of the word. See, e.g., Robinson (1966), p. 21. One should further note that, while Aristotle defines *syllogismos* (συλλογισμός) at the very beginning of his *Topics* as "an argument in which, certain things being laid down, something other than these necessarily comes about through them" (*Topics* I, 1, 100a 25), he also points out in *Posterior Analytics*, I, 1, 71a 7 that the premises are obtained "from men who grasp them." It would thus be wrong to think, merely because we were taught so, that Aristotle conceived the syllogism merely in terms of a solitary thinker; The definition also covers, and seems primarily intended to cover, dialectical situations.

(viii') **P** may not introduce any thesis *A*, **O** must first commit herself to any thesis *A*.[55]

This is because in dialogical logic the asymmetry is on the side of **P**, who is arguing for validity. By contrast, it becomes clear that dialectical games are symmetrically opposite: asymmetry is in favour of **O** who leads the game with his questions and can thus be said to *argue for inconsistency* on the part of **P**. Indeed, with rule (iv) Socrates, as **O**, never needs to assert anything. He just needs **P** to commit to the theses from which he will infer the contradiction. Strict obedience to formal rule (viii) thus guarantees that Socrates argues for inconsistency in the same way that, in dialogical logic, the formal rule (viii') guarantees that one argues for formal validity. This is a most remarkable feature of dialectical games, which will become important in the discussion of rule (ix), in a moment.

It is also linked with one of the most obvious, but often misunderstood, features of Socrates' philosophy: His repeated avowals of ignorance, most famously towards the beginning of *Apology* 21c, where Socrates tries to prove the oracle (who had claimed that he is the wisest) wrong. This should not mask the importance of avowals of ignorance in light of the fact that they occur in the context of dialectical games. There appears to be three contexts for such avowals: at the outset, during a play, and after the *elenchus*. The first case is ever so frequent since it is usually through an avowal of ignorance that Socrates tricks his adversary into committing himself to a claim, which then becomes the opening move of the play. Avowals also occur *during* the play, e.g., at *Lesser Hippias* 372b-e, when Socrates invokes his ignorance at the very moment when Hippias is forcing him to concede that he is committed to the view that those who commit injustice voluntarily are better than those who commit it involuntarily, a thesis he is known to disagree with (on this see, e.g., *Protagoras*, 351b and 359a). These avowals of ignorance, as they occur within the play, are crucial *to enforce rule (viii), without which Socrates could not argue for inconsistency*. It is, again, a testimony to the Greeks' depth of thinking about their games that this is exactly what Socrates portrayed himself as doing:

> "SOCRATES: You are the complete lover of discussion, Theodorus, and it is too good of you to think that I am a sort of bag of arguments, and can easily pick one which will show you that this theory is wrong. But you don't realize what is happening. The arguments never come from me; they always come from the person I am talking to. All that I know, such as it is, is how to take an argument from someone else - someone who *is* wise - and give it a fair reception. So, now, I propose to try to get our answer out of Theaetetus, not to make any contribution of my own."[56]

Another type of avowal of ignorance appears in discussions following the end of the play, for example in the *Gorgias*, where Socrates had been struggling a bit, while being put in the answerer's seat, before recovering and making mincemeat of Callicles in later stages of the dialogue:

---

[55]Rückert (2001), p.168.
[56]Plato, *Theaetetus*, 161a-b.

"These conclusions, at which we arrived earlier in our previous discussions are, I'd say, held down and bound by arguments of iron and adamant, even if it's rather rude to say so. So it would seem anyhow. And if you or someone more forceful than you won't undo them, then anyone who says anything other than what I'm now saying cannot be speaking well. And yet for my part my account is ever the same: I do not know how these things are, but no one I have ever met, as in this case, can say anything else without being ridiculous."[57]

Such avowals are of a different nature: They tell us that Socrates is aware that, even if A has failed the test of consistency many times and $\neg A$ passed it many times too, "held down and bound by arguments of iron and adamant," he could not say that he knows that $\neg A$, because the possibility is always opened that the next play for $\neg A$ will result in failure. Such avowals tell us, therefore, about what can be achieved and what cannot be achieved with dialectical games.

The set of rules would not be complete without a rule for winning:

(x) When **O** has driven **P** into an *elenchus*, the play ends with **O** winning; **P** wins by avoiding being driven into an *elenchus*.

This is implicit in Plato's dialogues, but explicit in the opening sentence of Aristotle's *Topics*:

"Our treatise proposes to find a line of inquiry whereby we shall be able to reason from reputable opinions about any subject presented to us, and also shall ourselves, when standing up to an argument, avoid saying anything contrary to it."[58]

One important caveat here is that **O** must get **P** to concede the contradiction, i.e., that **P** has committed himself to $\neg A$ when committed to further premises B and C. It is possible for **P** not to recognize the validity of this last deductive step, in which case the judgement of a moderator or of the audience could enter into play.

4 Vlastos 'refuted'

What rule (ix) throws into relief is the fact that a dialectical game for A appears to proceed through questions and answers in the midst of which **P** commits himself to further theses, say, B and C that are to be put in his 'commitment store' or 'scoreboard,' thus forming a set of premises $\{A, B, C\}$, from which in a final step, involving a logical inference, **O** is able to infer absurdity from that set. **O** might derive $\neg A$ from the further premises B and C, thus demonstrating that A, B and C are incompatible when taken together, i.e., that they form an inconsistent set. This is indeed perfectly in line with the above remark that in dialectical games **O** argues for inconsistency. In symbolic terms (with '$\bot$' standing for absurdity), **O** has shown that:

---

[57] Plato, *Gorgias* 508e-509a.
[58] Aristotle, *Topics* I, 1, 100a 18-20. Translation slightly modified.

$$(I) \quad A, B, C \vdash \bot$$

According to us, this *is*, strictly, the *elenchus*. The word is usually translated as 'refutation,' but, as we argued, the original meaning of 'test' remains and fits very well here: The compatibility of A with other premises, as well as the capacity of **P** to maintain consistency, has been tested by **O**.

This interpretation of the *elenchus* goes against Vlastos' 'standard *elenchus*.'[59] Vlastos' well-known presentation at Vlastos (1994), p. 11, can be adapted to our context as follows:

1. The proponent assert a thesis, A which Socrates - the opponent - considers false and targets for refutation.

2. Socrates secures (ad hoc) agreement to further premises, say B and C from which Socrates argues.

3. Socrates then argues, and the interlocutor agrees, that $B\&C$ entail $\neg A$.

4. Socrates then claims that he has shown that $\neg A$ is true and A false.

It is worth pointing out against Vlastos that (I) is merely a logical validity (one could say nowadays that it is true in every model, but models are not needed here). Applying the rule of the introduction of negation on the right-hand side of this sequent - a deduction rule which is incidentally also acceptable to the modern mathematical intuitionist -, one may derive from (I) the following sequent:

$$(II) \quad B, C \vdash \neg A.$$

If we are in a situation where B and C are true or simply more 'entrenched,' *then* this logical validity tells us that $\neg A$ is true or more 'entrenched,' assuming here that 'entrenchedness' is preserved by logical consequence. However, in absence of such information on B and C, or in a situation where A, B and C have the same degree of entrenchment, no such conclusion is available. This is illustrated by the fact that from (I), one can also derive, once again by negation introduction, the sequents:

$$(III) \quad A, C \vdash \neg B;$$
$$(IV) \quad A, B \vdash \neg C;$$

---

[59] Vlastos' interpretation is quite widespread and often simply taken for granted (see e.g., the references to it in footnote 20 above), but the number of notable authorities who have rejected it is surprising. See, e.g., Benson (1995), p. 47 n.6, for an extensive list, to which one could also add Frede (1992), pp. 210-211. At all events, we are not making an argument from authority but avoiding one: All that this shows is that Vlastos' interpretation is not 'authoritative,' and that it is perilous to take it for granted without first weighting the arguments for and against it.

That one should favour (II) over (III) or (IV) is an issue hardly ever raised by Socrates, except in the *Gorgias*, where Vlastos finds the only textual basis for his interpretation.[60] In the later stages of that dialogue, Socrates goes on arguing that, in our jargon $\neg A$ is in fact more entrenched than A. His interlocutor, Callicles, refuses to answer after a while. For that and other reasons, Socrates, who describes his own arguments as "of iron and adamant," nevertheless concludes with yet another avowal of ignorance, as we saw above. Even here, there is ultimately no claim that $\neg A$ is true and A false. An '*elenchus*,' therefore, concludes the game and Socrates' interlocutor is shamed (i.e., shown not to be the expert he pretends to be, etc.) for having been shown to hold an inconsistent set of premises $\{A, B, C\}$; He is also left contemplating the inconsistency of his beliefs, with no more than an indication that he should favour (II).

Vlastos' interpretation derived a good deal of its popularity among Socrates scholars from the fact that it opens the door to 'dogmatic' readings of Socrates and the early Platonic dialogues, according to which they are not truly *aporetic*, as Socrates is somehow putting forth some positive moral doctrines by refuting those of his interlocutors or arguing that some questions are in need of an answer. This interpretative task would be greatly simplified if one could show that in driving his interlocutor to an *elenchus*, Socrates is, each time, arguing for some truth by showing the falsehood of its contradictory. This is of course not a debate that we have to adjudicate here. Our claim here is that there is no 'standard *elenchus*' of the form (1)-(4), and that this avenue is not open for supporters of 'dogmatic' readings of the early dialogues. However, that Socrates does not show, by driving the proponent of A into an *elenchus*, the truth of $\neg A$ *is compatible with* the fact that he actually believed that $\neg A$, although he could not 'prove' it. His argument against Callicles in the later stages of the *Gorgias* is as clear an indication as any that he sometimes does. We are therefore not claiming that such readings are impossible, so that Socrates would not be driven by higher moral motives. We merely claim that this task will be harder than expected, because one cannot avail

---

[60]Vlastos' interpretation has a rather flimsy textual basis, namely a small set of remarks occurring in only one dialogue, at *Gorgias* 472b, 473b, 475e and 479e. It does not square with the extended 'meta-discussion' of dialectical games discussed at *Apology* 21b-23b (with the ensuing *elenchi* in the cross-examination of Meletus at 24d-28a) and *Sophist* 230b. In "The Dissolution of the Problem of the *elenchus*," Benson (1995), Hugh Benson has argued that one should look at the *Apology* for Socrates' own understanding of dialectical games, since it is the only place in the early dialogues where he is made to speak *of his own voice* and in a sustained way about these games. From an analysis of this passage, Benson draws a list of six features articulated by Socrates and examines five *elenchi* in *Euthyphro*, four in the *Laches* and eight in the *Charmides*, showing that they all possess these features while providing no compelling evidence that an *elenchus* is meant to prove the truth or falsehood of a given (moral) thesis, cf. Benson (1995), p.100. To this enormous weight of evidence, Benson provides some compelling remarks against the reading of *Gorgias* 472b, 473b, 475e and 479e, on which Vlastos based his interpretation, cf. Benson (1995), pp.100-109. However, discussing this last point would leave us too far afield; The reader is referred to Benson's paper, as we see no need to add to his arguments.

oneself of the 'standard *elenchus*' of the form (1)-(4), in order to infer that Socrates has either proven a moral truth $\neg A$, or simply believed that he did. Discarding A after it failed to pass the test is certainly a request to find some better candidate, but not a warrant to adopt $\neg A$.

It might also be replied that (II) was needed by Socrates in order to establish that (I), in which case, however, one only has a case of *reductio ad absurdum*. This corresponds, after all to the definition of 'refutation' in *Sophistical Refutations*:

> "a deduction to the contradictory of the thought that had been stated."[61]

We would like, however, to underline the fact that moves (I) and (II), i.e., respectively, pointing out the inconsistency of the set of premises $\{A, B, C\}$ and inferring $\neg A$ from B and C, are two *distinct* inferential moves. If one wishes to discuss the truth of the premises, one should then recall that valid arguments may have false premises - a fact already emphasized by Aristotle in *Topics* VIII, 11, 162a 10-11. Therefore, (II) might be valid, although B or C might be false. In order to go through with his interpretation, Vlastos must therefore show that, at least in some cases, the premises B and C are somehow proved true, in which case (II) can be read as (4), i.e., as proving the truth of $\neg A$. Our claim against this is simply that the truth of B and C is never established, but always conceded by the proponent, i.e., he commits himself but no one knows if he is entitled to them. One should note here that Vlastos is perfectly aware of this difficulty, and this is why he spoke about a 'problem of the Socratic *elenchus*:'

> "What Socrates in fact does in any given *elenchus* is to convict p of being a member of an inconsistent premise-set; and to do this is not to show that p is false but only that either p is false or that some or all of the premises are false. The question then becomes how Socrates can claim, as I shall be arguing he does claim in 'standard *elenchus*,' to have proved that the refutand is false, when all he has established is its inconsistency with premises whose truth he has not tried to establish in that argument: they have entered the argument simply as propositions on which he and the interlocutor are agreed. This is *the* problem of the Socratic *elenchus* [...]."[62]

Our point here is deceptively simple: There is no 'standard *elenchus*,' therefore no 'problem of the *elenchus*.' But Vlastos goes on arguing that Socrates must be making, in his own words, the "tremendous assumption" that "anyone who entertains the false (moral) belief A also entertains true (moral) beliefs B and C that entail $\neg A$."[63] This claim can hardly be sustained,[64] and Vlastos' discussion of

---

[61] Aristotle, *Sophistical Refutations* 1, 165a, 2-3.

[62] Vlastos (1994), pp. 3-4.

[63] Vlastos (1994), p. 25.

[64] This is labelled [A] by Vlastos in (1994), p. 25, who goes on arguing that [B] "Socrates' set of (moral) beliefs at any given time is consistent," so that one may infer that [C] "Socrates' set of (moral) beliefs at any given time is true" (Vlastos (1994), p. 28). Even if this were possible, it would be too farfetched. But it is not, as Brickhouse and Smith (1984) have shown further, that [A] and [B] could not entail [C]. See also Benson (1995), p. 48.

Socrates' attempt to solve the 'problem of the *elenchus*' seems too far-fetched to deserve further discussion in the context of this paper.

Instead of arguing for (*II*) on the basis of the alleged truth of B and C, one could salvage Vlastos' approach with the considerably weaker claim that B and C are more entrenched than A, so that it is better to give up A in the face of a contradiction. Since the game is played for A, there is perhaps a presumption that A is less entrenched. But, as we saw, this manoeuvre presupposes that 'entrenchedness' is preserved by the turnstile. Furthermore, one might argue here that the doxastic constraint gives us no reason to believe that any of the premises A, B or C is more entrenched than the others; even Vlastos recognizes this.[65] At all events, Socrates never carried on (except perhaps in the *Gorgias*), after an *elenchus*, pondering which of the premises, being less entrenched, ought to be given up, and more than once the premises introduced in the course of the play are on the face of it less plausible than the thesis being tested.

Since our dismissal of Vlastos' interpretation is based on a modern conception of logic not available to the Greeks, we should insist here that our use of logic makes for a charitable interpretation, not the reverse. From the point of view of modern logic, if Socrates merely pointed out to his interlocutor the inconsistency of the set his own premises $\{A, B, C\}$, then he is faultless. If, however, he thought he had 'refuted' A in the strict sense of having shown A false and $\neg A$ true, then his reasoning clearly comes out as deficient from our modern standpoint. This would just be Whig history all over again. One should note *en passant* that the last quotation above shows that Vlastos has some understanding of this and therefore that he is aware that under his interpretation Socrates comes out as an incompetent logician. Our interpretation is more *satisfaisante pour l'esprit*.

We should conclude our critical remarks, however, by pointing out that Vlastos considered his understanding of the *elenchus*, as "foundational" for his interpretation of Socrates.[66] This places him within a tradition that goes back, through Richard Robinson's *Plato's Earlier Dialectic* (1966) and Gilbert Ryle's *Plato's Progress* (1966), to Grote's *Plato and the other Companions of Sokrates* (1865).[67] Our disagreements notwithstanding, we concur with the fundamental importance of getting Plato on the dialectic games right for the study of his philosophy. (The same goes, of course for the role of dialectical games in the overall interpretation of Aristotle, see section 7.)

---

[65] Vlastos (1994), p. 13.

[66] Vlastos (1991), p. 111, n. 23.

[67] Of course, no unity of view is implied here, over and above the centrality given to the role of the *elenchus* in the interpretation of Socrates. Vlastos, for example, rejects Grote's understanding of the *elenchus*, which was that of 'ex-Vlastos,' at Vlastos (1994), p. 20. See also the remarks on Robinson's 'indirect *elenchus*' in Vlastos (1994), p. 12, n. 34. As for Richard Robinson, he misconstrues the *elenchus* as being of the form of a Modus Tollens: $A \rightarrow B, \neg B \vdash \neg A$. Alas, this is repeated in Kneale and Kneale (1962), p. 7.

5 More structural rules

Our modern dialogical games proceed in a finite number of steps, decomposing the assertion along the lines of its logical connectives, but not so here, as **O** must ask questions in order to obtain from **P** commitment to the theses that will serve for the *elenchus*. As **P** is well aware that **O** is trying to show an inconsistency in his beliefs, **O** has to proceed in a roundabout way, so that **P** would not suspect that he gives away key theses for the *elenchus*. Proceeding thus, the game might go on for quite some time. In order for the games to be playable, i.e., not just simply to avoid infinite games, but to keep them in a format manageable for their purpose (after all, more often than not they were being played in front of an audience), one needs some constraint on their length:

(xi) There is limit on the length of games.

Aristotle mentions a 'Time is up' or, in medieval parlance, a 'cedat tempus' rule in *Topics* VIII, 10, 161a10 and he also has at *Sophistical Refutations*, 34, 183a 25 a comment that implies that the time allowed might be set in advance by the players. The presence of such a rule makes sense: Under standard rules of dialogical logic, all plays are necessarily of finite length, but not here. A play cannot, however, go on indefinitely, e.g., when **O** seems repeatedly unable to drive **P** into an *elenchus*, so the game may be declared over, with **P** winning. Presumably, the game may be declared over by a umpire, the audience or even the players themselves.

The situation is slightly different for Plato, who insists in the *Theaetetus* on the importance of not having any time constraint, when describing the philosopher:

> "When he talks, he talks in peace and quiet, and his time is his own. It is so with us now: here we are beginning on our third new discussion; and he can do the same, if he is like us, and prefers the newcomer to the question in hand. It does not matter to such men whether they talk for a day or a year, if only they may hit upon that which is. But the other - the man of the law courts - is always in a hurry; he has to speak with one eye on the clock."[68]

This passage is of importance as it helps distinguishing the peculiar context of dialectical games from cross-examination in courts, where a clepsydra was used to limit the time of interventions, thus forcing speakers to resort to rhetorical ploys. As for the possible presence of an umpire, Hippias suggests that one be appointed at *Protagoras* 338b, in order that Protagoras and Socrates "steer a middle course" between the long speeches of the former, who "leaves the land behind to disappear into the Sea of Rhetoric" and the "excessively brief form of discussion" favoured by Socrates (338a). The suggestion is then rejected but there is an implicit recognition that questions or answers that are too lengthy are forbidden. This is not a 'cedat tempus' rule, but some control on excessive length, advocated here by

---

[68] Plato, *Theaetetus* 172d-e.

Socrates, is nevertheless enforced. Furthermore, as Ryle pointed out,[69] this passage suggests the prior existence of a practice of arbitration by an umpire. Even more interesting here is the idea that the umpire is to control the length of the moves in the game.

Closely related to (xi) is the following:

(xii) Delaying tactics are forbidden.

Aristotle's remarks at *Topics* VIII, 2, 158a 25–26 appear to be a forerunner of the 'no delay' rule in modern game semantics, which forbids delaying tactics such as repeating an attack or refusing, after conceding many instances, to agree to a generalisation already in play without providing a counterexample (see rule (xv) below). There is no explicit rule such as this in Plato's dialogues, but many instances where the players are asked to keep on track; perhaps the most striking example being when Socrates asks Protagoras to stop giving long-winded answers and keeps to shorts ones (*Protagoras*, 329a-b; see also *Gorgias*, 449b). As with (xi), delaying tactics could presumably be called off by an umpire, the audience or the players.

The audience appears at first blush to have a limited, albeit non-negligible role, e.g., at *Gorgias* 458b-c, or at *Protagoras* 339e, when Protagoras, after having driven Socrates into an *elenchus*, receives the "noisy round of applause" that destabilizes Socrates, who recounts: "I felt as if I had been hit by a good boxer;" He then tries to stall for time until he can think of a strategy to gain the upper hand again. The presence of an audience does not appear, however, to be a necessary condition, since there is none in *Euthyphro* and in *Crito*, if we forget the appeal to the laws (and the state) themselves, as a fictitious audience, from *Crito*, 50a onwards. We saw earlier, when discussing the principle of non-contradiction, that anyone contradicting themselves in front of an audience would lose credibility. What the absence of an audience indicates, therefore, is merely that the players themselves have *internalized* the third person viewpoint, and that he who contradicts himself merely in front of another player has lost credibility. This internalization does not stop at the players themselves, as everyone is free to do like Socrates and internalize the games themselves:

> "[…] not only now but at all times I am the kind of man who listens to nothing within me but the arguments that on reflection seems best to me."[70]

This internalization is already at work in the passages quoted at the end of section two, most notably *Gorgias* 482b-c and *Apology* 26e, where Socrates replies to Meletus, whom he just put in front of a self-contradiction: "You cannot be

---

[69] Ryle (1966), p. 115.
[70] Plato, *Crito* 46b.

believed, Meletus, *even, I think, by yourself*" (our emphasis). This process of internalization can be extended to account for the well-known definition of thought (*dianoia*, διάνοια) in the *Sophist*:

> "VISITOR: Aren't thought and speech the same, except that what we call thought is speech that occurs without the voice, inside the soul in conversation with itself?"[71]

For those reasons, we think that the very presence of an audience has implications that are lost within the currently popular 'existential' readings of the *elenchus* in terms of 'spiritual exercises,'[72] which transform dialectical games from a dynamic interaction between players into a 'struggle with oneself.'[73]

At all events, the result of an *elenchus* has a clear-cut result for the audience, whose role, in the Greek polity, was not merely that of spectators: They are also drawing their own judgement.[74] Dialectical games were played in public places such as the *Agora* and, provided a thesis A were successfully defended on a sufficient number of occasions, all members of the audience would be deferentially entitled to add it to their own deontic scoreboard;[75] if, on the other hand, proponents have repeatedly failed to defend it and members of the audience discover through such plays that A cannot be held consistently along with B, C, etc., which they consider to be more entrenched, it would be wiser for them to withdraw A from their scoreboard. Either way, the result becomes common knowledge (in a non-technical sense of this expression).

Finally, we should add a rule forbidding fouls of the 'eristic' or 'sophistic' kind:

(xiii) Fallacies are strictly forbidden.

As can be seen from the display of verbal tricks by Euthydemus and Dionosodorus in *Euthydemus*, the Greeks quickly realized that **O** can easily force **P** into what is only an apparent *elenchus* (*Laches* 196b). According to Plato, the *elenchus* would not be a real one because it would not be *believed* by **P** (*Theaetetus* 154d-e).[76] According to Aristotle, there is no real inference (*Topics* I, 1, 101a 3). Aristotle studied carefully these fallacious moves in *Sophistical Refutations*, a sort of companion volume to the *Topics*, whose set of examples actually overlaps those found in *Euthydemus*, which could also be used for pedagogical purposes. Following our discussion of Vlastos, one should note that, again, the distinction between 'apparent' and 'real' *elenchus* has nothing to do with truth and falsity, only with the possibility of an apparent logical inference as a step in the game.

---

[71] Plato, *Sophist* 263e.
[72] Hadot (1992).
[73] Davidson (1990), p. 476.
[74] On this point, see Ryle (1966), p. 198.
[75] The vocabulary here is taken, on purpose, from Brandom (1994), ch. 3.
[76] See Benson (1989) for a detailed discussion of this point. Indeed, the proponent needs to recognize the *elenchus* in order to feel shame.

As the well-known story goes, the words *eristike* (ἐριστική) and *sophistike* (σοφιστική) were used by Plato and Aristotle to cover cases where rule (xiii) was not enforced. Following them, we could indeed classify 'eristics' as games, but as *perversions* or *misuses* of dialectical games,[77] full of fouls and with no real *elenchus*. Plato tended to portray all Sophists as practitioners of 'eristics,' while Aristotle did the same for Megarians. This bias is misleading, not only because it is rather likely that fallacies crept in dialectical games before Sophists or Megarians - after all Aristotle was convinced that Zeno's arguments were sophistic and Zeno is the most often quoted 'Sophist' in *Sophistical Refutations* -,[78] but also because this accusation appears groundless in light of available evidence. As a matter of fact Aristotle hardly attributes a 'sophism' to a Sophist in his *Sophistical Refutations*. As Alexander Nehamas pointed out,[79] there is not a shred of evidence that Protagoras' *Techne Eristikon*, now lost, contained instructions for the use of fallacies, and it is even Protagoras who catches Socrates committing a fallacy at *Protagoras* 350c-351a! Even if we concede that some practiced the genre, it would be false to attribute it to all that went under the name of 'Sophist' (provided one argues, against the testimony of Aeschines, Isocrates and Aristophanes, that Socrates himself was not considered as a 'Sophist'). The original meaning of *eristike* was merely in reference to the (necessary) *agonistic* nature of dialectical games, and there is no reason to believe that the original *eristike* was just a game of fouls. Rehabilitating *eristike*, both in its Eleatic version and in the 'noble lineage' of the Sophists is indeed, in our eyes, a most important task.

The proscription of fallacies underscores an essential dimension of dialectical games, namely their embodiment of epistemic virtues. Of course, as games of consistency management, they already embody the requirement of consistency (and clarity in the theses to be discussed). But it is also rather significant that the very idea of fallacious forms of argument came about as a reflection on the proper conduct of those games. There are further more 'existential' facets here, related to 'self-control,' i.e., to the need to cultivate some traits of character in order to be a better player or even to be able to play. We have already seen some of this, for example the need to avoid anger but also the *courage* one needs to face the possible shame incurred by the *elenchus*, in Socrates' exhortation to Critias at *Charmides*, 166d-e, quoted above. As a matter of fact, that very dialogue seems as if written in order to teach its readers about the virtues of courage and moderation (as well as the advantages of cooperation through what is on the looks of it a purely agonistic game) as it is replete with discussions concerning these topics.[80] For that purpose, Plato even puts his character Socrates into a rather comical situation, when

---
[77] This is also Nehamas's position, Nehamas (1990), p. 8.
[78] See Dorion (1995), p. 54, n. 1.
[79] See Nehamas (1990), p. 7.
[80] On this, see Schmid (1998), Schmid (2002).

he looks inside Charmides' cloak, while everybody has their heads turned away:

"I saw inside his cloak and caught fire and was quite beside myself."[81]

As a result of his arousal, Socrates is momentarily unable to continue playing the game and he regains his composure and ability to play only after some time (at 156d). This can be construed as showing that control over one's impulses is a key to the proper playing of dialectical games. This facet, not to be confused with the 'spiritual exercises' of the 'existentialist' readings, is only thrown into relief through a proper understanding of the functioning of dialectical games, and the many remarks in Plato's dialogues that address it are certainly well worth a further investigation.

6 Some particle rules for dialectical games

As we said, dialectical games do not proceed by decomposing propositions along the lines of the logical connectives. Dialogical logic distinguishes between structural and particle rules, that define the meaning of the logical connectives, but, although there is, as we demonstrated amply, much thinking going on concerning the structural rules in meta-discussions embedded in Plato's dialogues and in Aristotle's *Topics*, there appears hardly to be any explicitly stated particle rules and no discussions about the connectives themselves (of the sort that one finds later on, e.g., Sextus Empiricus on the conditional). This does not mean that dialectical games do not involve implicitly rules for the connectives; After all they are presupposed in any discussion. According to the philosophy of the 'Erlangen School', that was developed in order to provide philosophical foundations to dialogical games, particle rules are to be abstracted from ordinary practice. A lot of ink has been spilt over this since the publication in 1967 of *Logical Propaedeutic. Pre-School of Reasonable Discourse*.[82] It has been suggested, with help of ideas from unusual suspects such as Hugo Dingler or Karl Bühler, that one looks for a starting point within something akin to Husserl's *Lebenswelt*.[83] Another suggestion would be simply to look at the regimented discussions of Ancient Greece, i.e., to look at dialectical games, not the *Lebenswelt*. Indeed, in these regimented discussions, rules for the connectives are *hardly ever made explicit, but always followed implicitly*, therefore, they are implicitly recognized as holding and one needs merely to make them explicit. An example, taken almost at random, should suffice to make the point. We chose a passage from *Ion*, where Socrates discusses with the rhapsode Ion his claim that

A A rhapsode speaks about poets "on the basis of knowledge or mastery."

---

[81] Plato, *Charmides* 155d.
[82] Kamlah and Lorenzen (1984).
[83] See Marion (2008) for a critical discussion and an entry into the literature.

Near the beginning of the dialogue, Socrates gets Ion to admit that one possible way to argue this is to claim that $A \rightarrow B$, with

B The rhapsode can tell that which of two poets is the best.

Ion having already committed himself to *A*, Socrates forces the play to move on to *B*, and then argues that $\neg B$, namely that Ion cannot tell which of two poets is best, because Ion had already conceded earlier on (*Ion* 531a) that he only knows about Homer - the *elenchus* happens at (*Ion* 532b). Here is, therefore, a snippet taken from *Ion* 531b exemplifying two implicit logical rules; it is taken from the middle of this discussion, at the stage were Socrates is trying to establish $A \rightarrow B$:

> "SOCRATES: Take all the places were two poets speak of divination, agreeing or not. Who would explain these better: a diviner or a rhapsode?"

Here Socrates asks Ion to choose: $C \vee D$. Ion chooses the former:

> "ION: A diviner."

And then Socrates suggests that Ion grants $C \rightarrow E$, as Ion grants *C*, thus forcing Ion to commit to *E*:

> "SOCRATES: Suppose that you are a diviner: then you would know how to explain where they agree and where they disagree.
> ION: That's clear."

The debate then goes on about *E*, with a series of questions raised by Socrates, but we need not go further: Implicit in this passage are the rules for negation, disjunction and implication.[84] Disjunction, like conjunction, follows the usual rules. This is not surprising as these connectives are usually not contested, but not so for negation and implication. Negation, is exemplified in our passage by Socrates undertaking to show that $\neg B$, once Ion has admitted to *B*. In this passage, it is the reverse from the usual rule in dialogical logic, where **O** would attack $\neg B$ by stating *B*, and seems to involve an exchange of roles, of sorts, since Socrates needs to show $\neg B$. The rule for implication seems to be:

(xiv) Once **P** has committed himself to *A* and $A \rightarrow B$, **O** can force **P** to defend his commitment to *B*.

---

[84]In *Ion*, Socrates drives Ion in a series of *elenchi* concerning a series of conditionals $A \rightarrow B$ in the manner outlined above for the *elenchus* at (*Ion* 532b). It is quite remarkable that Plato's 'Socratic' or 'aporetic' dialogues proceed as follows: Once Socrates has elicited commitment to an assertion *A* from his interlocutor, he usually elicits a further commitment to a conditional assertion of the form $A \rightarrow B$. Once the interlocutor has agreed to $A \rightarrow B$, Socrates actually concedes *A* to move against *B*, and the game goes on with a debate concerning *B*. Once the interlocutor has been driven to an *elenchus* concerning *B*, Socrates often goes on eliciting commitment from the proponent for another conditional assertion $A \rightarrow C$, from which a debate on *C* follows. This might explain why Robinson thought (see footnote 67 above) that the *elenchus* is of the form of a Modus Tollens.

This rule is closely related to the one found in dialogical logic, which displays a conception of the meaning of the conditional which differs from the truth-conditional one (where $A \rightarrow B$ is just equivalent to $\neg A \vee B$), namely that, once the **P** has committed himself to $A \rightarrow B$, **O** grants A, thus forcing **P** to defend his commitment to B: In order to debate on a conditional such as $A \rightarrow B$, the opponent must concede the antecedent A in order to force the proponent into debating the content of consequent B, which is after all where the semantic content of the conditional really lies. In dialectical games, however, **O** is not allowed to grant anything, let alone A, in order to play formally, and must therefore secure commitment from **P** first. Therefore, upon reflection one can see that the rules boil down to the same, the only difference being that in dialogical games **P** plays formally and in dialectical games it is **O** that plays formally.

Quantifiers are said to originate in the 19th century with C.S. Peirce (who already defined them in terms of games)[85] and Frege, so it is very surprising to find that Aristotle has an explicit rule at *Topics*, VIII, 8, 160b1:

(xv) If **O** has granted to **P** many instances but refuses to commit himself to the generalisation, then **P** may force **O** to provide a counter-argument.

Of course, in light of (x), such a rule is necessary, as granting instance after instance without granting the generalisation is a delaying tactic forbidden in (xi). This is not exactly equivalent to the rule in modern dialogical logic, but interestingly near enough.

7 Why playing dialectical games?

It may be useful to say a few words here about the point of playing dialectical games. For some classes of theses, e.g., cosmology, mathematical principles, ethical principles, statements in forensic, medical, or political contexts, etc. there is no possibility of a direct verification, and the Ancient Greeks appear to have considered the test of consistency offered by dialectical games as providing the best available warrant for a given thesis from one of these classes. It is therefore implicit in that practice that Ancient Greeks believed the best guarantee that the strongest available counterarguments were indeed marshalled against a given thesis A is a free and open discussion of A between all epistemic peers. Furthermore, they would clearly consider themselves safely entitled to A only when it was successfully defended against the strongest available counterarguments that can be levelled against it. To construct an argument in support of $\neg A$ does not provide the strongest possible epistemic test for A. This last point is recognized by Aristotle:

"Criticism of an argument when taken in itself and when presented in the form of questions, is not the same. For often failure to carry through the argument correctly in discussion is due to the

---

[85] See Hilpinen (1982).

person questioned, because he will not grant the steps of which a correct argument might have been made against his thesis; for it is not in the power of the one side only to effect properly a result that depends on both alike."[86]

And in order to marshal the strongest available counterarguments against *A*, one must defend *A* against an opponent that disagrees with *A*, hence the importance of the *agonistic* nature of those games. This serves to underscore the important of the third person point of view, which cannot be eliminated in favour of a 'struggle with oneself', unless the latter is an *internalization* of the practice of dialectical games. The point can be given further weight by pointing out that dialectical games presuppose not only that one faces an opponent, but also that it is not sufficient that one faces an opponent, *let alone an internalized one*, since that opponent, *especially if internalized*, might be particularly inapt at arguing against *A*: one must be in principle ready to answer the challenge from *any* opponent among one's epistemic peers.[87]

The idea seems to be that unsound theses can be progressively eliminated,[88] as they fail to pass the test of consistency, so dialectical games work as a *sieve*. Any thesis having successfully passed repeated tests, say, a given candidate for the principles of mathematics such as 'One is not divisible,'[89] can then be assumed as *hypothesis* or as one of the *homologemata* (ὁμολογήματα) of *Theaetetus* 155a-b. On the basis of such *hypotheses* one may derive further propositions, under the understanding that, *should they be found incorrect, all that was derived from them must be abandoned*. This is in accordance with Plato's well-known doctrine

---

[86] Aristotle, *Topics* VIII, 11, 161a 15-21. This point was recognized in the 19th century by George Grote's companion, John Stuart Mill, in *On Liberty*: "He who knows only his own side of the case, knows little of that. His reasons may be good, and no one may have been able to refute them, but if he is equally unable to refute the reasons on the opposite side; if he does not so much as know what they are, he has no ground from preferring either opinion. [...] Nor is it enough, that he should hear the arguments of adversaries from his own teachers, presented as they state them, and accompanied by what they offer as refutations. That is not the way to do justice to arguments, or to bring them in real contact with his own mind. He must be able to hear them from persons who actually believe them; who defend them in earnest, and do their utmost for them. He must know them in their most plausible and persuasive form; he must feel the whole force of the difficulty which the true view of the subject has to encounter and dispose of; else he will never really possess himself of the portion of truth which meets and removes that difficulty," Mill (1991), p. 42.

[87] This is, again, a point made by Mill (see the previous footnote) in this splendid passage (our emphasis): "There is the greatest difference between presuming an opinion to be true, because, with every opportunity for contesting it, it has not been refuted, and assuming its truth for the purpose of not permitting its refutation. *Complete liberty of contradicting and disproving our opinion is the very condition which justifies us in assuming its truth for purpose of action; and on no other terms can a being with human faculties have any rational assurance of being right*" (Mill (1991), p. 24). We should not be surprised that Mill ended up, like Grote, praising dialectical (and obligational) games, cf. Mill (1991), pp. 50-51.

[88] See, e.g., Plato, *Cratylus* 436d for a comment to that effect about geometric principles.

[89] On the role of dialectical games in testing candidates for the principles of mathematics, see Szabo (1978), especially Part 3.

concerning hypothetical reasoning in *The Republic*, where it is clear that the *hypothesis* is, at one stage, withdrawn from the dialectical test:

> "In order to avoid going through all these objections one by one and taking a long time to prove them all untrue, let's hypothesize that it is correct and carry on. But we agree that if it should ever be shown to be incorrect, all the consequences we've drawn from it will also be lost."[90]

This point is of great importance, considering the enormous influence in History of Plato's views about hypothetical reasoning.

As far as Aristotle is concerned, he was certainly ambiguous on this point, since he implies at times, e.g., at *Topics* VIII, 1, 155b 10-16, that the principles of science or *axiomata* (ἀξιώματα), might be withdrawn from the test of dialectical games by the philosopher. But Aristotle also says this, when describing the usefulness of his textbook:

> "For the study of philosophical sciences it is useful, because the ability to puzzle on both sides of a subject will make us detect more easily the truth or error about several points that arise. It has further use in relation to the principles used in several sciences. For it is impossible to discuss them at all from the principles proper to the particular science in hand, seeing that the principles are primitive in relation to everything else: it is through reputable opinions [*endoxa*, ἔνδοξα] about them that these have to be discussed, and this task belongs properly, or most appropriately, to dialectic; for dialectic is a process of criticism wherein lies the path to the principles of all inquiries."[91]

One should note here that the opinion of commentators has changed on this topic. Indeed, one used to believe with W. D. Ross that dialectical games belong to "a by-gone mode of thought" and that "his own *Analytics* [...] have made his *Topics* out of date,"[92] but today's tendency is to follow Jonathan Barnes and claim that Aristotle

> "[...] nowhere suggests that any other method will lead to results which conflict with or go beyond the results achieved by the method of *endoxa*."[93]

Of course, this is not an issue concerning the interpretation of Aristotle to be resolved here, although our approach is clearly in line with the more recent tendency.

8 From dialectic to syllogistic

To conclude, we would like to make some brief and tentative remarks about the origins of logic. Aristotle is credited for the codification of the rules of deduction

---

[90] Plato, *Republic* IV, 437a. See also *Cratylus* 436d on the necessity of an 'adequate examination' of first principles in order to uncover hidden inconsistencies.
[91] Aristotle, *Topics* I, 1, 101a 36-b4.
[92] Ross (1949), p. 59.
[93] Barnes (1980), p. 186. On this point, see the exchange between Bolton and Brunschwig in Bolton (1990) and Brunschwig (1990).

for some syllogisms in his *Prior Analytics*, which is usually considered, for a variety of reasons, the first book of logic.[94] He also happens to have written the only treatise on dialectical games handed down to us in complete form, the *Topics* (to which one should append, as we saw, the *Sophistical Refutations*). The chronology of the books composing the *Organon* is a matter of debate, but we assume here that the discovery of syllogistic was a late one and that the *Topics* is thus an earlier treatise than *Prior Analytics*. One may naturally ask: What prompted Aristotle to discover his syllogistic?

Our approach through parallels between dialectical and dialogical games suggests a possible answer. It has to do with a distinction, which is seldom fully grasped, between the level of *matches*, where anyone who has mastered the rules can go on playing, and the level of *strategies*, which involves the handling of some procedure. In other words, when the rules for dialectical games are set up and their knowledge is shared among epistemic peers, then anyone may play such games. Under these conditions, for a given thesis *A*, one play may result in a win, another in defeat, etc. One may go on with multiple plays, hoping for constant victory, in order to ascertain that one is entitled to *A*. If one remains at this unreflective level, one may win or lose by sheer luck, as one's opponent turns out to be either an imbecile or somehow naturally endowed with a superior ability at arguing. In other words, it is like playing chess having no clue about strategy. It is an entirely different thing to look at strategies, i.e., not only at the ability to play through simple mastery of the rules, but through reflection on games, i.e., study of tricks that will ensure a win. In dialogical logic, this is an important distinction, since it is at the level of strategies that one finds *validity*. There is evidence, e.g., from lists of titles in Diogenes Laertius, that at least at the time of Socrates, or even before and not only in Athens 'cram-books' were written to teach how to play dialectical games. Alas, only part of the *Dissoi Logoi* or *Double Arguments* has survived, this being our only available example of a 'cram-book.' Now, we suggest that the move from 'cram-books' such as the *Dissoi Logoi* to Aristotle's *Topics* and, further, *Prior Analytics* is explainable as an analogous move from the level of simple match-playing to the level of strategical thinking, through the discovery of winning 'combinations,' as in chess,[95] to reflection on the reasons why these 'combinations' work.

Jaakko Hintikka gave us a clue when he wrote about the difference between what we have called dialectical games and his own interrogative games (which include some purely deductive steps):

"In a sense, even the main formal difference between Plato's dialogical games and my interrogative ones had already been introduced by Aristotle. He was as competitive as the next Greek, and

---

[94]For example, someone could rightly insist on the introduction of the variable by Aristotle as a key moment in the history of logic.

[95]The analogy, a very good one, is Ryle's, Ryle (1966), p. 118.

hence was keenly interested in winning his questioning games. Now any competent trial lawyer knows what the most important feature of successful cross-examination is: being able to predict witnesses' answers. Aristotle quickly discovered that certain answers were indeed perfectly predictable. In our terminology, they are the answers that are logically implied by the witness' earlier responses. By studying such predictable answers in their own right in relation to their antecedents, Aristotle became the founder of deductive logic."[96]

We do not claim, however, that no meta-discussion or reflection at the strategy level really happened before Aristotle. On the contrary, our claim would be that there was a considerable lot going on, even before Socrates (Aristotle was not, to paraphrase Hintikka, the only competitive Greek), and that Aristotle capped this process only by operating more self-consciously and systematically at the level of strategies - after all, this is the very *raison d'être* of his *topoi*.

There are two steps involved here. The first one would be to learn through experience what series of moves will insure victory, i.e., what chess players would called 'combinations.' It is clear that a document like the *Dissoi Logoi*, or even Plato's dialogues in their own way, provide examples of such 'combinations,' and the numerous 'cram-books' that are now lost were probably all of this nature. At the very end of the *Sophistical Refutations*, Aristotle, aiming at Gorgias in particular, criticizes these 'cram-books' because they merely give recipes, but do not teach any 'know-how:'

> "[...] the training given by paid professors of contentious arguments was like the practice of Gorgias. For he used to hand out rhetorical speeches to be learned by heart, and they handed out speeches in the form of question and answer, which each supposed would cover most of the arguments on either side. And therefore the teaching they gave their pupils was rapid but unsystematic. For they used to suppose that they trained people by imparting to them not the art but its products, as though anyone professing that they would impart a form of knowledge to obviate any pain in the feet, were then not to teach a man the art of shoe-making or the sources whence he can acquire anything of the kind, but were to present him with several kinds of shoes of all sorts - for he has helped him to meet his need, but has not imparted an art to him."[97]

And immediately after this Aristotle goes on praising the novelty of his own work on that score, thereby implying that his syllogistic came about as reflection on strategies:

> "Moreover, on the subject of rethoric there exists much that has been said long ago, whereas on the subject of deduction we had absolutely nothing else of an earlier date to mention, but were kept at work for a long time in experimental researches."[98]

The 'experimental research' alluded to here probably consisted of repeated plays of dialectical games about a given thesis. Aristotle also gives in *Topics* concrete advice to students, which includes something like learning what chess players would called 'combinations:'

---

[96] Hintikka (2007), p. 3. This is not the first statement of that idea. See also Hintikka (1993), pp. 14-19.
[97] Aristotle, *Sophistical Refutations*, 34, 183b 36-184a8.
[98] Aristotle, *Sophistical Refutations*, 34, 185b 10-184a8-b5.

"It is best to know thoroughly argument upon those problems which are of most frequent occurrence, and particularly in regard to those theses which are primary; for in discussing these answerers frequently give up in despair. Moreover, get a good stock of definitions; and have those of reputable and primary ideas at your fingertips; for it is through these that deductions are effected. You should try, moreover, to master the heads under which other arguments mostly tend to fall. For just as in geometry it is useful to be practised in the elements, and in arithmetic having the multiplication table up to ten at one's fingers' end makes a great difference to one's knowledge of the multiples of other numbers too, likewise also in arguments it is a great advantage to be well up in regard to first principles."[99]

Following Slomkowski, we can see Aristotle's *topoi* in light of this as universal propositions or principles (often of the form of biconditionals) that can be used as premises of hypothetical syllogisms, knowledge of which helps one to win dialectical games, *via* a deduction.[100]

It would be wrong to infer from Aristotle's claim to novelty, above, that there was no thinking about 'combinations' before him, as the existence of 'cram-books' confirms this, but it is also interesting to note that Plato's dialogues can be seen in that context as useful for *pedagogical* purposes: They provide examples of 'combinations' and one can easily see how they could have been used in the Academy for teaching students how to play dialectical games. Ryle had already noticed that the same line of argument is found in the *Dissoi Logoi* and in *Protagoras*, where it is put in the mouth of Protagoras.[101] It is interesting to note further that, for obvious reasons given that he is out to debunk 'those reputed wise,' Socrates often uses a 'combination' surrounding skill or craft (*techne*, τέχνη). It consists in getting his adversary to commit himself to a link between the thesis debated and *techne*, and then use this to derive further consequences that lead to the contrary thesis, e.g., using the *techne* of the doctor and that of the captain in *Republic* 341d[102] to show, against Thrasymachus on 'Justice is the interest of the stronger,' that through his *techne* the stronger determines what is also of the interest of the weaker; in *Ion*, one finds a 'combination' around *techne* beginning at 532c, to undermine Ion's

---

[99] Aristotle, *Topics* VIII, 163b 17.
[100] Slomkowski (1997), p. 3.
[101] Ryle (1966), p. 116. See *Dissoi Logoi*, DK90, 6, 11 and Protagoras 327b-c.
[102] It is interesting to note that R.G. Collingwood referred to an earlier 'combination' on the very same terms, at *Republic* 333b, as exemplifying his claim that "By 'right' I do not mean 'true.' The 'right' answer to a question is the answer which enables us to get ahead with the process of questioning and answering." He goes explaining that "Cases are quite common in which the 'right' answer to a question is 'false;' for example, cases in which a thinker is following a false scent, either inadvertently or in order to construct a *reductio ad absurdum*. Thus, when Socrates asks (Plato, *Republic*, 333b) whether as your partner in a game of draughts you would prefer to have a just man or a man who knows how to play draughts, the answer which Polemarchus gives - 'a man who knows how to play draughts' - is the right answer. It is 'false,' because it presupposes that justice and ability to play draughts are comparable, each of them being a 'craft,' or specialized form of skill. But it is 'right,' because it constitutes a link, and a sound one, in the chain of questions and answers by which the falseness of that presupposition is made manifest," Collingwood (1939), pp. 37-38. Although we would object to calling the 'right' answer 'false,' it is clear that Collingwood understood very well the dynamics of dialectical games.

claim to possess the *techne* of the rhapsode, etc. This is another topic well worth further investigation.

Another very important feature of Aristotle's *Topics*, for which he claims novelty, is that he does not provide a textbook for **O** but for **P**, a change of perspective that fundamentally alters the nature of the game.[103] As we saw, the key moment of the dialectical game is the inferential step in rule (ix), when **O**, in possession of all elements, adds them up to infer the negation of the thesis at stake, thus driving **P** into an *elenchus*. The codification of this inferential step as, e.g., of the form of a *Modus Tollens*, does not just mark the birth of logic, it was for Aristotle providing the prospective *proponent* of dialectical games tools for winning. A deep philosophical question about logic, which almost never raised by philosophers is: Whence this or that rule but not another one? To explain the origin of Aristotle's syllogistic in terms of dialectical games, as we merely suggest here, might actually give us an important clue.

9 Acknowledgements

Some of the material included in this paper was presented by Benoît Castelnérac at annual congresses of the Canadian Philosophical Association in 2006, 2007, and 2008, and we would like to thank the commentators (François Renaud, Julien Villeneuve and John Thorpe) and the participants for their comments. The idea of an investigation of dialectical games from a dialogical standpoint was first expressed in Marion (2008), p. 18, where our paper was announced in a footnote under a different title. Earlier versions of this paper were presented at the Jockey Club, Department of Philosophy, Memorial University of Newfoundland in October 2008, at the Université Laval, Quebec City in the following month, and at the Montreal Interuniversity Workshop in the History of Philosophy in January 2009. We would like to thank participants, as well as Hugh Benson, Luc Brisson, Catarina Dutilh Novaes, Laurent Keiff, Neil Kennedy, Sara Magrin, Calvin Normore, Dario Perinetti, Shahid Rahman and, especially, Helge Rückert for their insightful comments. To all we are indebted, as their comments led to numerous improvements, from one version to another. We became acquainted with Rahman and Keiff (forthcoming), only at proof stage, and could not really take it into account. Their point of departure is also dialogical logic but they analyse the relationship between logic and rhetoric, mainly from Aristotle's standpoint. Nevertheless, there were too many points of contact with the present paper for us

---

[103] See Aristotle, *Topics* VIII, 5, 159a 34. This crucial change of perspective was, to our knowledge, first pointed out by Jacques Brunschwig (2007, pp. 278-279), who also correctly points out that it introduces deep changes in the nature of the dialectical games. However, the idea that it 'depersonalizes' dialectic – Dorion (1997) – is somehow exaggerated from our point of view; We have given reasons to believe that in Plato's dialogues too, one must give the answer one considers logically appropriate, not just the answer one 'believes.' It seems to us that the emphasis on **P** is rather of importance because it opens the door to the possibility of *arguing for validity*.

to take them into account at such a late stage.

BIBLIOGRAPHY

Aristotle. *The Complete Works of Aristotle*, 2 volumes, J. Barnes (ed.), Princeton University Press, Princeton NJ, 1984.

Barnes, J. (1980). "Aristotle and the Method of Ethics", *Revue internationale de philosophie*, vol. 34, pp. 490-511.

Benson, H. H. (1987). "The Problem of the *Elenchus* Reconsidered", *Ancient Philosophy*, vol. 7, pp. 67-85.

Benson, H. H. (1989). "A Note on Eristic and the Socratic *Elenchus*", *Journal of the History of Philosophy*, vol. 27, pp. 591-599.

Benson, H. H. (1995). "The Dissolution of the Problem of the *Elenchus*", *Oxford Studies in Ancient Philosophy*, vol. 13, pp. 45-112.

Bolton, R. (1990). "The Epistemological Basis of Aristotelian Dialectic", in D. Devereux & P. Pellegrin (eds.), *Biologie, logique et métaphysique chez Aristote*, pp.185-236, Éditions du C.N.R.S., Paris.

Brandom, R. (1994). *Making it Explicit. Reasoning, Representing, and Discursive Commitment*, Harvard University Press, Cambridge, Mass.

Brickhouse, T. C. and Smith, N.D. (1984). "Vlastos on the *Elenchus*", *Oxford Studies in Ancient Philosophy*, vol. 2, pp. 185-195.

Brunschwig, J. (1990). "Remarques sur la communication de Robert Bolton", in D. Devereux & P. Pellegrin (eds.), *Biologie, logique et métaphysique chez Aristote*, pp. 185-236, Éditions du C.N.R.S., Paris.

Brunschwig, J. (2007). *Aristote, Topiques, livres V-VIII*. Traduction, introduction et notes par J. Brunschwig, Les Belles Lettres, Paris.

Collingwood, R. G. (1939). *An Autobiography*, Clarendon Press, Oxford.

Davidson, A. I. (1990). "Spiritual Exercises and Ancient Philosophy: An Introduction to Pierre Hadot", *Critical Inquiry*, vol. 16, pp. 475-482.

Dissoi Logoi. *The Greek Sophists*, translation & introduction By J. Dillon & T. Gergel, pp. 318-333. Penguin, London 2003.

Dorion, L.A. (1990). "La subversion de l'elenchus juridique dans l'*Apologie de Socrate*", *Revue philosophique de Louvain*, vol. 88, pp. 311-344.

Dorion, L.A. (1995). *Aristote. Les réfutations sophistiques. Introduction, traduction et commentaire*, Vrin, Paris.

Dorion, L.A. (1997). "La 'dépersonnalisation' de la dialectique chez Aristote", *Archives de philosophie*, vol. 60, pp. 597-613.

Dutilh Novaes, C. (2007). *Formalizing Medieval logical Theories*, Logic, Epistemology and the Unity of Sciences, vol. 7, Springer, Dordrecht.

Fontaine, M. and J. Redmond (2008). *Logique dialogique: une introduction. Méthode de dialogique, règles et exercises vol. 1*, London, College Publications.

Frede, M. (1992). "Plato's Arguments and the Dialogue Form", in J. C. Klagge & N. D. Smith (eds.), *Methods of Interpreting Plato and his Dialogues*, pp. 201-219, Clarendon Press, Oxford.

Gorgias. *The Greek Sophists*, translation & introduction by J. Dillon & T. Gergel, pp. 43-97. Penguin, London 2003.

Grote, G. (1865). *Plato and the other Companions of Sokrates*, John Murray, London.

Hadot, P. (1992). *Exercices spirituels et philosophie antique*, 2nd ed., Albin Michel, Paris.

Hamblin, C. L. (1970). *Fallacies*, Vale Press, Newport News VA.

Hilpinen, R. (1982). "On C. S. Peirce's Theory of the Proposition: Peirce as a Precursor of Game-Theoretical Semantics", *The Monist*, vol. 62, pp. 182-189.

Hintikka, J. (1993). "Socratic Questioning, Logic and Rhetoric", *Revue internationale de philosophie*, vol. 47, pp. 5-30.

Hintikka, J. (2007). *Socratic Epistemology. Explorations of Knowledge-Seeking by Questioning*, Cambridge University Press, Cambridge.

Kamlah, W. and P. Lorenzen (1984). *Logical Propaedeutic. Pre-School of Reasonable Discourse*, University Press of America, Lanham NY.

Kapp, E. (1942). *Greek Foundations of Traditional Logic*, Columbia University Press, New York.

Kerferd, G. (1954). "Plato's Noble Art of Sophistry (*Sophist* 226a-231b)", *Classical Quarterly*, vol. 4, pp. 84-90.

Kneale, W. C. and Kneale, M. (1962). *The Development of Logic*, Clarendon Press, Oxford.

Krabbe, E. C. W. (2006). "Dialogue Logic", in D. M. Gabbay & J. Woods (eds.), *Handbook of the History of Logic*, vol. 7, pp. 665-704, Elsevier, Amsterdam.

Le Blond, J. M. (1939). *Logique et méthode chez Aristote*, Vrin, Paris.

Lesher, J. H. (1984). "Parmenides' Critique of Thinking. The *poluderis elenchus* of Fragment 7", *Oxford Studies in Ancient Philosophy*, vol. 2, pp.1-30.

Lesher, J. H. (2002). "Parmenidean *Elenchos*", in G. A. Scott (ed.), *Does Socrates have a Method?*, pp.19-35, Pennsylvania State University Press, University Park Pennsylvania.

Lewis, D. (1983). *Philosophical Papers. Volume 1*, Oxford University Press, Oxford.

Lorenz, K. (1973). "Rules versus Theorems", *Journal of Philosophical Logic*, vol. 2, pp.352-369.

Lorenzen, P. (1960). "Logik und Agon", *Atti del XII Congresso Internazionale di Filosofia*, vol. 4, pp. 187-194, Sansoni Editore, Florence.

Lorenzen, P. (1987). *Constructive Philosophy*, University of Massachusetts Press, Amherst.

Lorenzen, P. and K. Lorenz (1978). *Dialogische Logik.*, Wissenschaftliche Buchgesellschaft, Darmstadt.

Marion, M. (2008). "Why Play Logical Games?", in O. Majer, A.-V. Pietarinen and T. Tulenheimo (eds.), *Games: Unifying Logic, Language, and Philosophy*, pp. 3-26, Springer, Dordrecht.

Mill, J. S. (1991). *On Liberty and other Essays*, Oxford University Press, Oxford.

Moraux, P. (1968). "La joute dialectique d'après le huitième livre des *Topiques*", in G. E. L. Owen (ed.), *Aristotle on Dialectic*, pp. 277-311, Clarendon Press, Oxford.

Nehamas, A. (1990). "Eristic, Antilogic, Sophistic, Dialectic: Plato's Demarcation of Philosophy from Sophistry", *History of Philosophy Quarterly*, vol. 7, 3-16.

Plato. *Complete works*, J. M. Cooper & D.S. Hutchinson (eds.), Hackett, Indianapolis IN, 1997.

Rahman, S., and L. Keiff (2004). "How to be a Dialogician", in D. Vanderveken (ed.), *Logic, Thought and Action*, pp. 359-408, Dordrecht, Springer.

Rahman, S., and L. Keiff (forthcoming). "La dialectique, entre logique et rhétorique", *Revue de métaphysique et de morale*.

Robinson, R. (1966). *Plato's Earlier Dialectic*, Clarendon Press, Oxford.

Ross, W. D. (1949). *Aristotle*, 5th ed., Methuen, London.

Rückert, H. (2001). "Why Dialogical Logic?", in H. Wansing (ed.), *Essays on Non-Classical Logic*, pp. 165-185, World Scientific, Singapore.

Ryle, G. (1966). *Plato's Progress*, Cambridge University Press, Cambridge.

Schmid, W. T. (1998). *Plato's Charmides and the Socratic Ideal of Rationality*, SUNY Press, Albany NY.

Schmid, W. T. (2002). "Socratic Dialectic in the *Charmides*", in G. A. Scott (ed.), *Does Socrates have a Method?*, pp. 235-251, Pennsylvania State University Press, University Park Pennsylvania.

Slomkowski, P. (1997). *Aristotle's Topics*, Brill, Leiden.

Szabo, A. (1978). *The Beginnings of Greek Mathematics*, D. Reidel, Dordrecht.

Trevaskis, J. R. (1955) "The Sophistry of Noble Lineage", *Phronesis*, vol. 1, pp. 36-49.

Vlastos, G. (1991). *Socrates: Ironist and Moral Philosopher*, Cornell University Press, Ithaca NY.

Vlastos, G. (1994). *Socratic Studies*, Cambridge University Press, Cambridge.

Logic, Act and Product

JACQUES P. DUBUCS, WIOLETTA MISKIEWICZ

ABSTRACT. Logic and psychology overlap in judgment, inference and proof. The problems raised by this commonality are notoriously difficult, both from a historical and from a philosophical point of view. Sundholm has for a long time addressed these issues. He begins his beautiful piece of work (2002) by summarizing the main difficulty in the usual provocative manner of the author: one can start, he says, by the act of knowledge to go to the object, as the Idealist does; one can also start by the object to go to the act, in the Realist mood; never the two shall meet. He is himself inclined to accept the first perspective as the right one and he has eventually developed an original version of antirealism which starts, not from considerations about the publicity of meaning, in the manner of Dummett, but from an epistemic standpoint, trying to search in a non-Fregean tradition of analysis of judgement and cognate notions a way of founding constructivist semantics. The present paper ploughes the same field. We concentrate on the significance, for Sundholm's program, of the perspective that has been opened by Twardowski in his important essay on acts and products Twardowski (1912a).

1 Problems for logic in the intentionalist framework

Judgment can be conceived from two different standpoints: the subjective perspective that takes it as an actual episode in the mental life of the judger, and the objective perspective that considers it as related to real or ideal entities whose conformation makes the judgment correct or wrong in a determinate way.

On the objectivist side, one finds philosophers - Frege could be taken as a paragon of this view - who claim that, to make logic objective, there is no other means than to extrude from the mind the contents of the judgements and to consider them as stable and mind-independent entities. Dummett has convincingly shown that this anti-psychologist move, in its original Fregean form, leads to an 'ontological mythology'[1] - one would like to say an 'epistemological mythology' too: by extruding thoughts from the mind, one has to assume that they populate a 'third realm' of reality, distinct of both the physical and the mental worlds, and one has therefore to solve the new problem that results in this way, namely that of explaining how we could cognitively access these thoughts. According

---

[1] Dummett (1996), p.25.

to Dummett, the 'linguistic turn' of philosophy, which is to him the characteristic mark of analytic philosophy, just results from the need of providing a version of anti-psychologism that were free from these mythologies and oddities. The intercalation of words and sentences between the judger and the contents of her judgments preserves the objectivity of judgement while avoiding mythology, by locating meanings outside the mind but firmly in the spatio-temporal world:

> "One in this position has therefore to look about him and to find something non-mythological but objective and external to the individual mind to embody the thoughts which the individual subject grasp and may assent to or reject. Where better to find it than in the institution of a common language?".[2]

This interpretation of the linguistic turn, as well as the question whether it should be really considered as characteristic of analytical philosophy, have been much discussed in recent philosophical literature, and we will leave it untouched. Rather, our aim is to discuss a similar inflection that the partisans of the subjective approach of judgment have envisaged at the same time to solve symmetric problems raised by the primitive formulation of their theory, to examine the reasons why they eventually renounce to such a linguistic turn, and to argue that they were right in doing so. The starting point of the subjectivist conception of judgment is a general analysis of mental life, which simply conceives judgment as one of the moods, *inter alia*, of intentionality, beside presentation (*Vorstellung*), emotion, volition and so on. Judgment enjoys no special status at all: conscious presentation of something is taken as the privileged intentional relation, and judgment is said to follow just the same way as other conscious activities. In the most often quoted passage of his writings, Brentano expresses the point as follows:

> "Every mental phenomenon is characterized by what the Scholastics of the Middle Ages called the intentional (or mental) inexistence (*Inexistenz*) of an object, and what we might call, though not wholly unambiguously, relationship (*Beziehung*) to a content (*Inhalt*), direction toward an object (which is not here to be understood as meaning a reality (*Realität*)), or immanente objectity (*Gegenständlichkeit*). Every mental phenomenon includes (*enthält*) something as an object within itself, although they not all do so in the same way. In presentation something is presented, in judgment something is asserted (*anerkannt*) or rejected (*verworfen*), in love loved, in the hate hated, in desire desired and so on."[3]

In the Brentanian perspective, propositional attitudes (e.g. judgments) are therefore treated in the same way as non-propositional attitudes as love or hate and the last ones are taken as fundamental.[4] The objects of the judgments are considered as no more structured than the intentional objects involved in other mental phenomena. As a consequence, judgments can even no longer be viewed, in the

---

[2]Dummett (1996), p.25.
[3]Brentano (1995), p.88. The translation of *Beziehung* by 'reference', as in the English edition quoted, is misleading; 'relationship' should be definitely preferred.
[4]Cf. Brandl (1996b), p.263.

Aristotelian mood, as referring to conglomerates of elements as subjects and predicates: according to Brentano, every judgment is judgment of existence. According to the 'ideogenic' theory he defended, the object of a judgment is always a thing the existence of which is affirmed or denied (e.g. when someone is judging that S is P she is actually judging that some S that is P exists): the whole judgment consists in the approval (or denial) of the existence of the presentation you have:

> "It can be shown with utmost clarity that every categorical proposition can be translated without any change of meaning into an existential proposition, and in that event the 'is' or 'is not' of the existential proposition takes the place of the copula.
>
> I want to prove this with some examples.
>
> (...) The categorical proposition 'No stone is living' means the same as the existential proposition 'A living stone does not exist' or 'There is no living stone'."[5]

This intentionalist perspective on judgment poses two kinds of problems for logic: the question of the unstructuredness of their contents - judgment is just affirmation or denial of a single presentation -, and the question of their volatility, namely that of their confinement in the episodes of the mental life of individual subjects.

1.1 Lack of structure

Logic and psychology jointly contribute to the analysis of judgment. According to a widely shared consensus, psychology deals with the mental act of judging, while logic primarily deals with the content of judgment, and with the act only in an indirect way. Logical constants (negation, conjunction, etc) belong to the content and give it its logical structure. The first duty of logic is to tell, given this structure, what the truth-value of the whole content is, supposing the truth-values of the constituent parts that are articulated in the structure be already known. As a second duty, logic has, of course, to tell also which judgments are correct, but this second task is a trivial by-product of the first one: a judgment will be considered as correct if it consists in the assertion of a true content or in the denial of a false one, and as incorrect in the other cases. That amounts to say that, should be no logical structure in the content of judgments, logic would have nothing to say about judgments. To the logicians, the first, formidable, difficulty of the intentionalist approach of judgment is this: as intentionalists consider the contents of judgments as simple and devoid of any logical structure, they have nothing to do but keep mute about judgment if the intentionalist is right.

This worrying situation could have been a motive, for logic-minded intentionalists, to bring some changes in the Brentanian credo by putting forward, in a way or another, structured entities - linguistic contents - that were intimately associated to the judging-acts.

---

[5]Brentano (1995), p.213-214.

One has however to be careful about the exact nature of the problem for which a solution was sought by intentionalists. It has been sometimes said,[6] that 'Brentano's problem' consisted in explaining intentionality - taken as a relation of the subject to propositions, as it is apparently the case in belief, desire and other mental states - in a naturalistic frame. In other words, Brentano is interpreted as having raised, and declared to be unsolvable in a materialist setting, the question how it is possible, to a bodily creature, to be related to a certain object while no causal transaction between the creature and the object can be plausibly envisaged as a support or conveyor for the relation. To *this* enigma, a familar scheme of answer consists in decomposing the intentional relation (let us say x's belief that A) into two different parts. The first one relates x to a token of a sentence p meaning that A - this first relation is innocuous from a naturalist standpoint, as both relata are physical entities (e.g. x is ready to sincerely and overtly assent to p when suitably interrogated, or x has some neural code for p in her belief-box, or something else) -, and the second one relates the sentence-type p to its meaning, namely the proposition that A. This second relation is taken as naturalistically innocent as the first one, for it amounts to the familiar relation of linguistic meaning that connects words and things in absence of any physical relation relating them. In short, one solves the intentionality enigma in putting all the weight of the problem on the allegedly unproblematic relation of linguistic meaning. Words, as it were, are in charge of the travel.

For several reasons, *this* 'linguistic turn' has never been seriously envisaged in the intentionalist tradition. First, one has to keep in mind the deep difference between contemporary analytic philosophy of knowledge and mind, and the scene where Brentano, Twardowski and others were playing: as it has been convincingly established by Haldane (1989), the intentionalist tradition was not at all interested in the contemporary 'enigma' of the incorporation of intentional mental states into a physicalist account of the world. Second, the basic items of intentionalist conception are things we do or perform - mental *acts* -, not things we have or live in - mental *states*. This utterly different conception is much less suitable for a 'linguistic turn' in the way Field envisages. Third, Brentano did not assume any strong correspondence between acts of judging and sentences that might be intimate to them. He defends a non-propositional theory according to which, it is sufficient, for two judgments being the same, that their matter (the presentation involved in them) and their form (affirmative or positive) are the same. Nothing linguistic at all is involved in these identity conditions. Moreover, he considers that linguistic utterances - including 'inner' utterances in the silent speech of the judger - are not compulsory ingredients of judging acts for another reason yet, namely that, to him, a judging activity can be performed even in the absence of conceptual content, as for example when we accept a perceptual experience of something cyan, while we

---

[6]Field (1978), pp.9 and following, is the *locus classicus* in this respect.

do not possess or do not apply the concept of 'cyan', or when we express our acknowledging of a painful experience by simply crying rather than in an articulate way. Four and not least, intentionalists were well-conscious of the necessity of making room for logic in the analysis of judgment, and therefore of taking it as equipped with a logical structure but, as we are going to develop in the last part of this paper, they considered that this structure should be put on the act-side, not on the content-side.

1.2 Volatility

Intentional existence, or rather inexistence, existence in the improper (*nicht wirklich*) sense, is as transitory and personal as the mental acts that host it. The objects to which mental acts refer to just inhabit (*einwohnen*) the individual mind while it acts so and so, and judgment does not make an exception to this rule. What could be added to that? Brentano was the first one to be not interested in the question of the 'ontological status' of intentional objects (an expression that he, incidentally, did not use very often), but he became more and more concerned with a misinterpretation of his doctrine in terms of *entia rationis* - his phrase for Platonic entities - and eventually decided to cut short this reading, which he considered as parasitic and mistaken.

An appealing strategy, to deal with a problematic relatum of a relation, is *adverbialization*: one ceases to consider the suspect as a genuine relatum, and one transforms it into a mere *modifier* of the relation, which has therefore one argument less than before. In another context, speaking of propositional attitudes instead of judgments, Quine has envisaged favorably this strategy in a well-known passage of *Word and Object*. After having canvassed various construals of belief-objects, he writes:

> "A final alternative I find as appealing as any is simply to dispense with the objects of the propositional attitudes. We can continue to formulate (...) the propositional attitudes with the help of the notations of intensional abstraction (...) but just cease to see these notations as singular terms referring ot objects. This means viewing 'Tom believes [Cicero denounced Cataline]', no longer as of the form '*Fab*' with a = Tom and b = [Cicero denounced Cataline], but rather of the form '*Fa*' with 'a = Tom and complex '*F*'."[7]

Brentano made a similar move at the end of his carreer, when he proposed a purely adverbial theory of judgment - as well as of mental acts and life in general -, in which the question of the ontological status of intentional objects was radically emptied of any sense: to mentally refer to an object ω eventually became an accidental and transitory monadic state of the subject, a 'ω-thinking', comparable in this regard to a 'sitting' or to a 'singing'. The price for such an adverbialization is high. First, by making the contents of judgment *mute*, one endangers the architecture of cognitive life: the judgment that *Not A* - more exactly, the judgment that the

---

[7] Quine (1964), p.216.

man in the street calls in this way - should be more, and more specifically, related to the judgment that *A* than sitting and singing are related. Second, one fails to see, in absence of stable relata, what the identity conditions of judgments could be, and what the guarantee that judgments could be made another time by the same individual on another occasion, *a fortiori* how they could be made, with certainty of dealing with the same judgment, by another individual. Now, such iterability and sharability seem mandatory for argumentation, rational discussion, and logic. How to conceive individual and dated acts of judging as token of judgment-types that can be realized by other tokens, in other circumstances or by other individuals? Here again, there could have been another motive, for the intentionalist tradition, to make adopt a 'linguistic turn.'

Roughly expressed, the stabilization problem is this. Brentano-style psychology has to do with instantaneous acts of mind, while logic is, *prima facie*, concerned with more stable items. Should this last assumption - logic has necessarily to do with something stable - be taken as granted, the development of a logic in accordance to the intentionalist principles ought to overcome the momentariness of mental acts by proposing a conception that associates to them stable entities. This stabilization can be sought in two directions, either in considering mental acts as intermittent effects of stable dispositions, or in considering them, or their products, as tokens of general types. Let us consider these two strategies in turn.

The dispositionalist strategy, all things considered, is hardly relevant to the intentionalist case. For, either we take 'disposition' in a weakly causal sense, alluding to the fact that the mental life should enjoy some minimal stability - *once a first judgment has been performed*, it is uncommon that the judger randomly deviates -, or we take it in a strongly causal acception, which has to do with a *direct* explanation of series of transitory judging-acts by deeper and more stable doxastic propensities that were just revealed by the acts they cause. Dispositions, weakly conceived, are of few use: they leave us just with the question we were dealing with, namely that of giving sense to the idea of the repetition of a judgment. The strong sense of 'disposition' amounts to considering human beings, as it were, as *driven* to judge intermittently but repeatedly in the same way by an underlying force, the same way radio-active nucleus are bound to lose energy by periodically emitting ionizing particles; but this account does not the job either, for the analogy rests on a mistake. Propensity to radio-active decay gives raise, at each time of its actualization, to a particle emission from which a punctual decay of radio-activity actually results. By contrast, propensity to associate presentations - which is the way dispositionalist theorists (Hume, Bain, Mill, ...) conceive dispositional beliefs - simply does not result in judging acts, but rather in actual linkages between presentations, what is throughly distinct.[8] Thus, dispositional beliefs, because they

---

[8] Brentano (1995), II, VII, §2. That is not to say, of course, that psychologists near to Brentano remain indifferent to the question of mental habits or psychological tendencies. Quite the opposite, some

are not dispositions to judge, are not a suitable way of stabilizing judging acts.

The other strategy for stabilizing judging-acts, namely by considering them as tokens of some types, looks much more promising, but it has to overcome a certain number of difficulties:

> a) As the most attested and apparently unproblematic range of application of the pair type/token is the domain of linguistic items (*grosso modo*, there is no insuperable difficulty in asserting that there are two tokens of the type 'there' that are enclosed in this parenthesis), there is a strong temptation to take a 'linguistic turn,' namely to embedd mental acts into a world of words and to define act-types as acts directed toward sentence-types. Nonetheless, this move is just unfaithful to one of main Brentano's motto, namely that phrases as '*x* judges that *A*' can be at most tolerated as neutral, minimal and non-committing ways of speaking of mental acts, or of reporting them, in the common idiom, but that they can never be taken at face value, because the actual content of judging-acts is not propositional.
>
> b) The type/token distinction can be constructed in different ways, which are not equivalent with respect to the scope and significance of the intentionalist enterprise. On the one hand, types can be conceived in an ideal way as entities enjoying some ontological priority relative to the entities that tokenize them (this priority might go so far as to consider as possible, at least conceptually, types that were not exemplified at all); on the other hand, one can think of types as merely resulting by an abstraction process from classes of actual or potential items. Each option has its own advantages and inconveniences. The downward strategy is certainly more appropriate to insure the 'purity' of logic, while the ascending one is largely more comfortable to guarantee that the standards of logic can be actually met.
>
> c) The types and tokens that are at stake in the present discussion are classified according, not to some unproblematic equivalence relation as typographic equiformity, but to equivalence relations that present much more difficulties, as equivalence in meaning. One has therefore to be prepared to envisage various construals of typification, depending for example on what analysis is provided for the very notion of meaning.

2 Intentionalist ways out

2.1 Stabilization

As said just above, various strategies of stabilization can be envisaged, and have been in fact developed, in the intentionalist tradition.

---

of them had strong interest in these dispositions, falling under the umbrella of the physical determinants of psychic activity. Cf. Höfler (1897), §§12 and 25.

Husserl (1900-01) is often taken as having found the Holy Grail: a whole conception of mental life, meaning and logic which meets the opposite requirements of psychic accessibility and objective ideality of the contents. To insure that we are speaking of the same meaning across repeating judging acts, Husserl assumes that acts of judgment have a transcendent ideal content beyond their immanent real one. These timeless and non-individual contents, ideal conditions of possibility of acts, are nevertheless conceived as displaying themselves before the judger, and therefore as able to be accessed in individual transitory acts of judgement. A detailed discussion of how and whether this kind of conciliation between logic and psychology is possible is largely beyond the scope of the present paper. We prefer to focus the discussion on the alternative analysis presented by Twardowski (1912a), which differs from Husserl's conception on several significant points.[9]

After careful examination of the intricacies of the relationship between mental acts and their linguistic expression, Twardowski canvasses an original and detailed theory of the objectification of judgments, which tries to prevent the mistakes that arise from indiscriminate reliance on their public expression. He firstly considers some wrong ways of stabilizing mental activities by means of language. The most pernicious one, he says, lies in confounding mental acts with their 'external' linguistic expression, as when one equates a singular act of judgment with the affirmative sentence that the judger is disposed to utter at the same moment. The verbal utterance that sometimes accompanies the mental act is in itself unproblematic and hardly worth of mention, as a neutral, minimal and non-committing way for the judger of conveying some information about her own mental life in the colloquial idiom. However, this *enunciatio* is often approximative and in relaxed style, just determinate enough to put the hearer who shares the same situation in a position of getting an idea of the act that has been performed. The tendency of taking this sentence as a perfectly reliable guide to grasp the act leads to major inconveniences, just because the reusability of the sentence does not fit with the momentariness of the act. As early as 1900, Twardowski had shown that this mistake is at the root of the relativism of truth, namely of the doctrine according to which the truth-value of the judgments is subject to variations depending on circumstances. The relativist

---

[9]Twardowski, a pupil of Brentano, was the founder of the Polish analytical School of philosophy. He was not himself a logician in the technical sense of the term. Not to mention Frege, he had no deep acquaintance with mathematics as, for example, Husserl could have had. While he was convinced of the importance of Brentano's contribution to logic, he renounced eventually to publish the systematic exposition he had in mind. His manuscript Twardowski (1924), which contains significant developments on Brentano's reducibility of any judgment to existential form, stopped after a striking comparison between Brentano and Bolzano. Nevertheless, Twardowski never stopped to insist, even against some of his own fellows, on the validity of the 'ideogenic' theory of judgments. He wrote a whole paper in 1907 to defend it, while he recognized that he was at the time in Poland, as it were, the last of the Mohicans: "Only my text adopts the idiogenetic perspective with respect to the essence of a judgment, whereas all of the others subscribe to one of the allogenetic theories", Twardowski (1907), p.99.

wrongly assumes, on the basis of the invariance of the linguistic by-products of the judgments, that the judgments themselves do not vary according to circumstances. The use of indexical terms by the judger are a typical occasion for such mistakes. The correct view is that the sentence 'It's raining now' does not express the same judgment at any time. Twardowski stresses that the element that varies according to the circumstances is not the truth-value of the judgment, which remains fixed, but the judgment expressed by the sentence:

> "Although there is a very intimate connection between a judgment, on the one hand, and a statement, which is the external expression of the judgment, on the other, the statement is nonetheless not identical with the judgment (...). The relativists, however, do not take this distinction into account, and only because of this lack of rigour are they in a position of adduce examples of judgments which apparently support their theory concerning the existence of relative truths (...). In the absence of appropriate indications, or as a result of inattentiveness, one may be misled to believe in such cases that an invariance in the expression is accompanied by an invariance in the product bound up to it. Upon closer examination it turns out that the invariance in the mental product was merely illusory, and that in fact what we have is an identity of external expression for two different mental products."[10]

To recover the true content of the judgments from the sentences that are uttered on the occasion of their performance, the replacement of the indexicals by absolute spatio-temporal coordinates is however not sufficient. We should also grasp the whole way the judger uses the words in these sentences and what he really intends to mean by them. To that, it is not enough to belong to the same linguistic community as him, for the relevant notion of meaning is less that of *linguistic* meaning than that of *speaker's* meaning, namely a notion of meaning that involves the communicative intentions of the speaker. As Twardowski notices, the cause of the fact that "the same statement can express different judgments is to be found (...) in the manner in which we employ speech for expressing our judgments," as it is plainly evident by the fact that we may indifferently answer by 'Yes ! I have' to the two questions whether we have read Sienkiewicz's *Quo vadis* and whether we have ever been married (Twardowski (1900), p.150.).

Twardowski's thesis is not, however, that some judgments might be properly ineffable. The external expression of judgment is indeed in most cases able to "accomplish its objective perfectly in colloquial speech," for the resulting statements are generally intelligible to those around the speaker, since "the circumstancies accompanying the speaker's words fill in for what the words do not express"

---

[10] Twardowski (1900), p.149. At the time of this article, Twardowski has not yet established his distinction between actions and products, to which Twardowski (1912a) is entirely devoted. In Wartenberg's German translation checked by Twardowski (see Twardowski (1902), p.39) one simply separates the 'sentence', which expresses a judgment, from the judgment itself as a 'mental activity' (*psychische Tätigkeit*). Generally speaking, Twardowski was a passionate and competent philologist, attentive to cross-linguistic evidence and trying to find the most rigorous way of expressing his philosophical views in several languages (Polish, French, German, English). That is the reason why, to interpret his writings, one should keep in mind all the versions and translations of his texts. The systematic trilingual edition Twardowski (1912b) of his seminal work has been settled to meet this requirement.

(Twardowski (1900), p.150). But his analysis shows that the recourse to the very sentences uttered by the speaker can be hardly considered as a suitable way to stabilize the content of his mental acts beyond the particular circumstances of its performance, for thoroughly different judgments may be expressed by the same words. Linguistic utterances are a reliable guide toward the mental activity of the speaker only insofar as the hearer shares the same communicative situation. For other people, especially for remote readers, the quotational report of theses utterances is of little help to decide with certainty what mental act has been performed. Thus, the worst way of referring to a judgment is the paratactic way ('x judges: it's raining'): by the very nature of the judgment, acts of judging cannot be assimilated to acts of saying. This condemnation extends, *a fortiori*, to the elliptic way that takes the mere *enunciatio* for the judgment itself:

> "Some use 'judgments' to refer to precisely that we here refer to with 'sentence'. Prof Łukasiewicz does so, among others, defining a judgment as a 'sequence of words or other signs which state that some object has or does not has a particular attribute' (Łukasiewicz (1971), p.497). But in treating a judgment as a sequence of words or other signs Dr Łukasiewicz must distinguish from this sequence of words or other signs what constitutes its meaning. As a matter of fact, Dr Łukasiewicz also speaks of 'meaning-equivalent jugdments', defining them as judgments that 'express the same thought in different words' (Łukasiewicz (1971), p. 500). Now, this thought, expressed in those words, is obviously nothing other than a judgment in the sense of a product of an action of judging, thus, if the word 'judgment' is made to serve for designating a 'sequence of words or other signs' that express this sort of thought, an expression will then be lacking for designating such a thought."[11]

The problem raised by Twardowski goes far beyond the 'practical' difficulty of recovering the thought from the linguistic sentence that accompanies its appearance, or even the 'conceptual' problem of introducing a suitable distinction between the sentence and the thought. It lies in the fact that, if one explains or individuates a thought by a certain relationship to the meaning of a sentence p, one will be also committed to explain the secondary thought that the sentence p has just the right meaning, or that the utterance of p is just the most suitable, in context, to guide the hearer to the content of the primitive thought.

Twardowski's gives to his decoupling of the acts of judgment from their linguistic externalisation an original form by means of his distinction between activities (*czynnościach*) and products (*wytworach*). He claims that the statements that accompany judging acts cannot be considered as their result or products - these products are the judgments themselves -, but rather as derived 'psychophysical' products, namely as the products of a concomitant physical activity (e.g. in the case of a lying act, the psychical product is the lie and the psychophysical product it the sentence that has been insincerely uttered). After quoting approbatively Bergmann, who stresses the transitoriness and the momentariness of the products

---

[11] Twardowski (1912a), pp.129-130, fn.56.

of the judging activity[12], he states in clear terms the *stabilization problem*:

> "To be sure, we also say that certain beliefs have persisted through the ages, and that the thoughts of a sage can outlive him. But in these cases the issue is not the continued actual existence of products independent of the actions that produced them; it is, rather, a matter of repeating through a succession of generations actions and products that are similar to those that have occurred in preceding generations, or in that sage. (...) Hence, when we speak on the enduring existence of products of this sort, it is either a matter of the same kind of actions and products repeating themselves, or of their *potential* existence."[13]

In other words, the judgments produced by judging-acts are not more enduring as these acts: generally speaking, "there is (...) no place for mental products within the domain of enduring products."[14] Some momentary acts, of course, generate enduring products, in such a way that these products are ontologically distinct of the acts that give birth to them. The act of building a house, for example, belongs to this category, for the house persists in existing beyond the end of the building act. Not so, however, for mental acts as judgments, whose products instantaneously vanish and cannot survive, except by being regenerated and brought into renewed existence by new mental acts of the same kind. To achieve this process in another way than by mere coincidence, language is required. For the linguistic *by-products* that usually accompany the judgmental activities can continue to exist long after the judging-acts that correspond them if their physical substratum is suitable for that, for example if they consist in written words. That does not mean, however, that these words are the fixed content of the judgments considered. As there is no mental endurance beyond renewal of similar mental activity, these words have a mere rôle of *stabilizers*, which is actually played only in the case where they are read, understood and taken as the support and occasion of a new act of the same type, by the same judger who remembers her previous act in this way and iterates it, or by another judger who is lead in this way to consider the presentation involved in the original judgment and to assent it once again on her side:

> "The issue is not the continued actual existence of products independent of the actions that produced them; it is, rather, a matter of repeating through a succession of generations actions and products that are similar to those that have occurred in preceding generations, or in that sage (...). When we speak of the enduring existence of products of this sort, it is, either a matter of the same kind of actions and products repeating themselves, or of their *potential* existence.[15] It is indeed for this reason that these products may be termed non-enduring, namely, in the sense that they do not endure in the mode of *actual* existence any longer than the action by means of which they originate."[16]

---

[12] Bergmann (1879), p.38: "The judgment (*Urteil*) is what emerges simultaneously with the judging (*Urteilen*), and is the immediately vanishing product of the latter."

[13] Twardowski (1912a), p.116.

[14] Twardowski (1912a), p.119.

[15] This potentiality may be conceived within the broadest possible scope, as when we speak, say, of the 'existence' of truths which no one knows yet, i.e. of the 'existence' of true judgments that no one has ever passed. Obviously, what is involved here is the capability of passing these judgments, and that which exists is not the judgments but the capacity to pass them.

[16] Twardowski (1912a), p.116.

Judgments cannot be stabilized beyond the potential reiteration of the judging-acts. In particular, one is not allowed to consider that stability could be actually achieved by means of language, as if a judging-act consisted in being transitorily related, in the mood of assent or of denial, to a fixed propositional content regularly meant or designated by a certain sentence. Quite at the opposite, in saying that a judging-act is merely *accompanied* by a sentence, or creates a token of it as a mere psychophysical by-product, Twardowski considers that the judgment should be considered as the *meaning* of the sentence. This acceptation of 'meaning' is uneasy to grasp,[17], but it is certainly a modest notion, by large not reaching as far as the Husserlian notion of 'irrevocably identifiable' objective meaning. To Twardowski, judging-acts do not refer, with approbative or rejective force, to sentences. In a diametrically opposite way, it is the sentences that express, or refer to, judgments, considered as products of judging acts:

> "In regard to a mental product that expresses itself *in* a certain psychophysical product - i.e. when that psychophysical product is an expression of the mental product - we occasionally say that the psychophysical product *signifies this mental product*, to wit - that the mental product is signified by means of the psychophysical product. But we only speak that way under quite specific circumstances, namely, when the psychophysical product in which some mental product expresses itself can itself become the partial cause for the subsequent emergence of the same or a similar mental product, and when it plays this role of a partial cause by eliciting the same or similar mental action as that which gave rise to the given psychophysical product."[18]

To sum up: the sentence 'p' means the mental product J iff it is both uttered as a by-product of the mental act of producing J by the judger and taken as an occasion of a similar mental act by the reader.

That characterization deserves explanation.

a) The question raised by Twardowski is not that of the *effability* of thoughts, namely that whether the mental activity is capable of being suitably expressed by the sentence that means its product. For a psychophysical product may express, in this sense, a mental act, without being taken as the support of a new similar act (the second condition for meaning is compulsory). As an example, a scream expresses the pain of the subject without meaning it, for it does not give rise to a comparable pain in the hearer.

b) For the same reason, an utterance cannot be taken as a mere *indication* of the underlying judgment, as smoke indicates the fire that causes it. Intimation (*Kundgebung*), namely the voluntary act of making known to the others which judgment one has performed, is not even sufficient. To mean a judgment, a sentence ought to be a *successful* intimation or command, to the reader, of performing the same judgment. This original feature of Twardowski's conception insures the potential convergence of the judgers without

---

[17] Cf. Buczyńska-Garewicz (1977) and Smith (1989).
[18] Twardowski (1912a), p.121.

achieving it by means of the assumption of ideal meanings or of pre-existing contents waiting for being grasped in a way or another by different individuals: meaning is at the very start defined by the fact that speakers and writers should be able to evoke and provoke in their hearers or readers mental acts that are suitably analogue to theirs own acts. In that way, the meaning of a sentence can be characterized as the equivalence class of the singular mental acts activated in a community by its audition or reading.

c) In defining meaning in this way, Twardowski expresses a considerable skepticism toward any conception that relies on language and meaning to characterize mental activities. For him, there is no room for a 'linguistic turn.' The correct understanding of the others' words is viewed as fragile as such an extent that, to insure it, one has to define meaning in terms that already involve mutual comprehension and possible interaction. As a surprising by-product of this view of meaning, the linguistic expression of the judgment is pushed on the side of illocutionary acts, to which, in principle, it does not belong. Judgment, as usually conceived, does not make on others any comparable effect as it promises, even in the case when the judging-act is accompanied by the production of voiced or written items. Somebody may be afraid by my threat of punishing her, not at any extent, at least in a so direct way, by hearing my judgment about a related topic. Not so in Twardowskian view, which puts the potential efficacy on the mental life of others among the correctness conditions of the utterances of the judger.

d) Twardowski's road toward stabilization is bottom-up, starting from various individual judgments to come by *abstraction* to a class of similar judgments, instead of starting from hypothetical ideal contents and judgments-types directed toward these contents to come then to individual realizations which were tokenizing those ideal types in the mental life of singular judgers. To be plainly rigorous, this abstracting way of achieving generality should settle in clear terms the similarity conditions of individual judgments. By the very nature of the case, one cannot account for these conditions by invoking the conjunction of sameness of propositional content and sameness of attitude (affirmation or denial) toward that content. The similarity should be characterized in psychological terms, by keeping on the side of mental acts. One has to say that Twardowski is less concerned by such a characterization than by establishing that the similarity that is in need of definition can be achieved in a very plausible way by virtue of the bodily, cognitive, linguistic and cultural similarity of the judgers:

> "If an enduring psychophysical product elicits the mental product expressed in it, whether successively in one and the same person, or successively or simultaneously in different persons, then it obviously elicits not just one product but as many as there are actions

> that produce them. Now, these products will not be completely identical, but will differ from each other to a greater or lesser extent. Suffice it to recall how varied are the mental products that arise in different persons who are affected by the same picture or sentence. However, insofar as we regard that psychophysical product as a product that signifies some sort of mental product, the disparity among the mental products elicited by it dare not go too far - there must a group of common attributes in these individual mental products. And it is precisely these common attributes (in which these individual products accord) that we ordinarily regard as the meaning of the psychophysical product, as the content inherent in it, provided of course that these common attributes correspond to the intent with which that psychophysical product was utilized as a sign. This is also why we say that some sentence arouses 'the same' thought in different individuals, although strictly speaking it arouses as many thoughts as there are persons, since these thoughts are not identical to each other. But we abstract away from what makes these thoughts differ, and consider as the thought that comprises the meaning of the sentence only those of the thought's constituents that are in accord with each other and with the corresponding constituents of the thoughts of the person who makes use of that sentence. Thus, we speak of only a single meaning of a sign - barring cases of ambiguity - and not of as many meanings as there are mental products that are aroused, or capable of being aroused, by the sign in the persons on whom it acts. Now, a meaning conceived in this manner is no longer a concrete mental product, but something at which we arrive by way of an abstraction performed on concrete products."[19]

e) Judgement is not a random guess, but a cognitive act on the basis of available evidence. Thus, there is some rationale in the fact that the judgments which have been performed, once they find an adequate expression in meaningful statements, are able to provoke in turn other similar judgments. Of course, the intersubjective network of judgments that arise in this way is not to be understood in a merely causal way: linguistic communication does not amount to telepathy, and the public expression of my judgments just provides you with an occasion of following me by making in turn a similar mental act you are entirely responsible for. Nevertheless, one can wonder whether the condition, stated by Twardowski, of provoking the hearer to perform the same judging-act as the speaker is not a bit too strong to define the meaningfulness of a sentence. For it is commonsensical that one may very well understand a sentence without to approve it and grasp what the others think without being in agreement with them. There is here some risk, to the intentionalist, of an infelicitous confusion between mutual understanding and mutual agreement. The rival perspective, which considers judgments as individual acts of approbation or denial directed toward objective, pre-existing propositional contents, is *prima facie* more comfortable in dealing with the problem, for it clearly separates the act of grasping the meaning of the sentence from that of judging whether the sentence, understood in this way, is right or wrong. To solve the problem, it is not enough to insist that, in Twardowskian conception, a sentence can have the meaning it has, while its utterance remains causally inert in some, or even in most, of the people

---

[19]Twardowski (1912a), p.p.127-128.

who hear or read it. For the point is that meaningfulness is only equated with *potential* provocation to act mentally in the same way as the writer, by which Twardowski means that the utterance should be at least efficient on some readers. Therefore, these readers, say the competent readers of the sentence, are just supposed to testify to its meaningfulness by performing the same mental act as the author: nothing has been gained in this way, unless one assumes some intentionalist counterpart of the verificationist theory of meaning, in posing that an adequate and truly meaningful expression of a judgment should put its competent reader in a position of being able, in principle, of recognizing the well-foundedness of the judgment. Twardowski, who never flirted with such extreme verificationism, canvasses an original way of escaping the difficulty, by putting forward a notion which is, in some respect, germane to the propositionalist notion of unasserted content, namely that of *unperformed judgment*. In the first place, such an inauthentic judgment is not a product without a producing act - which were a mere *contradictio in adjecto* indeed -, but an effect that has, as its origin, some unusual or non-standard source. That happens, for example, in lies, where the insincere statement may have, and is intended to have, as an effect, some mental act in the hearer, while it has not its standard source, namely a comparable act in speaker's mind. Twardowski calls 'artificial statements' (Polish '*sztuczne powiedzenia*'; German '*künstliche Aussagen*'), or simply 'artifacts' such products, that are of course materially undiscernable from the authentic products they imitate:

> "Such artifacts occur frequently within the realm of psychophysical products. Ample use of them is made, e.g., by an actor who assumes a demeanor through which some feeling is to expressed. As a rule, however, the actor actually just pretends to have this feeling, so that this demeanor, that psychophysical product, does not emerge as the result of a genuine feeling that ordinarily expresses itself in such a demeanor, but as the result of a representation of a feeling, that is to say as the result of a represented [i.e. imagined] feeling. (...) It is possible to form artificial, surrogate sentences that are not expressions of actual judgments, but rather expressions of articifical products that substitute for actual judgments, namely, merely represented judgments. Hence, the meaning of these artificial sentences will also not be passed judgments (which is to say, actual judgments) but merely represented judgments - that is to say, the representations of judgments."[20]

Twardowski gets in this way just what he needs to distinguish between misunderstanding and disagreement: the mental activity that a meaningful statement sets in motion in its reader can be limited to the act of *presentation* of the judgment in question, without reaching the very repetition of the judgement in her own mind. The linguistic traces of the mental activity of others are potentially unfaithful, in the sense that they may leave the hearer uncer-

---

[20]Twardowski (1912a), pp.129-130.

tain whether the activity has been actually performed or merely envisaged, not to say fainted. One would say, in the traditional vocabulary, that the force, rather than the content, is specifically vulnerable to communicative indeterminacy. Lie is an ever open possibility: there is no physical feature of the utterance which could be taken as an undefeasible warrant that the sentence has been seriously voiced and that an act of assertion has been actually performed. Expressed in Twardowskian language, that means that there is no undefeasible evidence, on the hearer-side, allowing to decide definitely between the standard interpretation (a judging act has been actually performed by the speaker) and the artifactual one (she has just considered the judgment). The undiscernability of the artifactual and the standard or 'charged' utterances *qua* physical products, as well as the reasonable idea of the superevenience of the psychophysical properties of a product upon its physical properties (same speech sounds or written traces, same mental activation in a given hearer/reader at a specified time and place), recommend that we take, as a condition of the understanding of the utterance, the weakest requirement that ought to be satisfied either in the standard or in the artifactual case, namely that we only require, as a condition of understanding, the performance of an act of presentation of the speaker's underlying judgment. As it is impossible to do regularly better, it is enough, to understand in the right way the linguistic traces of the judicative activity of the others, to form in oneself a presentation of what is like to judge as they do: as regards the very replication of this activity, it is, however, not mandatory.

f) Unperformed judgments play sensibly the same rôle, in the intentionalist framework, as unasserted contents do in the propositionalist perspective, insomuch that Twardowski sometimes expresses himself as if a judging-act, instead of being the approval or denial of a single presentation, were the act of accepting or rejecting a previously unperformed judgment, namely as if a judging-act consisted in giving, at a second stage, the green light for actual status to unperformed judgments that were candidates merely contemplated at a first stage:

> "[Artifactual sentences] are not psychophysical products that express actual - that is to say, passed - judgments. They express merely 'represented judgments,' and these represented judgments only substitute for passed, i.e., actual judgments, just as those sentences substitute for actual sentences, i.e., ones that express actual judgments. In this case, a psychophysical artifact expresses a mental artifact. Preserved surrogate products of this sort present the most extreme case of making mental products independent of the actions owing to which alone they can truly (actually) exist. Operating with such surrogates in both science and everyday life makes it all the easier to slip into operating with non-surrogate products independently of the actions producing them, especially since it frequently happens that actual and surrogate products appear interchangeable, as, e.g., when we eventually pass a judgment which at first we had contemplated with disbelief

(*niedowierzanie*)."[21]

On this basis, Twardowski develops a *demythologized* interpretation of the Bolzanian notion of 'proposition in itself' (*Satz an sich*), which he conceives, not as a mind-independent entity, but as a judgment 'that has been rendered independent from the action of judging,'[22] namely as 'a product that is taken as independent from the activity that normally produces it.'[23]

To summarize, language participates to the stabilization of the judgments, not by playing an *intermediation* rôle between individual judgers and Fregean thoughts enjoying mind-independence and eternity, but by participating as physical substratum in the process of exchanging, presenting and remembering individual judgments that eventually appear as independent from the activity that produced them. At this stage, unperformed judgments may very well be taken as unasserted contents in intentionalist clothes. As we will argue now, a deep difference, however, separates the two notions as soon as logic appears on the scene.

2.2 Logical structure

People who have the slightest tincture of formal logic are taken aback by hearing that Brentano is sometimes credited for a new foundation of the field. Abruptly expressed, one could say that one is at most indebted to him for an analysis of 'It's raining,' not even extendable, to his own avowal, to 'Socrates is running.' His attack against the Aristotelian thesis of the reducibility of sentences to the subject-predicate form was not, by large, unprecedented, and many other offensives are more convincing, as the one that stresses the unlikelihood of the reducibility of every relation to a monadic one. To sum up, Brentano did not measure up, at any degree, to competition with Frege, who was definitely the founder of modern logic.

Moreover, few of Brentano's many followers, admirers, thurifers and commentators have perceived in the right way, and tried to defend as a plausible line research *in logic*, his main original idea, namely that of moving the logical structure from the content of judgment to the very act of judging, in short of taking judging acts as the very stuff of logic and of impulsing a dynamic turn for that discipline.

---

[21] Twardowski (1912a), pp.130-131. Twardowski's own German version adds that frequent *oscillation* between performing and not performing, which is characteristic of intellectual life, reinforces the misleading impression that judgment is approval or denial of mind-independent prior contents:

"Und da wir sowohl in der Wissenschaft als auch im täglichen Leben fort und fort mit solchen unabhängig erscheinenden stellvertretenden Gebilden operieren, umso mehr, als wir sehr oft bald das echte, bald das stellvertretenden Gebilde erzeugen, z.B. in dem Falle wo wir zunächst ein Urteil uns bloß vorstellen, dann es als wahr annehmen also fällen, dann aber wieder an ihm irre werden und es uns aufs neue bloß vorstellen." (Twardowski (1912b), p.121)

[22] Twardowski (1912a), p.131, fn.61.
[23] Twardowski (1912b), p.122.

One has to say that the very founding father of contemporary logic, Frege, had declared a *fatwa* against any such move. He argued lenghty that the logical constants ought to considered as a part of the content of the judgment, not as a modification of the act, and that they are contradictory contents rather than opposite judgements. The topical locus is Frege (1966), but he had stated, as early as Frege (1969), that he considered that principle as one of his highest achievements. There is little doubt that this opinion, coming from someone as Frege, had impressed, not to say intimidated the partisans of intentionalism.

As a matter of fact, the difficulty of moving the logical constants from contents to acts is unequal. For some of them, e.g. conjunction, no real difficulty is raised. These constants are tailored for structuring propositional contents, but they can be acclimated in a smooth and easy way in the realm of judging acts. For example, conjunction makes easily sense for mental acts, for 'x judges that A *and* B' may be truthfully transformed into 'x judges that A and x judges that B.' Not so for other logical constants, especially for negation: clearly, 'x judges that *Not A*' does not amount to 'x does not judge that A.' It is the reason why the treatment of the negation focuses the controversy.

To the partisan of propositional content in Frege-style, the problem of negation is easily solved by internalizing the denial to the current content, namely by prefixing it with a negation sign. She is then in a position of satisfying a certain exchangeability requirement, according to which the denial of A (whatever A could be) ought to be correct if and only if the acceptance of A is wrong: as *Not A* is true iff A is false, the denial of A, once rephrased in the assertion of *Not A*, is correct iff *Not A* is true, namely iff A is false, namely iff the assertion of A is wrong.

The same move is not open to the intentionalist, who cannot invoke the mutual play of negation and truth-values at the propositional content level. Since it certainly belongs to the intentionalist's duties to provide a *logical* explanation of the link between acceptation and denial - these varieties of judgment cannot be simply taken as different, unrelated *species* of the judging activity -, she has to find her own way on the basis of the intentionalist conception of truth and evidence. As we will see, the crucial rôle of unperformed judgments - as distinct of unasserted contents - surfaces at this place.

The intentionalist way out rests on the principle that, to analyse judgments, the right notion is that of evidence rather that of truth, which is only a derivative notion in this field. Brentano's thesis, which simply equates truth with knowledgability, is this, *expressis verbis*:

> "The judgment of someone is correct if she judges in the way, someone would judge it as being evident."[24]

Thus, one can judge truly by guess, or without adequate evidence, provided it is

---

[24] Brentano (1974), p.139.

possible, to somebody else, to judge in the same way with evidence. Returning to the problem of indicating a suitable logical link between affirmation and denial, the question at stake is clearly that of defining the conditions for the *correctness* of these acts. Now, an asymmetry between the two species immediately appears, which has no counterpart in the propositionalist framework. Evidence for *A* entitles to affirm that *A*, while the lack of such an evidence does not entitles to deny the same. Sketchily expressed, the lack of evidence is not a variety of evidence, while the lack of truth is a variety of truth-value. The lack of truth of a propositional content guarantees the correctness of the assertion of the negated content, but the lack of evidence of an unperformed judgment is not a sufficient reason to pass the opposite judgment.

To clarify the point, let us suppose that we have just been informed that someone has performed such or such judgment, and let us keep aside any doubt about her performance. As we are not connected with her by other means than language, we do not know, usually, what kind of evidence she had to judge in this way. Suppose however, first, that she has left, not only written traces of her judgment, but also traces of her evidence for this judgment, for example a formal derivation of a theorem if it is a question of mathematics. We have firstly to check the conformity of these traces to some formal standards, just as if we were evaluating the accordance of whatever physical product with an agreed level of quality, then, in case of positive verdict, to try in our turn to perform the mental act of proof that is required, taking the written by-product as guideline ('real' proofs are not signs in the books of mathematics, but mental activities). Let us now take the most frequent case, when the judgment is communicated without justification. As it has been performed on the basis of an evidence that is hidden to us, we cannot do better than to ask ourselves what is like *in general* to perform such a judgment, and what is like *in general* to have evidence for it. The adequate evidence for a denial cannot be limited to a negative verdict concerning her presumptive reasons to judge as she does, or concerning the reasons that present themselves in her own mind. Actually, an adequate evidence for denial cannot be weaker than a negative review of *any* putative evidence favoring the judgment, result that can be for example achieved by convincing oneself that any 'evidence' of this kind would suffice to establish the opposite of something very simple we have adequate evidence for. To sum up, a denial is, to the evident judger, a *stratified* act, which takes as its basis the presentation of an unperformed judgment and which transforms any putative evidence for it into evidence for another unperformed judgment, known as incorrect.[25]

The decision of putting the logical structure on the act-side rather than on the content-side, not only forces the adoption of a non-classical semantics, but raises the vexed issue of psychologism. Anyway, Twardowski was convinced that Brentano's perspective, understood in this manner, cleared up this infamy in the

---

[25] The analysis above is consonant with the conclusions of Schaar (1997).

right way:

> "The differentiation of mental actions and products, as well as the differentiation of the various types of mental products, may render no modest service (...). Indeed, a rigorous demarcation of products from actions has already contributed enormously to liberating logic from psychological accretions."[26]

As the respective domains of psychology and logic encompass, according to Twardowski (Twardowski (1912a), p.111), the judgment as an act and the judgment as product, this *satisfecit* is threatened by a sort of dilemma. Either one distinguishes *really* between acts and products, for example in considering that judgments as products inhabit a 'third realm,' distinct of both the mental and the physical domains, and then the whole doctrine becomes hardly discernable from Dummett's 'platonician mythology;' or one distinguishes between them in a merely *conceptual* manner, as for example the activity of walking may be distinguished from its concomitant product, the walk, and then the 'rigorous demarcation' between psychology and logic seems to vanish. Actually, there is few doubt that Twardowski chose the second option. The only reservation one can express in this respect has to do with his apparent confidence that *classical* logic could survive the new, dynamic perspective opened in this way.

BIBLIOGRAPHY

Bergmann, J. (1879). *Allgemeine Logik, Volume I, Reine Logik*, Ernst Siegfried Mittler und Sohn, Berlin.

Brandl, J.L., (1996a). "Kazimierz Twardowski Über Funktionen und Gebilde: Einleitung zu einem Text aus dem Nachlass", *Conceptus*, 29, n.75, pp.145–156, Academia Publisher.

Brandl, J.L., (1996b). *Intentionality*, in Albertazzi, L., Libardi, M. and Poli, R. (eds.), *The School of Franz Brentano*, pp.261-284, Springer Verlag, Nijhoff International Philosophy Series.

Brandl, J. and Wolenski, J. (1999). *Kazimierz Twardowski. On Actions, Products and Other Topics in Philosophy*, Series *Poznan Studies in the Philosophy of the Sciences and the Humanities*, Rodopi, Amsterdam and Atlanta.

Brentano, F. (1903). *The Origin of the Knowledge of Right and Wrong*, English Translation by C. Hague of: *Vom Ursprung sittlicher Erkenntnis*, 1889, Archibald Constable Publisher, Westminster.

Brentano, F. (1956). *Die Lehre vom richtigen Urteil* Edited by F. Mayer-Hildebrand, Francke, Bern.

Brentano, F. (1995). *Psychology from an Empirical Standpoint*, International Library of Philosophy, Routledge and Keagan Paul, New York. English translation by A.C. Rancurello, D.B. Terrell and L. L. McAllster of: *Psychologie vom empirischen Standpunkt*, Verlag von Duncke und Humblot, Leipzig, 1874.

---

[26]Twardowski (1912a), p.132.

Brentano, F. (1974). *Wahrheit und Evidenz*, edited by O. Kraus, Felix Meiner Verlag, Hamburg, 1974.

Buczyńska-Garewicz, H. (1977). "Twardowskis Bedeutungslehre", *Semiosis*, 7(2), pp.55-66.

Dummett, M.A.E. (1996). *Origins of Analytical Philosophy*, Harvard University Press.

Field, H. (1978). "Mental Representation", *Erkenntnis*, 13, pp.9-61.

Frege, G. (1966). *Die Verneinung. Eine logische Untersuchung (1918-1919)*, in Patzig, G. (ed.), *Logische Untersuchungen*, pp.213-218, Vandenhoeck and Ruprecht, Göttingen.

Frege, G. (1969). *Kurze übersicht meiner logischen Lehren* (unpublished manuscript of 1906), in Hermes, H., Kambartel, F. and Kaulbach, F. (eds.), *Nachgelassene Schriften*, pp.213-218, Felix Meiner Verlag, Hamburg.

Haldane, J. (1989). "Naturalism and the Problem of Intentionality", *Inquiry*, 32(3), pp.305-322.

Höfler, A. (1897). *Psychologie*, Verlag von F. Tempsky, Wien. 2d edition, Adamant Media Corporation, 1930.

Husserl, E. (1900-01). *Logische Untersuchungen*, Max Niemeyer, Halle. In *Husserliana XVIII* and *XIX/1-2*, M. Nijhoff Publ., The Hague, Boston and London, 1975 and 1984.

Łukasiewicz, J. (1971). "On the Principle of Contradiction in Aristotle", *The Review of Metaphysics*, pp.485-509, Philosophy Education Society, Inc. The Catholic University of America. English translation by V. Vedin of: "O zasadzie sprzeczności u Arystotelesa", *Studium krytyczne*, Kraków, 1910.

Miskiewicz, W. (2004). "'L'affaire Zimmermann'. À propos des influences bolzaniennes dans l'École de Lvov et de Varsovie", in Thouard, D. (ed.), *Aristote au XIXème siècle*, Series *Cahiers de Philosophie. Apparat critique*, vol. XXI, pp.377-394, Presses Universitaires du Septentrion, Lille.

Quine, W.v.O. (1964). *Word and Object* MIT Press.

van der Schaar, M. (1997). "Judgment and Negation", in Hahn, L.E. (ed.), *The Philosophy of Roderick M. Chisholm*, Series *Library of Living Philosophers*, vol.XXV, chapter XIII, pp.291-318, Open Court Publications Company, Chicago.

Smith, B. (1989). "Kasimir Twardowski", in Szaniawski, K. (ed.), *The Vienna Circle and the Philosophy of the Lvov-Warsaw School*, pp.313-373, Kluwer Academic Publisher, Dordrecht.

Sundholm, B.G. (2002). "A Century of Inference: 1837-1936", in Gärdenfors, P., Wolenski, J. and Kijania-Placek, K. (eds.), *The Scope of Logic, Methodology and Philosophy of Science*, vol.II, pp.565-580, Kluwer Academic Publisher, Dordrecht.

Twardowski, K. (1900). "On So-Called Relative Truths", English translation by A. Szylewicz of: "O tak zwanych prawdach względnych", in Brandl and Wolenski (1999), pp.147-169.

Twardowski, K. (1902). "Über sogenannte relative Wahrheiten", *Archiv für systematische Philosophie*, vol. VIII, pp.415-447, German translation by M. Wartenberg (with the help of K. Twardowski) of: "O tak zwanych prawdach względnych"; repr. in Pearce, D. and Wolenski, J. (eds.), 1989, *Logischer Rationalismus. Philosophische Schriften der Lemberg-Warschauer Schule*, pp. 38-58, Athenäum, Frankfurt am Main.

Twardowski, K. (1907). "On Idio- and Allogenetic Theories of Judgement", English translation by A. Szylewicz of: "On idio- i allogenetycznych teoriach sądu", Polish original text, together with French and German translations by K. Twardowski himself; in Brandl and Wolenski (1999), pp.99-101.

Twardowski, K. (1912a). "Actions and Products. Comments on the Border Area of Psychology, Grammar and Logic", English translation by Arthur Szylewicz, in Brandl and Wolenski (1999), pp.103-132.

Twardowski, K. (1912b). "O czynnościach i wytworach", Polish original text, together with French and German translations by K. Twardowski himself; http://www.elv-akt.net/ressources/archives.php?id_archive=1.

Twardowski, K. (1924). *Franciszek Brentano*, Unpublished manuscript available on the website of the ELV-AKT project http://www.elv-akt.net/ressources/archives.php?id_archive=1.

Bolzano and (Early) Husserl on Intentionality

WOLFGANG KÜNNE

ABSTRACT. This paper is about the way Bolzano in 1837 and Husserl around 1900 conceive of the contents of mental acts and states (and of illocutionary acts). After some historical preliminaries I show how the the framework of Bolzano's theory of propositions (*Sätze an sich*) and notions (*Vorstellungen an sich*) is assimilated in Husserl's *Logical Investigations*, explain the early Husserl's species conception of these abstract entities and reconstruct his criticism of Bolzano's account of questions.

1 Introduction

The first volume of Edmund Husserl's *Logical Investigations*, entitled 'Prolegomena to Pure Logic', came out in 1900. It was perhaps the most successful book he ever published. In this book Husserl praised the first two volumes of Bernard Bolzano's *Wissenschaftslehre* of 1837 as

> "a work which [...] far surpasses everything that world-literature has to offer in the way of systematic contributions to logic."[1]

In my lectures I will talk about the early Husserl as a reader of Bolzano's masterpiece, hoping thereby to contribute to a proper understanding of certain aspects of Bolzano's and Husserl's theories and of what those theories are about. I shall concentrate on the question how Bolzano in 1837 and Husserl around 1900 conceive of the contents of mental acts and states.

The heroes of my lectures were both born in the same state of the Habsburg Empire. Bolzano spent all his life in *Bohemia*, which occupies the western two thirds of what is now the Czech Republic. He was born in Prague, he was a Catholic priest and a mathematician whose work was to impress Cantor, Dedekind and

---

[1] Husserl (LU) I, 225 / LI 1, 142: "[E]in Werk, das [...] alles weit zurückläßt, was die Weltliteratur an systematischen Entwürfen der Logik darbietet." In 1911 the key is a bit lower: these volumes, he now says, occupy "the highest rank in the logical world-literature of the 19th century," Husserl (BW) vol. 7, 97. Page references are always to the 2nd edition of (LI). After the oblique stroke I refer to the 2001 paperback edition [!] of Findlay's translation, but here as in all other cases translations from German are my own.

Peirce. He held the chair for the philosophy of religion at the university of his home-town, until the Emperor in Vienna personally saw to it that he was sacked. He spent his philosophically most productive years in a small village in Southern Bohemia, and he died in Prague in 1848, highly respected by both his German-speaking and his Czech-speaking compatriots. Husserl, like Mach, Freud and Gödel, was born in *Moravia*, which occupies the eastern third of what is now the Czech Republic. He, too, was a mathematician by training. He studied philosophy in Vienna with Brentano, but then he spent his entire academic life in Germany, where he became an ardent German patriot (who always preserved his Austrian intonation, as Gadamer reports). He became Privatdozent in Halle and held a chair of philosophy first in Göttingen, then in Freiburg, until the government of the Austrian would-be painter in Berlin saw to it that he was sacked. He died in Freiburg in 1938, highly *un*respected by most of his former German colleagues.

2 The Rediscovery of Bolzano in the School of Brentano

It was half a century after Bolzano's *Wissenschaftslehre* had fallen dead-born from the press that three philosophers came to appreciate the treasures it contains: Benno Kerry, Kasimir Twardowski and Edmund Husserl, all of them pupils of Franz Brentano in Vienna.[2] Interestingly, Brentano himself did not encourage them to read the WL, on the contrary. One rather gets the impression that for his pupils studying that book was a way of emancipating themselves from the overwhelming influence of their teacher. The first among those rebels is nowadays almost entirely forgotten. Benno Kerry made Bolzano as mathematician and as logician the hero of several papers in a very long series called "On Intuition and Its Psychological Processing" (*"Über Anschauung und ihre psychische Verarbeitung"*). It was published between 1885 and 1891 in a, if not the, leading German journal of philosophy of that time.[3] But it is not because of his rediscovery of Bolzano that Kerry acquired a kind of secondary immortality in the philosophical community but because of the fact that his criticism of Frege's *Grundlagen der Arithmetik* in the very same series of papers[4] was deemed worthy of a reply: Frege's famous essay "On Concept and Object" was triggered by this criticism.[5]

In 1894 the Viennese *Habilitationsschrift* of Brentano's Polish pupil Kazimierz Twardowski was published: "On the Content and Object of Presentations" (*"Zur Lehre vom Inhalt und Gegenstand der Vorstellungen"*).[6] This book contains the first detailed discussion of several Bolzanian concepts and doctrines that ever appeared in the Habsburg monarchy (and as we shall see, it was to provoke a vigorous

---

[2] Cf. Künne (1997), in Künne (2008) ch. 9, §4.
[3] Kerry (1885-1891), essays I - III & VIII. For details see Künne (1997), §5.
[4] Frege (1884), Kerry (1885-1891), essays II & IV.
[5] Frege (1892).
[6] Twardowski (1894).

response from Husserl). At the centre of Twardowski's inquiries is the distinction between acts of presenting (*Vorstellen*), their contents and their objects. As he points out:

> "Bolzano consistently sticked to this distinction [...] What I call 'content of a presentation' he called 'objective presentation' or 'presentation in itself', and he distinguished from this both the object and the 'presentation somebody has' or the 'subjective presentation' by which he means the act of presenting."[7]

Let me fix my own terminology at this point and register a harmless terminological mismatch between Bolzano and Twardowski. In order to save breath (and space) I shall call Bolzano's 'presentations-in-themselves (*Vorstellungen an sich*)' or 'objective presentations' *notions*,[8] just as I shall call his 'sentences-in-themselves (*Sätze an sich*)' or 'objective sentences' *propositions*. As for the terminological mismatch, Bolzano uses '*Inhalt* (content)' in such a way that thinking of a person as the *brightest* daughter of the *silliest* father in Barcelona and thinking of a person as the *silliest* daughter of the *brightest* father in Barcelona are two acts that have the same content but not the same '*Stoff* (matter).'[9] Similarly, he uses 'content' in such a way that the judgement that Mallorca is larger than Menorca and the judgement that Menorca is larger than Mallorca are two acts which have the same content but not the same matter. The Bolzanian content of an act consists of the same components as its matter, but the content, unlike the matter, is indifferent as to the mode of composition: You can specify it by a list. When Twardowski says 'content' he wants to capture what Bolzano calls 'matter.'

Twardowski leaves no doubt about the question to whom he owes his interest in Bolzano's book. Around 1926, when the Lemberg-Warsaw School of Analytic Philosophy whose founding-father Twardowski had become was flourishing, he said in his "Self-Portrait" about his *Habilitationsschrift*:

> "I endeavoured to write it in the spirit of Franz Brentano and – Bernard Bolzano, whose *Wissenschaftslehre* I studied carefully ever since Kerry's papers "On Intuition and Its Psychological Processing" had awakened my interest in that book."[10]

The one point at which Twardowski takes himself to disagree strongly with Bolzano concerns the question of empty presentations. According to Bolzano,

---

[7]Twardowski (1894), p.17: "Bolzano hat mit grosser Consequenz an diesem Unterschiede festgehalten [...]. Bolzano gebraucht statt des Ausdruckes 'Inhalt einer Vorstellung' die Bezeichnung 'objective' Vorstellung, 'Vorstellung an sich' und unterscheidet von ihr einerseits den Gegenstand, andererseits die 'gehabte' oder 'subjective' Vorstellung, worunter er den Act des Vorstellens versteht."

[8]My rendering of '*Vorstellung an sich*' by 'notion' needs to be protected against a possible misunderstanding. In the context of this paper this term must not be understood as just another word for 'concept,' for Bolzano distinguishes within the class of notions 'intuitions (*Anschauungen an sich*)' from pure 'concepts (*Begriffe an sich*).' The former are atomic notions under which exactly one object falls, the latter are notions which neither are nor contain any intuitions, see Bolzano (WL) I, 325-31. In this paper Bolzano (WL) is always quoted by volume and page number.

[9]Bolzano (WL) I, 244.

[10]Twardowski (1926), p.11.

some acts of presenting and the notions that are their matter lack an object: They are, as he puts it, 'object-less (*gegenstandlos*).' To use an example that was not yet available to Bolzano, the notion expressed by the definite description 'the planet between Mercury and the Sun' is such that no object is picked out by it, in spite of the fact that Le Verrier and many other astronomers performed acts of presentation that had this notion for their matter. Kerry had followed Bolzano in this respect, and Husserl was to follow suit. But Twardowski contradicted Bolzano at this point:

> "For each presentation there is an object that is presented by it, whether this object exists or not, just as for each name there is an object whose name it is, no matter whether this object exists or not. Although it was correct to maintain that the objects of certain presentations do not exist, one went too far in maintaining that [...] such presentations do not have any object, that they are objectless."[11]

With this kind of reasoning, Twardowski prepared the ground for the Theory of Objects (*Gegenstandstheorie*) which was soon to be developed by another of Brentano's unfaithful students, Alexius von Meinong. One of the corner-stones of Meinong's Theory of Objects, or perhaps rather its slippery ground, is 'the Principle of the Independence of Being-thus-and-so from Being (*das Prinzip der Unabhängigkeit des Soseins vom Sein*):' An object can have all sorts of properties even though it has no being, that is, although it neither exists nor subsists (*besteht*). Thus, the intra-mercurian planet Vulcan has the property of being a planet, although astronomers were bound to look in vain for this planet.[12] Bolzano, on the other hand, always maintains:

> "as the old principle has it – of what there isn't there are no properties."[13]

In 1900, in the *Prolegomena* to his LI, Husserl used Bolzanian weapons in his battle against what he called psychologism in the philosophy of logic. Bolzano had insisted that:

> "it is obviously a shift of the proper point of view (a distortion of the right perspective) if one pretends to be dealing with the most general laws of *thinking* when one is actually concerned with the most general conditions of *truth* itself."[14]

---

[11]Twardowski (1894), p.24: "[Es wird] durch jede Vorstellung ein Gegenstand vorgestellt, mag er existieren oder nicht, ebenso wie jeder Name einen Gegenstand nennt, ohne Rücksicht darauf, ob dieser existiert oder nicht. War man also auch im Recht, wenn man behauptete, die Gegenstände gewisser Vorstellungen existieren nicht, so sagte man doch zu viel, wenn man behauptete, unter solche Vorstellungen falle kein Gegenstand, solche Vorstellungen hätten keinen Gegenstand, sie seien gegenstandslose Vorstellungen."

[12]Cf. Meinong (1904). On Meinong as a reader of Bolzano see Künne (1997), in Künne (2008) ch. 9, §10.

[13]Bolzano (1838), p.293: "wie schon der alte Kanon besagt – non entis nullae sunt affectiones;" cf. Bolzano (WL) II, 213. As to the "Kanon" cf. Descartes, *Principia philosophiae* I, §52; Leibniz, *Essais de théodicée*, III, S 387.

[14]Bolzano (WL), I, 65: "[Es ist] offenbar eine Verschiebung des rechten Gesichtspunkts, wenn man

Husserl enthusiastically embraced this, and (following Frege's footsteps[15]) he turned it against Sigwart, Lipps, Wundt, Erdmann *e tutti quanti*, just as Bolzano had turned it against the tradition of the logic of Port-Royal. Apparently Brentano had heard some rumours that he might be one of the inofficial targets of Husserl's attack against psychologism. In winter 1904, living in Florence and gradually losing his eye-sight, Brentano wrote to his former pupil: "I would be very grateful if you could refer to one important point where you take yourself to have moved away from me and beyond me."[16] In his long reply of January 1905 Husserl said:

> "The 1st volume of my LI fights against 'psychologism' in logic and epistemology, that is, against what I regard as a very harmful overestimation of psychology as allegedly providing the basis for philosophy as a whole, hence also for pure logic and epistemology [...] I discovered extremely fruitful contributions to pure logic in Bolzano's WL which I admire very much."[17]

That was a message his old teacher did not like at all:

> "[So your pure logic] is not to be based upon psychological insights, [...] and in this respect you praise Bolzano as your teacher and guide. I cannot deny that there is quite a bit here that worries me. [...] I dare say that the realm of intelligibilia (*Gedankendinge*) in which unfortunately even a respectable thinker like Bolzano got lost can be shown to be absurd. But in any case, I congratulate you because of your spiritual contact with this noble and serious thinker. From such thinkers one can learn something even where they are in error."[18]

In his reply Husserl was very keen to pacify his old master:

> "[Bolzano's views about notions (*Vorstellungen an sich*) and propositions (*Sätze an sich*) and] the fact that the first two volumes of WL contain valuable contributions to logic that are independent of empirical psychology [...] had a strong impact upon me, and so did Lotze's reinterpretation of Plato's theory of forms. And yet I could not call Bolzano a 'teacher' and 'guide' as regards those things my LI are about. I still regard myself, and I still call myself, *your* pupil."[19]

---

dort von den allgemeinen Gesetzen des *Denkens* zu handeln vorgibt, wo man im Grunde die allgemeinsten Bedingungen der *Wahrheit* selbst aufstellt." Husserl underlined several such passages in his copy of WL, e.g., I, 47, lines 23-27. I am very grateful to my former student Guillaume Fréchette for providing me with an exact description of Husserl's annotated copy in Louvain.

[15] Cf. the discussion of psycholgism and of Frege's anti-psycholgism in Künne (2009), ch. 1, §8.

[16] Husserl (BW), vol.1, p.24.

[17] Husserl (BW), vol.1, pp.27, 29: "Der 1. Band der Log. U. streitet gegen den 'Psychologismus' in der Logik u. Erkenntnistheorie, d.i. gegen eine m.E. sehr schädliche Überschätzung der Psychologie als vermeintlicher Fundamentaldisciplin für die gesammte Philosophie, u. somit auch für die reine Logik u. Erkenntniskritik [...]. Außerordentlich fruchtbare Ansätze zur Behandlung einer reinen Logik bietet die von mir sehr bewunderte Wissenschaftslehre Bolzano's."

[18] Husserl (BW), vol.1, pp.31, 34.

[19] Husserl (BW), vol.1, p.39 and 40, n.*: "[Bolzanos] Conceptionen [der] Vorstellungen und Sätze an sich und der Umstand, dass in den beiden ersten Bänden der Wissenschaftslehre wertvolle logische Darstellungen vorliegen, die von empirischer Psychologie unabhängig sind, [...] haben auf mich stark gewirkt, ebenso Lotze's Umdeutung der Platonischen Ideenlehre. Indessen als 'Lehrer' und 'Führer' kann ich Bolzano hinsichtlich dessen, was ich in meinen Log. Unters. dargestellt habe, doch nicht nennen [...]. Ich fühle und nenne mich nach wie vor als *Ihren* Schüler."

In 1909 a young philosopher in Prague, Hugo Bergmann (a pupil of Brentano's faithful pupil Anton Marty), had begun working on a book on the greatest philosopher of his home-town. In a letter he told Brentano, somewhat unwisely, of this project, whereupon he received a none too gracious reply from Florence:

> "As regards Bolzano, when I came to Austria [in 1874] I drew attention to him who had almost entirely fallen into oblivion, and I spoke of him with praise. [...] I had [...] to encourage young people in Austria to study philosophy. What had thus far been achieved in this country was not so very much [...]. Only Bolzano proved to be outstanding in several respects. In an age of extreme decay [i.e., in the heyday of German Idealism] he got clear about its character; he did not allow himself to be impressed by the name of Kant. He preferred Leibniz and was right in doing so [...]. So you will understand that I found no cause to enter into a criticism of Bolzano's weaknesses. As a result, [...] Twardowski, Husserl and Kerry immersed themselves in the study of Bolzano's work. What was wrong therein they did not recognize as such [...] I may be allowed not to accept any responsibility at all for the many odd and absurd views Husserl came to maintain by paying attention to Bolzano."[20]

So from Brentano's point of view the growth of interest in Bolzano's WL among some of his pupils is the history of an apostasy.

3 Objectless Presentations: Husserl's Defence of Bolzano against Twardowski

Husserl rejected Twardowski's objection to Bolzano when Twardowski's ink was hardly dry, and he also disagreed with one aspect of his interpretation of WL.[21] In a manuscript of 1894 he defended Bolzano's contention "that not very presentation is such that there is an object presented by it," against Twardowski's objection that had just been published.[22] The pseudo-division of objects of presentations into those that exist (like the planet between Saturn and Neptune) and those that don't (the planet between Mercury and the Sun) really is a distinction between two kinds of *presentations*: between objectual and non-objectual presentations. Husserl appeals here to an enlightening analogy between Twardowski's (and Meinong's) distinction and the 'pseudo-division of objects into determinate and indeterminate ones (*Quasi-Einteilung der Gegenstände in bestimmte und unbestimmte*).' If a man is longing for a woman then there might be a particular woman he is longing for: It has got to be Carmen, nobody else would do. But sometimes, or so I have

---

[20] Bergmann (1946), pp.125-127: "[...] [D]ie Verantwortung für so vieles Absonderliche und Absurde, wozu [...] Husserl unter Berücksichtigung von Bolzano gelangt [ist], darf ich doch vollständig ablehnen." Bergmann (1909) was the first monograph on Bolzano's philosophy, and for a century it was to remain the only one: cf. Künne (1997), in Künne (2008), ch. 9, §§11-12. In his letter Brentano also claims that Meinong got deeply involved with Bolzano's WL. That is an error. See Künne (1997), in Künne (2008), §10.

[21] The vehemence of these reactions shows that Husserl did not become a reader of the *Wissenschaftslehre* by studying Twardowski's book. He discussed Kerry along with Frege in his first book, *Philosophie der Arithmetik* (1891), and as a mathematician he had come across Bolzano already as a student for Weierstrass. For details see Künne (1997), in Künne (2008), §8.

[22] Husserl (1894), p.303 ff/ Husserl (1894b): "daßnicht jeder Vorstellung ein Gegenstand entspricht."

been told, he just wants release from womanlessness: It hasn't got to be a particular woman. So are there women who are determinate and others who are indeterminate, just as some women are dark and some are fair? "Of course, the answer will be: each object is in itself a determinate object."[23] The pseudo-distinction between determinate and indeterminate women really is a distinction between those longings for a woman that are individually focussed and those that are not. Similarly, Husserl maintains, Bolzano did well to opt for a distinction between objectual and non-objectual presentations rather than for a distinction between existent and non-existent objects. In Husserl's Brentanoesque language:

> "[The statement that] such-and-such an object is a 'merely intentional' object [...] means: the intention, the [state of] 'having in mind' exists, but not the object. Whereas, if the intentional object exists then not only the intention, not only the [state of] 'having in mind' exists, but also the object that one has in mind."[24]

What prima facie looks like a distinction between those objects that are merely intentional and those that are not, really is a distinction between two kinds of intentional acts: those which do not satisfy the condition that there is an object such that it is what the agent of the act has in mind, and those which *do* satisfy this condition.

And what about Twardowski's interesting analogy between presentations and names ("each presentation presents an object, just as each name names an object")? Well, Bolzano and Husserl might have replied:– It is simply not true that each name names an object. What *is* true is that every instance of the equivalence schema for names,

For all x, '$N.N.$' names x iff $x = N.N.$,

expresses a truth, provided we assume a negative free logic. According to such a logic, a sentence of the form '$N.N. = N.N.$' expresses a falsehood if and only if the name substituted for '$N.N.$' has no bearer. So no matter whether we substitute 'Mercury' or 'Vulcan,' we obtain a truth from the equivalence schema. Once again, what is needed is not a distinction between existent and non-existent name-bearers, but only a distinction between two kinds of names: those that do have a bearer and hence permit existential generalization and those that lack a bearer and hence do not permit existential generalization. – At least for Bolzano it can be shown that he takes the side of the advocates of negative free logic.[25] (None of this amounts to a

---

[23] Husserl (1894), p.313/ Husserl (1894b), p.148: "Man wird natürlich antworten: Jeder Gegenstand ist in sich bestimmt."

[24] Husserl (LU), II/1, p.425; (LI), 2, p.127: "Der Gegenstand ist ein 'bloß intentionaler', [...] heißt: die Intention, das einen so beschaffenen Gegenstand 'Meinen' existiert, aber nicht der Gegenstand. Existiert andererseits der intentionale Gegenstand, so existiert nicht bloßdie Intention, das Meinen, sondern auch das Gemeinte."

[25] Cf. Künne (2006), in Künne (2008), ch.8.

refutation of the Twardowski-Meinong view concerning names and presentations. It is only meant to clarify the Bolzano-Husserl stance.)

In his 1896 review of Twardowki's book Hussel emphasized that in a crucial respect the author entirely misunderstood Bolzano.[26] Twardowski takes himself to be in agreement with Bolzano when he maintains that the content of an act of presenting an object X is a *mental* object that comes into being in the course of the act, that mediates the reference of the act to its 'primary' object X and that is itself the 'secondary' object of this act.[27] Husserl protested that the content of an act of presenting, conceived along Twardowskian lines, is categorially different from what Bolzano regards as the content, or rather, the matter of such an act, for a Bolzanian notion is not any kind of mental entity but rather an abstract object. This objection hits the nail on its head. But in order to make this clear I must at last try to elucidate Bolzano's conception of propositions and of notions.

4 A Sketch of Bolzano's Theory of Propositions and Notions

Since Bolzano assigns to propositions (and acts of judgment) conceptual priority over notions (and acts of presenting), let us start with his account of propositions. He took the concept of a proposition (*Satz an sich*) to resist analysis or conceptual decomposition (*Erklärung*) but he always stressed that there are, and that there have to be, other ways of helping one's addressee to comprehend a concept. He calls them *Verständigungen* (elucidations).[28] Consider a report of the following type: 'Johanna claimed that Plato's most important teacher was witty, Juanita asserted the same thing, though in different words, and Joan believes what they said.' Here a that-clause is used to single out something that is

1. asserted by different speakers,

2. distinct from the linguistic vehicles used for saying it,

3. believed by somebody.

"Now, this is the sort of thing I mean by proposition," Bolzano could say. What is said and thought in such sayings and thinkings is a proposition. As we all know, sometimes what somebody said would have better remained unsaid, but some sayables and thinkables, for better or worse, *do* remain unsaid and unthought forever. (At least, Bolzano would feel obliged to add as a theist, they remain unthought by cognitively finite thinkers.) Thus, for example, there are ever so many

---

[26] Husserl (1896), p.353, about Twardowski (1894), p.17. Three decades later Roman Ingarden provided Twardowski with a copy of this review that had remained unpublished: Husserl (BW) vol. 1, 183.
[27] Twardowski (1894), §4.
[28] Cf. Bolzano (WL), IV, 243-5, 488-90, 542-5, 547.

decidable questions nobody ever bothers to ask, let alone to answer:[29] the question how often the letter A occurred in the last 999 issues of El País would have been such a question, I guess, had I not wasted a bit of breath by formulating it. The one true answer, and the many false answers, to such a tedious question are sayables and thinkables Bolzano wants to be subsumed under the concept of a proposition.

The sayings and thinkings of Johanna, Juanita and Joan are causally efficacious, but like all other sayables and thinkables, what our three graces said and thought is causally inert, or as Bolzano puts it, it is not *wirklich* (actual) in the sense of *wirksam* (capable of acting upon something).[30] Bolzano also has another way of putting this that can easily mislead: He is prone to say that propositions don't exist or have no existence. In his mouth that doesn't mean that there are no propositions. He treats the verb 'exist' as a first-order predicate that applies to all and only those objects that are capable of acting upon something. The phrases 'there is' or 'there are,' on the other hand, Bolzano takes to be second-order predicates that serve to ascribe to a notion the property of being 'objectual (*gegenständlich*),' i.e., of applying to at least one object.[31] So in Bolzano's mouth the statement "There is something that does not exist" is not paradoxical at all.

Now this may again be misleading, I am afraid, because to some ears it makes him sound like Meinong. So let me immediately insert another warning. Consider the statements

1. There are golden mountains

2. Golden mountains exist.

According to Frege and most contemporary philosophers, these are two ways of saying the same thing, and that thing is false. Bolzano and Meinong are agreed that these are not two ways of saying the same thing, but here their agreement ends. Meinong takes (1) to be true and (2) to be false, whereas Bolzano, like the Fregeans, regards both as false, though, unlike the Fregeans, for different reasons: (1) is false because no object whatsoever falls under the notion of a golden mountain, and (2) is false because no object *that is capable of acting upon something* falls under that notion.

Judgings are those mental acts, and believings are those mental states, in which a proposition is acknowledged as true.[32] Bolzano's official view is that a thinking with a propositional matter is always an act of judging or a state of believing, for according to his classification of mental events and states, if a thinking is not

---

[29] Bolzano (WL), p.218.
[30] Bolzano (1851), p.2; cf. Bolzano (1838), p.85; Bolzano (WL) I, pp.362-366, III, 16, and at many other places.
[31] Bolzano (WL) II, pp.52-54, pp.214-218.
[32] Bolzano (WL) I, p.78, p.154; III, p.108, p.199 f.

a judging or believing then it has got to be a thinking of an object, a presenting (*vorstellen*).³³ This view, which is well-entrenched in the philosophical tradition, has odd consequences. If you merely entertain the thought that p, without committing yourself as to its truth-value, then the matter of your thinking is not the proposition that p but, as in all cases of presenting something, a notion. In this case, it is a notion under which exactly one proposition falls.³⁴ But a notion does not have a truth-value.³⁵ So the matter of one's merely entertaining the thought that p is not in the truth-line of business at all. Isn't this strange? Or consider a *modus ponens* argument. Somebody first judges that if p then q, then she acquires the conviction that p, and finally she concludes that q. Now at first she does not yet judge that p, that's something she only does when she takes the second step. But then, according to Bolzano's official theory, it is not the case that the proposition that p is first *part* of the matter of her thinking and then, when it comes to be acknowledged as true, its complete matter. So why is the argument valid? All this is very counter-intuitive, and Bolzano, I am happy to say, does not always stick to it.³⁶

Which stance does Husserl take here? In his LI he devotes a lot of space to showing that the noun '*Vorstellung* (presentation)' and the verb '*vorstellen* (presenting)' as used by German-speaking philosophers ever since Christian Wolff made it their common coin is multiply ambiguous.³⁷ If only the noun had to be translated into English, 'idea' would have been optimal, but alas, no verb corresponds to it. That's why translators of Twardowski and Husserl use the somewhat stilted terms 'presentation' and 'presenting,' and with little enthusiasm I have followed them. Now in one usage of '*vorstellen*', Husserl points out, it is followed by a 'that'-clause ("*sich vorstellen, dass es sich so-und-so verhält*") and means: merely entertaining the thought that things are thus-and-so.³⁸ About this usage Husserl says:

> "No one would question that (it holds a priori, in essential generality that) *for every judgement there is a presentation* [a non-committal thinking] *endowed with the same matter*, and therefore presenting the same thing in exactly the same manner, as the judgement judges about it. To the judgement, e.g., 'The earth's mass is about 1/325,000 of the sun's mass,' corresponds, as 'mere' presentation [as non-committal thinking], the act performed by someone who hears and understands this utterance, but sees no reason to take a stand on the issue."³⁹

---

³³Bolzano (1838), p.25; Bolzano (WL) II, pp.67-68; Bolzano (1849) §11.

³⁴Bolzano (WL) I, p.99, p.155, p.157.

³⁵Bolzano (WL) I, pp.238-243.

³⁶Cf. Bolzano (WL) I, p.78; II, p.4.

³⁷(LI), 5th, §§22-45).

³⁸Husserl (LU) II/1, p.499; (LI) 2, 171: No. 2.

³⁹Husserl (LU) II/1, p.446; (LI) 2, 139-140 (my emphasis): "Daß es nun *zu jedem Urteil* (a priori, in Wesensallgemeinheit gesprochen) *eine Vorstellung gibt, die mit ihm die Materie gemeinsam hat*, und die also genau dasselbe in genau entsprechender Weise vorstellt, wie das Urteil es urteilt, wird niemand bezweifeln. So entspricht beispielsweise dem Urteil 'die Erdmasse ist ungefähr 1/325.000

So Husserl acknowledges that merely entertaining the thought that p has exactly the same matter as judging, or believing, that p.[40] Apparently, he forgot what I called Bolzano's official doctrine, for otherwise he would not have claimed that "no one would question [this]."[41] But is his general statement correct? Is it really true that to each judgement to the effect that p there corresponds an act of merely entertaining the thought that p?[42] Can one merely entertain the thought that a rose is a rose without seeing any reason to take a stand on the issue? A proposition that is self-evidently true cannot be the complete matter of a *non-committal* propositional thinking. (Of course, it can be a proper *part* of the matter of such an act, e.g., when you mereley entertain the thought that it is not the case that a rose is a rose.)

Of all the many uses of '*Vorstellung*' and '*vorstellen*' Husserl carefully distinguishes, only one is relevant for Bolzano's theory (and Twardowski's): 'presentation as a nominal act (*Vorstellung als nominaler Akt*)'.[43] (A mental act-or-state is *nominal* just in case it is not *propositional*, and it is propositional just in case its explicit 'intimation [*Kundgabe*]' is the utterance of a complete sentence.[44]) I think we can get fairly close to the sense of 'presentation' that is pertinent here if we think of presenting something as: having something in mind, adverting to something, thinking of something.

Bolzano's explanation of '*subjektive Vorstellung*' takes for granted that we already know what a judgement is. This is in itself remarkable, for it reverses the traditional order of explanation. According to Bolzano,

(SUBJPRES)
x is an act-or-state of presenting (*subjektive Vorstellung*) iff:
x is a mental act-or-state & x is not a judgement-or-belief &
x could be an ingredient of a judgement-or-belief.[45]

If Juanita judges or believes that Socrates was witty, then she has Socrates and the property of being witty in mind. One can advert to Socrates without making a

---

der Sonnenmasse' als die ihm zugehörige 'bloße' Vorstellung der Akt, den jemand vollzieht, der diesen Ausspruch hört, versteht, aber kein Motiv findet, sich urteilend zu entscheiden."

[40] And so does Frege, e.g., in Frege (1918), pp.62-63. On the stance he takes in this regard, see Künne (2009), ch. 1, §7 and ch. 2, §3.

[41] There is no doubt that he knew that doctrine. Of the three passages mentioned above in which Bolzano characterizes merely entertaining the thought that p as a 'subjective presentation' of the proposition that p, Husserl marked two in his copy of Bolzano (WL) (I, p.99, lines 25-29, and p.155, *sub* d), and he wrote a question mark in the margin next to the first passage. (Fréchette's observation.).

[42] Cf. Stepanians (1998), pp.231-239.

[43] Husserl (LU) II/1, 500; (LI) 2, 171: No. 3.

[44] In Husserl's mouth, as in that of most medieval philosophers and of Peter Geach, 'propositional' means *sentential*, so when he calls a mental act propositional he is not using this word in its primary sense.

[45] Bolzano (WL) I, 161; III, 5-6.

judgement about him, for example by seeing, hearing, or mentally picturing him. Under favourable circumstances any of these acts can be involved in somebody's making a judgement about Socrates. But as each of *our* judgements about Socrates shows, one can have Socrates in mind without perceiving him, and as most of our judgements about him also show, one can have him in mind without producing a mental image of him.

Let me insert here an aside on a widespread prejudice among analytical philosophers. John Perry is a typical example: Having said that "many mental states and activities exhibit the feature of intentionality: being directed at objects," he goes on to explain this as involving two claims: Intentional mental acts or states are, firstly, directed at propositions, and, secondly, directed at the objects, properties and relations these propositions are about.[46] As to the first point, Bolzano would hardly accept, and Husserl would definitely reject, the claim that judgings and believings are directed at those propositions which are their matter. (I shall come back to this issue in §4.) The second point is currently more important: Perry takes for granted that all intentional mental acts-or-states are ascribable in the style of 'N.N. [VERB]s that *p*.' Sometimes an ascription of an intentional act-or-state that does not have this format, e.g., 'Schliemann was looking for Troy,' can be expanded to show a sentence within a sentence, thus 'Schliemann tried to bring it about that he himself finds Troy.' Quine, to whom this suggestion is due, was aware of the fact that there are other examples that cannot so easily be "accommodated under the propositional attitudes;" thus 'Schliemann is thinking of Troy,' 'is imagining Troy.' "But perhaps they can with some torturing."[47] (Actually, he did not to tell us how to torture the latter two sentences.) Dummett is also inclined to believe that all recalcitrant cases can be propositionally tamed, but he frankly admits: "this is no more than a hope."[48] From Bolzano to Husserl, all continental philosophers who brooded over intentionality saw no good reason for such a hope, and I think they were right. For all I know, nobody ever succeeded in showing that, say, admiring Socrates (or Ulysses for that matter), liking him, hating him, despising him, or pitying him are at bottom *propositional* acts or states.[49] Since you can 'do' none of these things without having Socrates in mind, without adverting to him, withouth thinking of him, *vorstellen* may very well the most basic of all non-propositional acts or states that exhibit intentionality. At any rate, it is the only one Bolzano saw

---

[46] Perry (1994), pp.386-388.

[47] Quine (1960), pp.179-180, cf. 152 ff.

[48] Dummett (1973), p.269; cf. pp.187, 264, 290. (Searle does not claim that all intentional acts-or-states "contain an entire propositional content," but as regards those which don't he maintains that "in general, they require the presence of beliefs and desires," Searle (1994), p.383; cf. Searle (1983), pp.6-7, 34-35. Even this non-reductivist claim is dubious: Which belief, and which desire, has got to be present when somebody sees a tree? But perhaps the prefix 'in general' is meant to shield off counterexamples.)

[49] For a recent survey of failed attempts see Montague (2007).

reason to discuss in a book entitled *Wissenschaftslehre*.

Now what is a *Vorstellung an sich*, a notion? If Juanita comes to believe that Plato's most important teacher was witty, whereas Juan comes to believe that the philosopher who drank the hemlock was witty, then they both have Socrates in mind, but he is not "thought under the same attributes (*unter denselben Merkmalen gedacht*):"[50] Juanita's and Juan's subjective presentations have the same object, but their acts differ as regards their matter. This matter is what Bolzano calls a notion. He pursues two strategies when it comes to explaining this concept. The first one relies on the above explanation of presenting and on the concept of being the matter of a mental act or state (which Bolzano takes to be indefinable[51]):

(NOTION1)
x is a notion (*Vorstellung an sich*) if and only if
there could be an act-or-state of presenting
of which x is the complete matter.[52]

Why '*could* be'? Consider the singular term 'the number of occurrences of the letter A in the last 999 issues of El País.' Presumably, if I had not produced a specimen of this term for consideration, nobody would ever have performed an act of presentation the matter of which is expressed by this singular term: Nevertheless, at some time or other this notion *could* become the matter of some human thinker's thinking.[53] Furthermore, our cognitive capacities are limited. None of us, I presume, is able to perform an act of thinking-of-a- person, the matter of which is expressed by the singular term that is the result of prefixing thousand times the phrase 'The mother of the father of' to the name 'Socrates.'[54] But in some possible world there is a thinker who is able to perform such cognitive feats,[55] and since Bolzano is a theist he is convinced that actually there is at least one such thinker.

In (NOTION1) Bolzano is careful to say '*complete* matter,' since sometimes a proposition is *part* of the matter of an act-or-state of presenting. For example, if you judge that the theorem that (in any right triangle the square on the hypotenuse is equal to the sum of the squares on the other two sides) is unknown to many people, then you advert to the Pythagorean Theorem, and this theorem is also *part* of the matter of your judgement, but of course it isn't its complete matter, for if it were then your judgement would have been that $c^2 = a^2 + b^2$. Only a notion can

---

[50]Bolzano (WL), III, p.14.

[51]Bolzano (1935), 18.12.1834, pp.84 f.

[52]Bolzano (WL), III, 5 f.

[53]Cf. Bolzano (WL), I, p.217 f.

[54]Cf. Bolzano (WL), I, pp.283, 356.

[55]Husserl concurs. Talking about notions and propositions as what can be expressed in words, i.e., as 'meanings (*Bedeutungen*),' he says: "There are [...] countless meanings which, in the common, relational sense, are merely possible ones, since they are never expressed, and since they can, owing to the limits of man's cognitive powers, never be expressed," Husserl (LU) II/1, p.105; (LI), 1, p.233.

be the complete matter of an act-or-state of presenting, just as only a proposition can be the complete matter of a judgement or belief.

Bolzano prefers his second account of what a notion is. For him it is a matter of course that propositions resemble sentences in being structured wholes:

> "It seems indisputable to me that every, even the simplest, proposition is composed of certain parts, and that parts do not [...] merely appear [...] in the verbal expression of a proposition, but are contained already in the proposition."[56]

Invoking the conception of propositions as structured wholes, Bolzano explains 'notion' as follows:

(NOTION2)
x is a notion (*Vorstellung an sich*) if and only if
x is a non-propositional part of a proposition.[57]

Bolzano is careful to say '*non-propositional* part,' since sometimes a part of a proposition is itself a proposition (as in the above attribution of geometrical ignorance). What Bolzano seems to have overlooked is that we need to impose some restriction on the concept of Non-Propositional Part of a Proposition if this account of a notion is to be extensionally equivalent with its predecessor. Consider the proposition that 7 is prime or 7 is not prime. Part of this proposition is expressed by the following fragment of a sentence: '( ) is prime or [ ].' Is this part of our proposition really a notion? Could there really by a subjective presentation of which this strange item would be the complete matter?

If both Johanna and Juanita judge that the philosopher who drank the hemlock was witty, then there occur two mental episodes with the same matter. Part of this matter is the notion expressed by 'the philosopher who drank the hemlock.' Notions inherit their ontological status from the entities of which they are parts. Hence they are not what Bolzano calls 'actual (*wirklich, wirksam*),' they are causally inert.[58] Neither propositions nor notions are waves in anyone's stream of consciousness. So Husserl was justified in accusing Twardowski of partially misrepresenting Bolzano's views on notions.

As regards identity conditions for notions and propositions, Bolzano is a hyperintensionalist: The fact that two sub-sentential expressions (sentences) have the same intension, i.e., the same extension (truth-value) in all possible worlds, does

---

[56] Bolzano (WL), I, 222: "Es däucht mir [...] unwidersprechlich, daß jeder auch noch so einfache Satz [sc. an sich] aus gewissen Theilen zusammengesetzt sey; daß sich nicht etwa [...] nur in dem wörtlichen Ausdrucke eines Satzes [sc. an sich] erst gewisse Theile [...] hervorthun, sondern daß diese Theile schon in dem Satze an sich enthalten sind."

[57] Bolzano (WL) I, p.216; II, p.18.

[58] Bolzano (WL) II, p.222.

not guarantee that they express the same notion (proposition). The notion expressed by 'equilateral triangle' differs from that expressed by 'equiangular triangle' in spite of the fact that the two expressions are intensionally equivalent. You can think of an object as an equilateral triangle withouth thereby thinking of it as equiangular: The notion expressed by 'equiangular' is not part of the notion of an equilateral triangle.[59] Suppose some notions are atomic (*einfach*),[60] and suppose furthermore that the notion expressed by 'something' is one of them.[61] As to the latter supposition, think of the use of 'something' not as a quantifier but as a general term that has the same sense as 'object.'[62] (We use this word in this manner when we say: 'She caught sight of a black, slimy, weird looking something, and she screamed.') Under these two assumptions Bolzano can prove that it cannot be true that the notion of a property necessarily had by all S is always a part of the notion of an S:

> "If each property of an object, which it must have in order to be the object of a certain notion, had to occur as a component of this notion, then all simple notions would have to contain the notion of a something. But in order to remain simple, they could contain no other notions besides. Consequently, [...] there would be only one simple notion, namely that of a something. Who does not see that this is absurd?"[63]

Now if a complex notion X is not only intensionally equivalent with notion Y but also consists of the very same notions as Y, doesn't that suffice for their being one notion? No, it doesn't, replies Bolzano and offers the following example:[64] The notion expressed by '4^2' is composed of the notions expressed by '4,' 'to the power of' ('x multiplied y-times by x') and '2,' the notion expressed by '2^4' is built up from the very same components, and necessarily, $4^2 = 2^4$. Nevertheless, the two expressions flanking the identity operator express different notions, two different ways of thinking of 16. Now in this case, the components are the same (hence the Bolzanian 'content' is the same, too), but the mode of composition is not. So Bolzano suggests the following condition of identity for complex notions:

> "If not only the components which are to compose a notion are specified but also the mode of composition, then the whole notion is obviously already completely determined. This implies

---

[59]Bolzano (WL), vol. I, pp.269-281, pp.287-296, p.302.

[60]Bolzano thinks he can prove that there ara atomic notions: The alleged proof is to be found in Bolzano (WL), vol. I, pp.263-265.

[61]Cf. Bolzano (WL) vol. I, pp.219, 330, 360, 416, 447, 459; Bolzano (WL), vol. III, pp.20, 194; Bolzano (WL), vol. IV, p.335; Bolzano (1935), 23.08.1833, 25; 14.01.1834, 39.

[62]Bolzano (WL), vol. I, pp.459, 461 f. Cf. Husserl (LU), II/1, pp.296-297 / (LI), 2, p.50.

[63]Bolzano (WL), vol. I, p.276: "Müßte [...] jede Eigenschaft eines Gegenstandes, die ihm mit Nothwendigkeit zukommt, sobald er der Gegenstand einer gewissen Vorstellung seyn soll, auch als Bestandtheil in ihr selbst vorkommen: so müßten alle einfachen [gegenständlichen] Vorstellungen die Vorstellung Etwas enthalten. Um aber einfach zu bleiben, müßten sie dann nebst dieser einen sonst keine andere enthalten. Mithin [...] gäbe es nur eine einzige einfache Vorstellung, nämlich die Vorstellung Etwas. Wer sieht nicht, daß dieses ungereimt sey?"

[64]Bolzano (WL), vol. I, p.446.

that there cannot be two notions which do not only have the same components but also the same mode of composition."[65]

Unsurprisingly, Bolzano takes a proposition P to be identical with proposition Q just in case P and Q are composed of the same notions in the same way.[66] So the proposition that ABC is an equiangular triangle differs from the proposition that ABC is an equilateral triangle, and the proposition that $4^2 > 15$ is not identical with the proposition that $2^4 > 15$.

The proposition expressed in an utterance of a declarative sentence does not coincide with the conventional linguistic meaning of that sentence, or at least it does not always do so. Sentence-synonymy is not a *necessary* condition of propositional identity: the sentences 'The *WL* is a long book' and 'The *WL* really is a long book' don't have the same conventional linguistic meaning, but, Bolzano maintains, their utterances express the same proposition.[67] Sameness of conventional linguistic meaning is not a *sufficient* condition of propositional identity either: my Monday utterance of 'I am feeling sick today' expresses a proposition that differs from that expressed by my Tuesday utterance, and from all propositions expressed in your utterances, of this sentence.[68]

The propositions expressed by other-person and other-time utterances of 'I am feeling sick today' are variants of the proposition that my utterance of this sentence now expresses, and some of those variants are true, some false. When making this claim we keep part of a proposition constant and observe which effect the exchange of certain notions by others has on the truth-value of the proposition.[69] It is by considering the result of systematically varying notions in propositions that Bolzano obtains a unified account of concepts such as universal validity, logico-analytic truth and deducibility, which is his most important contribution to *logic*. Let me try to convey the spirit of his account with the help of a few examples. Rough as it is, this sketch may suffice for the brief comparison with Husserl's account of analyticity which I shall attempt at the close of the next section. Consider the true proposition that is expressed by my utterance of

(*P*)   If the cover of my copy of *LI* is red then it is coloured.

This proposition has the interesting property that all its variants, which differ from it only in that the notion expressed in that utterance by 'the cover of my copy

---

[65] Bolzano (WL), vol. I, p.434: "Wenn aber nebst den Bestandtheilen, aus welchen eine Vorstellung zusammengesetzt werden soll, auch noch die Art ihrer Verbindung angegeben würde: so wäre offenbar schon die ganze Vorstellung bestimmt. Daraus folgt denn, daß es nie zwei objective Vorstellungen gebe, die bei den nämlichen Bestandtheilen auch noch die nämliche Verbindungsart unter denselben haben."
[66] Bolzano (WL), vol. I, pp.147, 434; vol. IV p.338 f.
[67] Bolzano (WL), vol. I, pp.123, 197, 246 ff.
[68] Cf. Bolzano (WL), vol. II, pp.77-78.
[69] Bolzano (WL), vol. II, p.78.

of *LI*' is replaced by a different (non-empty) notion, are also true. In virtue of this feature that proposition is, as Bolzano puts it, 'universally valid (*allgemeingültig*)' with respect to this notion.[70] Now consider the truths that are expressed by my next two utterances:

(*Q*)   If the cover of my copy of *LI* is red and rough then it is red.

(*R*)   If the cover of my copy of *LI* is red then it is not the case that it is not red.

These propositions are universally valid not only with respect to the notion expressed in my mouth by 'the cover of my copy of LI' but with respect to all the non-logical notions they contain: Every non-logical notion in these truths can be varied *salva veritate*, and that makes them *logico-analytic* in Bolzano's sense. His explanation of this concept[71] can be provisionally codified as follows:[72]:

($\approx$)
x is a logico-analytic truth iff
as regards the non-logical notions contained in x,
x is universally valid with respect to each of them.

Bolzano does not claim to have an absolutely firm grasp on the distinction between logical and non-logical notions:

> The concepts which form the invariable part of these propositions all belong to logic [...]. Admittedly, the whole domain of concepts belonging to logic is not so sharply circumscribed that controversies could never arise.[73]

Bolzano's own use of the title 'logical concept' is very broad indeed. It covers not only the notions that are expressed by logical constants, such as 'not,' 'or,' and 'there is,' and by the word 'has' when used to signify exemplification of a property, but also concepts of formal ontology, as Husserl would call them, e.g.,

---

[70] Bolzano (WL), vol II, pp.77-82.

[71] Bolzano (WL), vol. II, pp.83-89.

[72] For the caveats that are required cf. the account of Bolzano's conception of analyticity and its Quinean echo in Künne (2006), in Künne (2008), ch. 8.

[73] Bolzano (WL), vol.II, p.84: "[D]ie Begriffe, welche den unveränderlichen Theil in diesen Sätzen bilden, [gehören] alle der Logik [an] [...] [Freilich ist] das Gebiet der Begriffe, die in die Logik gehören, nicht so scharf begrenzt [...], daß sich darüber niemals einiger Streit erheben ließe." This attitude is rather close to Tarski's in 1936 concerning the difference between logical and extra-logical terms: "No objective grounds are known to me which permit us to draw a sharp boundary between the two groups of terms (Mir sind keine objektiven Gründe bekannt, die eine scharfe Grenze zwischen beiden Gruppen von Termini zu ziehen gestatten)," Tarski (1936), p.10. On Tarski's later position see Simons (1992), pp.13-40.

object, property, collection, part, and meta-logical concepts, as one might call them, such as *proposition, notion, truth, validity, deducibility, objectuality.*[74]

Note that the Bolzanian concepts of universal validity and of logico-analytic truth apply to propositions, not to sentences. This should come as no surprise, for Bolzano takes truth itself to be a property of propositions. The importance of this observation jumps into the eye as soon as you ask yourself: what about the proposition expressed by

(S)   If something is a vixen then it is female,

is it also a logico-analytic truth? Yes, it is, answers Bolzano, since it is identical with the proposition that if something is both a fox and female then it is female. So Bolzano's account covers the paradigm cases that Kant used when he tried to explain what he meant by 'analytic,' but of course, it covers far more, and it is far more precise.

5   The Assimilation of Bolzano's Framework in Husserl's *Logical Investigations*

Let me repeat part of a statement Husserl made in a letter to Brentano: "Bolzano's views about notions and propositions [...] had a strong impact upon me [...]." As a matter of fact, the Bolzanian framework I have sketched in the last section shines through Husserl's theory of intentionality in the 5th *Logical Investigation*, entitled *"On Intentional Experiences and their 'Contents' (Über intentionale Erlebnisse und ihre 'Inhalte')."* Since most of the secondary literature on Husserl is written in complete ignorance of Bolzano, there prevails a deafenig silence about this fact. Let me outline the essential features of Husserl's theory by parading in alphabetical order five Spanish thinkers:

(A) Antonio judges that Plato's teacher Socrates once drew an approximately equilateral triangle in the sand.

(B) Belinda also judges that Plato's teacher Socrates once drew an approximately equilateral triangle in the sand.

(C) Carlota judges that *Xanthippe's husband* Socrates once drew an approximately equilateral triangle in the sand.

(D) Dolores judges that *Xanthippe's husband* Socrates once drew an approximately equi*angular* triangle in the sand.

---

[74]Cf. Bolzano (WL), vol. II, pp.84, 392; vol. III p.240. As to the nomenclature used for the first group, compare Husserl (LU), I, pp.242-246 / (LI), 1, pp.152-155; vol.II/1, pp.225-226, pp.251-254 / (LI), 2, p.435, pp.455-457.

(E) Emilio, habitually cautious, merely entertains the thought that Plato's teacher Socrates once drew an approximately equilateral triangle in the sand.

In §20 of the 5th *Logical Investigation* one finds the key concepts that are employed in the Husserlian description of these acts. All five thinkers perform 'propositional acts,' acts that are standardly reported as above, in the style of 'N.N. [VERB]s that *p*.' If we want to know which 'quality (*Qualität*)' an act has, we have to look at the verb in 'N.N. [VERB]s that *p*.' If we want to know which 'matter (*Materie*) or interpretive sense (*Auffassungssinn*)' an act has, we have to inspect the embedded sentence in 'N.N. [VERB]s that *p*.' And if we want to know which 'object (*Gegenstand*)' a propositional act has (if it has any), once again the embedded sentence is what we have to look at.

Since the verb in the reports of the first four acts is the same and since this verb has the same sense in all four reports, we may safely assume that the quality (whatever that may be) of those four acts is the same. All five propositional acts have two ingredients Husserl calls 'nominal acts (*nominale Akte*).' One of them is an act of identifying that Husserls calls 'subject-act,' the other one is an act of ascribing that Husserl calls a 'predicate-act.' (Neither of them is a *speech*-act.)

Now in the case of *A*'s, *B*'s and *E*'s propositional acts the ingredient nominal acts both have the same matter and, consequently, their propositional acts have the same matter: Not only do *A*, *B* and *E* identify the same object and ascribe to it the same property, they also conceive of that object and of that property in the same way. *C*'s act differs from all others as regards its matter, and the same holds for *D*'s act. *C*'s act differs from those of *A*, *B* and *E* in that its *subject*-act, though having the same object as its counterparts, conceives of it in a different way. *D*'s act differs from those of *A*, *B* and *E* in that its *predicate*-act, too, though ascribing the same property, conceives of it in a different way. Like Bolzano, Husserl is a hyper-intensionalist. Using Bolzano's standard example, he writes:

> "The presentations *the equilateral triangle* and the *equiangular triangle* differ in content, and yet they are directed [...] to the same object: they present the same object but in a 'different way.' Obviously, the same hold *mutatis mutandis* [...] for statements which, though otherwise having the same sense, differ only in such 'equivalent' concepts."[75]

Husserl's nominal acts are what Bolzano calls subjective presentations. Their Husserlian *Materie* is their Bolzanian *Stoff* by another name,[76] and as you will

---

[75]Husserl (LU), II/1, pp.414 f. / (LI) 2, p.122:"So sind die Vorstellungen *das gleichseitige Dreieck* und *das gleichwinklige Dreieck* inhaltlich verschieden, und doch sind sie beide [...] auf denselben Gegenstand gerichtet. Sie stellen denselben Gegenstand, aber noch in 'verschiedener Weise' vor. Ähnliches gilt [...] selbstverständlich dann auch für Aussagen, welche, im übrigen bedeutungsidentisch, sich nur durch solche 'äquivalente' Begriffe unterscheiden."

[76]Cf. Bolzano (WL), vol.III, p.10.

have noticed a while ago, even the difference in name evaporates when one is talking English.

Only E's act differs from the others with respect to its quality. As regards act-quality, there are just two possibilities: An act is either 'positing (*setzend*),' or it is not.[77] In the case of a propositional act, a [VERB]ing that p is positing just in case it involves a commitment as to the truth of the proposition that p. So under this reading of 'act-quality' there is no 'qualitative' difference between, say, seeing that p, recalling that p, expecting that p, being pleased that p, and being sorry that p, though they are vastly different as regards what Searle calls their 'psychological mode.'[78] Clearly, E's act is not positing, but it has the very same matter as A's and B's acts. As I anticipated a while ago, this is a point where Husserl does not embrace Bolzano's official doctrine.

And there is a further point of difference which becomes visible as soon as we ask: Does a propositional act as a whole also have an object, over and above the objects of the ingredient nominal acts? Bolzano's implicit answer is, No. Husserl's explicit answer is, Yes (5th *Logical Investigation*, §17). In order to get hold of the concepts that are employed in Husserl's answer let us have a closer look at the subject-acts in the five propositional acts we just considered. The object of those acts is always good old Socrates. But each subject-act also contains a subservient act which provides it with an auxiliary object: sometimes Plato is also adverted to, sometimes Xanthippe. Husserl calls the object of the nominal act as a whole its 'primary object' and the object of the subservient act the 'secondary object' (of the whole act).[79] Clearly, a nominal act can also have an object of the third (etc.) degree: When you are thinking of Sarah as the mother of the father of Jacob, then Isaac is the secondary object of this act, and Jacob is its tertiary object. Normally, primary objects are not identical with the objects of subservient acts, but there are exceptions: Thus when you are thinking of the third positive integer as the product of 3 and 1, or of the *Wissenschaftslehre* as the most important work of the author of the *Wissenschaftslehre*, your acts contain subservient acts whose object is identical with that of the whole act. Now as for propositional acts performed by our five thinkers, Husserl's view is that their subject-acts and predicates-acts are subservient acts, and that the object of the whole act is a state of affairs (*Sachverhalt*).

States of affairs are more coarsely individuated than propositional matters. Let us remove in our examples (1 – 5) the appositions which bring in auxiliary objects, and let us assume, as I did above, that being F is the same *property* as being G just in case necessarily, everything that is F is G, and *vice versa* (and that the same holds *mutatis mutandis* for relations). Perhaps (it is hard to be sure) Husserl

---

[77] Cf. Husserl (LU) II/1, pp.411-413 / (LI) 2, pp.119-121; Husserl (LU) II/1, pp.479-485 / (LI) 2, pp.159-162; Husserl (LU) II/2, p.64 /(LI) 2, p.226.

[78] Searle (1983), pp.4-13; Searle (1994), pp.380-383.

[79] For Moore's reflections on this see Künne (1991), §six.

would accept the following suggestion: If two propositional acts are such that the same property is ascribed to the same object and no auxiliary object is thought of in either act, then they are directed at the same state of affairs. So, perhaps, one and the same state of affairs is the primary object of both Dolores's judging that Socrates once drew an approximately equi*lateral* triangle in the sand and of Emilio's merely entertaining the thought that Socrates once drew an approximately equi*angular* triangle in the sand. Similarly, Frasquita's fear that *Hesperus* is uninhabitable may be directed at the same state of affairs as Gabriel's hope that *Phosphorus* is uninhabitable, and Isabel's mere surmise that this bowl of rice (*B*, for short) *needs more common salt than* that plate of spaghetti (*P*, for short) and Joaquin's strong conviction that *B needs more sodium chloride than P* may both be directed at the same state of affairs. (But note that the fact that a correct report of an intentional act does not mention any auxiliary object in its content clause does not imply that the act that is reported does not contain any subservient act that brings in an auxiliary object.) In any case, I suggest that we formally represent elementary Husserlian states of affairs as ordered pairs containing an n-place property and a non-empty ordered set of n individuals. So in the case of Isabel's mere surmise and Joaquin's strong conviction, the *intentional object* of both mental states can be represented equally well as {needing more common salt than; {bowl *B*; plate *P*}} and as {needing more sodium chloride than; {bowl *B*; plate *P*}}, for these are the same sequence.[80] By contrast, using the same format the *matter* of her surmise can be represented as an ordered pair the first element of which is the *sense* of 'needs more common salt than,' whereas the matter of his conviction is represented as an ordered pair the first element of which is different, namely the sense of 'needs more sodium chloride than.' In the case of Frasquita's fear and Gabriel's hope, the intentional object could be represented equally well as {being uninhabitable; {Phosphorus}} and as {being uninhabitable; {Hesperus}}, since these sequences are identical. Using the same format, the *matter* of her fear can be represented as {the sense of 'is uninhabitable;'{the sense of 'Hesperus'}}, and the matter of his hope as {the sense of 'is uninhabitable;' {the sense of 'Phosphorus'}}. It is a delicate question whether the two co-referential names differ in sense, but it is not a delicate question whether the sequence representing the propositional matter of, say, Frasquita's fear is identical with the sequence representing the state of affairs her fear is directed at. Being uninhabitable is a property of various regions in the universe, but is the sense of 'is uninhabitable' a property of any region in the universe? Furthermore, in Bolzano and Husserl senses of predicates are more finely individuated than properties.

---

[80]Not any old ordered pair containing an n-place property and a non-empty ordered set of n individuals represents a state of affairs. Thus the sequence {*being the negative square root of*; {Socrates; Plato}} does not; for the concatenation of words 'Socrates is the square of Plato' makes no sense, so it cannot serve to specify the intentional object of an act of propositional thinking.

Husserl complains:

"In Bolzano the contrasting juxtaposition of proposition and states of affairs was missing."[81]

There is no reason to suspect that Bolzano *confused* propositions and states of affairs: The latter category is simply absent from his ontology. Whether the absence of the latter category marks a real deficiency in Bolzano's theory remains to be shown. At any rate, its absence does not prevent him from realizing that some judgements have objects they are *about*. Whenever the subject-notion in the propositional matter of a singular judgement is objectual, the object picked out by that notion is what the judgement is about. So by Bolzano's lights the one and only object somebody's act of judging that Socrates is witty is directed at is the man it is about. Consequentially, someone's judging that Vulcan is inhabitable is not directed at anything: Since the subject-notion is empty, there is nothing this judgement could be said to be about.[82] Is such a judgement nevertheless directed at a state of affairs? I don't think Husserl posed this question anywhere in the *Logical Investigations* (but because of the unsurveyability, for cognitively finite beings, of Husserl's *Opera omnia* I do not dare to claim that he never ever posed it). The way I introduced the formal representation of states of affairs commits me to a negative answer, for I stipulated that the sequence of objects be non-empty. I find the negative answer very intuitive: no *Sachen*, no *Sachverhalt*. There *is* such a thing as the state of affairs that Hesperus is inhabitable, but for all I know this state of affairs isn't a fact, it does not obtain. But Vulcan's being inhabitable is not even a state of affairs that does not obtain.

Husserl and Bolzano are agreed that the matter of an act-or-state is not what the latter is directed at. As regards Husserl, there cannot be any serious doubt that he subscribes to this claim. In the case of Bolzano one can get confused by the fact that he sometimes uses 'y grasps x [y *fasst x auf*]' as a stylistic variant of (the converse of) 'x is the matter of y.'[83] This should not be taken to imply that he takes the matter of an act to be its intentional object. If an act x has an object at all, as in the case of a subjective presentation of the moon, then this object is, as Bolzano strongly emphasizes, different from the matter of the act.[84]. About that object one could say (though Bolzano doesn't) that the act is directed at it, but one could not sensibly say this about the matter of the act. If the act is a judging then the only thing Bolzano could agree to call its object is the object, if any, that falls under the subject-notion in the proposition which is the act's matter. Once again, the only candidate for the title 'what the act is directed at' that Bolzano reckons with is vastly different from the matter of the act.

---

[81] Husserl (1913), p.130, in Husserl (GW), p.299: "[Bei Bolzano] fehlte die Gegenüberstellung von Satz [an sich] und Sachverhalt." Cf. Beyer (1996), pp.124, 166 f.

[82] Bolzano (WL), vol.II, pp.25, 77.

[83] Bolzano (WL), vol. III, p.6; Bolzano (1935), 22.11.1834, p.65; 18.12.34, p.84 f.

[84] Bolzano (WL), vol.I, pp.218-220.

Husserl and Bolzano are also agreed that claim that the matter of a propositional act-or-state is not what the latter is directed at does not imply that such acts or states are never *about* propositions. Normally they are not, of course. If Leon judges that

(L)   in any right triangle the square on the hypotenuse is equal to the sum of the squares on the other two sides,

($c^2 = a^2 + b^2$, for short), then the Pythagorean Theorem is not what he is thinking about, but the matter of his thinking. But if Manuela judges that

(M)   the Pythagorean Theorem is unknown to many people,

then this theorem is not the matter of her thinking but the object her thinking is about. As we know from Bolzano, there are also mixed cases: Sometimes a proposition is both an object to which a thinker adverts and *part of* the matter of her judgement. If Nina judges that

(N)   the theorem that $c^2 = a^2 + b^2$ is unknown to many people,

then she adverts to the Pythagorean Theorem by entertaining the thought that $c^2 = a^2 + b^2$. Nevertheless, in this mixed case as in its predecessors, the proposition which is the matter of her judgement is different from the object her judgement is about. If you first think that (L) and then that (N), the modification of what was the matter of the first act is what Husserl calls 'the operation of nominalization.'[85] This is a grammatical metaphor, of course, and he might as well have called this process 'the operation of objectification' if he had not put this term already to a different use. Presumably he shied away from calling it 'reification' because to his ears that sounds too much like something he abhors: 'metaphysical hypostatization.'[86]

At the beginning of this section, I cut Husserl short. Here is a bit more of what he let his indignant former teacher know:

> "Bolzano's views about notions and propositions *and the fact that the first two volumes of WL contain valuable contributions to logic* that are independent of empirical psychology [...] had a strong impact upon me, [...]."

Or in an earlier letter: "I discovered extremely fruitful contributions to pure logic in Bolzano's *Wissenschaftslehre* which I admire very much."[87] The impact as regards logic was by no means confined to the anti-psychologistic spirit of the first two volumes of WL. One of Bolzano's most valuable contributions to logic was his account of logico-analytic truth that I outlined at the end of the last section. I

---

[85] Husserl (LU), II/1, pp.482-484 / (LI) 2, pp.161-162.
[86] Husserl (LU) II/1, p.101 / (LI) 1, p.230.
[87] For references see §1 above. Cf. also Husserl (1913), p.129, in Husserl (GW), p.298.

shall close this section with a consideration of Husserl's account of analyticity in §12 of the 3rd *Logical Investigation*.[88]

Husserl agrees with Bolzano that Kant's attempts at explaining 'analytic' "do not deserve to be called 'classical'." He distinguishes two kinds of analytic truths: those those that contain non-formal (material, *sachhaltige*) concepts, and those that don't. (Sentences do not contain concepts. So when Husserl says '*analytische Sätze*', what he means are analytical propositions. The word '*Satz*' is used here in the same way as it is used in '*der Satz des Pythagoras* [the Pythagorean theorem]' or in '*der Energieerhaltungssatz* [the law of conservation of energy].') Analytic truths that contain material concepts are called 'analytic necessities,' those that don't are dubbed 'analytic laws.' Elsewhere Husserl suggests a more telling phraseology: 'impure analytic truths' versus 'pure *(reine)* analytic truths.'[89] Impure analytic truths are said to be 'formalizable *salva veritate*.' Formalization *salva veritate* is an operation that gets us from a truth that contains material concepts to a truth that contains only formal concepts. The upshot of formalization *salva veritate* cannot be a skeleton of a proposition, for otherwise *veritas* would never be preserved: after all, schemata aren't true. Rather, the upshot of that operation is a pure analytic truth.

Here is an example of what Husserl is driving at:[90] the proposition expressed by my utterance of

(R)   If the cover of my copy of *Logical Investigations* is red, it is not the case that it is not red

is formalizable *salva veritate*, for the material concepts this impure analytic truth contains can *salva veritate* be replaced by formal concepts. The result of this operation is the general truth expressed by

(R*)   If something is somehow then it is not the case that it is not thus,

or, if you prefer symbols and are not afraid of second-order quantification, it is expressed by

(R†) $\forall x \forall f (fx \rightarrow \neg\neg fx)$.

---

[88]Husserl (LU) II/1, pp.254-256 / (LI) 2, pp.20-22.

[89]Husserl (1903/04), p.195.

[90]The example Husserl himself gives is, in an interesting way, rather unfortunate. He declares the proposition [If this house is red then redness belongs to this house] to be formalizable *salva veritate* and hence analytic. He doesn't spell out the corresponding 'analytic law,' but obviously it would have to be the proposition [If something is somehow then the property of being thus belongs to it], i.e., [$\forall x \forall f$ (x is $f \rightarrow$ the property of being f belongs to x)]. The property variant of Russell's paradox shows that this alleged law is false.

The concept of being formalizable *salva veritate* is Husserl's attempt at catching Bolzano's concept of being a logico-analytic truth.[91] The proposition expressed by

(S)   If something is a vixen then it is female

is analytic by Husserl's lights, too. Being identical with the proposition that if something is both a fox and female then it is female, it is formalizable *salva veritate*:

(S†) $\forall x \forall f \forall g((fx \& gx) \to gx)$

When it comes to Husserl's pure analytic truths or 'analytic laws,' however, Bolzano's conception diverges sharply from Husserl's.[92] *Per definitionem* such truths are truths that contain only formal concepts. Bolzano does not regard every truth that consists of nothing but logical notions as analytic. For example, the proposition that there is at least one notion, and the proposition that some notions are complex, are not universally valid with respect to any notion they contain, so they are synthetic,[93] and yet they contain nothing but logical notions (in the Bolzanian acceptation of this term).[94] If Husserl were to deny that the concept of a proposition and that of a notion are 'formal concepts,' as presumably most logicians nowadays would, Bolzano would not be silenced. Consider the propositions expressed by

(T*)   There is at least one object

and by its closest approximation in the symbolism of the predicate calculus,

(T†) $\exists x(x = x)$

Neither of them is excluded from the realm of pure analytic truths by Husserl's claim that such truths are free from 'individual existential import (*individuelle Existenzsetzung*)' such as that carried by the indexical definite description in my utterance of (R). But neither of these propositions is universally valid with respect

---

[91] In the margins of Bolzano (WL) II, pp.77-89, where this notion is introduced, there are many *Nota bene* signs in Husserl's copy of the *Wissenschaftslehre*. (Fréchette's observation.)

[92] ([...] and my interpretation of Bolzano and Husserl from that in Textor (2000).)

[93] Bolzano (WL), III, p.240.

[94] In Bolzano (WL), III, p.240, sub No. 2), the propositions just mentioned are classified as logical propositions. Cf. Bolzano (VZL), p.127. The critical remark about Kant in Husserl (LU) II/2, p.203 / (LI) 2, p.319 strongly suggests that Husserl regards *all* logical laws as (pure) analytic propositions, which is exactly what Bolzano denies in Bolzano (WL), III, p.240. From the margins in Husserl's copy it is clear that he read that page, but what he deems worthy of a NB sign is No. 3) on that page. (Fréchette's observation.)

to any notion it contains, so by Bolzano's lights, they are both synthetic.[95] It is worth recalling here that Russell found it embarrassing that $(T\dagger)$ turns out to be a theorem in the system of *Principia Mathematica*.[96]

6 Husserl's *Species* Conception of Notions and Propositions

It's high time that I let Husserl finish his sentence (in the letter to Brentano I keep on referring to):

> "Bolzano's views about notions and propositions and the fact that the first two volumes of WL contain valuable contributions to logic [...] had a strong impact upon me, *and so did Lotze's reinterpretation of Plato's theory of forms*."

Elsewhere Husserl reports that his reading of the chapter "The World of Forms (*Die Ideenwelt*)" in Hermann Lotze's *Logik* of 1874 helped him to de-mythologize Bolzano's realm of propositions and their parts.[97] It is not easy to see how Lotze's (rather wild) reinterpretation of Plato's theory of forms could have been helpful here. According to Lotze, the realm of Platonic forms should be conceived of as a realm of propositions, and the being of propositions should be regarded as their *Geltung*. Now firstly, Platonic forms, such Beauty and Courage, Identity and Difference, obviously don't have a propositional structure, so if their realm is occupied by propositions then they can only survive the Lotzean transformation of Plato's metaphysics by being parts of propositions. So let us suppose that this is what Lotze is driving at. But then, how can the being of propositions consist in their *Gelten*? After all, *Geltung*, standardly translated as 'validity,' comes down here to truth. Since quite a few propositions are false (roughly half of them), and since Husserl is as much convinced as Bolzano that falsehoods as well as truths are indispensable for an adequate philosophy of mind and of logic, he cannot possibly endorse Lotze's claim that for propositions *esse est verum esse*. So what on earth is it that impressed Husserl? It is Lotze's suggestion that propositions are universals that are instantiated in the mental realm.[98] In his forceful reply to the very hostile criticism of both Bolzano's *Wissenschaftslehre* and Husserl's *Prolegomena* by the Hungarian philosopher Melchior Palagyi, Husserl wrote in 1903:

> "It was only by reflecting on Lotze's not completely transparent argumentation that I got hold of the key to [...] Bolzano's conceptions and to the treasures of his WL [...] Bolzano's doctrine that propositions are objects but lack 'existence' now becomes easily understandable as the claim that they have the 'ideal' being of 'general objects' [species] [...] rather than the real being of things or of dependent moments of things – of temporal particulars in general."[99]

---

[95] Cf. Bolzano (WL), II, p.375.
[96] Russell (1919), pp.203 and fn. Russell almost wrote a review of (LI); Künne (1991), §i.
[97] Lotze (1874), §§313-321.
[98] See Beyer (1996), pp.131-152.
[99] Husserl (1903), pp.156 f.: "Erst die innere Verarbeitung der [...] nicht völlig abgeklärten Gedanken Lotzes gab mir den Schlüssel zu den [...] Konzeptionen Bolzanos und zu den Schätzen

Now if propositions (and notions) are species, what are the members of these species? Husserl's answer is: particularized properties of mental acts-or-states.[100] This answer invokes an Aristotelian framework, so let us try to get some illumination from the Ontological Square in *Categories*, chapter 2:

(1) Socrates this rose	(3) Socrate's pallor the redness of this rose
(2) the kind, Man the kind, Rose	(4) the property of being pale the property of being red

According to Aristotle, the entities in (1) are individuals in the category of Substance, while the entities in (2) are Kinds in that category; on the right-hand side, the entities in (3) are indiduals in the category of Quality, whereas those in (4) are Kinds in this category. So just as the items in (1) are members of the Kinds in (2), so the items in (3) are members of the Kinds in (4). In Husserl's language, the items in (1) and (3) are 'real' ("i.e., objects of a possible sense-perception"[101]) and 'temporal particulars (*zeitliche Einzelheiten*),' while those in (2) and (4) are 'ideal' and 'general objects / universals (*allgemeine Gegenstände*).'

Type-(3) entities inhere in type-(1) entities. According to the traditional reading of Aristotle's text,[102] inherence is a relation that satisfies the following conditions:

(A)   for all x, for all y, if x inheres in y, then x cannot exist at any time at which y does not also exist, and

(B)   for all x, for all y, for all z, if x inheres in y & x inheres in z, then y must be identical with z.

There is a problem with condition (B) which has been noted in the literature and which actually also threatens (A).[103] No rose is identical with its blossom, but isn't

---

seiner WL [...] Bolzanos Lehre, dass Sätze [an sich] Gegenstände sind, aber doch keine 'Existenz' haben, [gewinnt] die leicht verständliche Bedeutung, dass ihnen das 'ideale' Sein [...] 'allgemeiner Gegenstände' zukomme [...], nicht aber das reale Sein von Dingen oder unselbständigen dinglichen Momenten, von zeitlichen Einzelheiten überhaupt." For Palagyi's criticisms see Künne (1997) in Künne (2008) ch. 9, §8.

[100] See Husserl (LU), II/1, pp.100-103 / (LI) 1, pp.230-232; Husserl (1908), pp.31-35.

[101] Husserl (LU) II/1, p.280; II/2, pp.139, 151 / (LI) 1, p.479, 2, pp.782, 791.

[102] The reading of *Categories*, ch. 2 that was standard from Porphyrios to Ross, Anscombe and Ackrill was challenged by G.E.L. Owen (in: *Phronesis* 19, 1965) and M. Frede (in: *Antike und Abendland* 24, 1978) who plead for taking the items in (3) to be *infimae species*.

[103] See Schnieder (2006).

the redness of a red rose nevertheless identical with the redness of its blossom? After all, its stalk, its thorns and its leaves are all green. (The same holds *mutatis mutandis* for Socrates's pallor, of course.) We must either replace 'identical with' by 'not wholly distinct from' (and explain that notion carefully), or we must say that, strictly speaking, the redness of a rose does not inhere in that flower but in its *blossom*. In order to preserve condition (A), the latter escape route is preferable. When you come to think of it, what we tend to call the red rose in Carmen's hair is no longer a rose: it is just the most beautiful part of what *was* a rose, and the redness of its petals survived the death of that flower. No wonder that referring to a type-(3) item as the redness of a particular *rose* turns out to be a case of *totum pro parte*, semantically in the same boat as the politically less innocent habit of saying 'America' when one is talking about the U.S.A. Generally, the F-ness of a inheres in that part p of a which is such that a is F because p is F.

Various titles have been bestowed upon type-(3) entities, as traditionally conceived. For many centuries they were called 'individual accidents' or 'modes.' In the last decade of the 20th century, Australian philosophers gave currency to a dreadful misnomer (of American, sorry, of U.S. origin), 'tropes.' Bolzano calls the items in (3) 'adherences (*Adhärenzen*).'[104] About one of the examples in (3), he writes:

> "*This* red (*numero idem*) cannot be found in any other rose. The red that is found in a second rose can be *similar, very similar* to that found in the first, but it cannot be *the same*, precisely because it is not the same rose; for *two* roses, *two* reds are required."[105]

As Bolzano's case shows, you can coherently believe in type-(3) entities *without* taking properties to be species thereof. Husserl refers to the items in (3) as individual moments, and he does adopt the species conception of properties. Here is what he says (somewhat convolutedly, as usual) about the same example:

> "[A] red thing [is] a thing containing a case of red. [...] The primitive relation between Species and Instance emerges: it becomes possible to look over and compare a manifold of instances, and perhaps to judge with self-evidence: In all these cases the individual moments differ, but in each the same Species is realized: this red is the same as that – specifically treated it is the same colour – and yet again this red differs from that one – i.e., individually treated, it is a different individual feature of an object."[106]

---

[104] See especially Bolzano (1838), pp.21-27, discussed in Künne (1998) und Schnieder (2002).

[105] Bolzano (1935), 23.08.1833, pp.32-33: :"*Dieß* Roth (*numero idem*) kann sich an keiner zweyten Rose finden. Das Roth, das sich an einer zweyten Rose findet, kann jenem, wenn Sie wollen, *gleich, sehr gleich* kommen, aber *dasselbe* kann es nicht seyn, eben weil es nicht dieselbe Rose ist; zu *zwey* Rosen werden *zwey* Röthen erfordert."

[106] Husserl (LU) II/1, p.109 / (LI) check [2nd Inv., §1].: "Ein Rotes [ist ein] einen Fall von Rot in sich Habendes. [...] Es tritt das primitive Verhältnis zwischen Spezies und Einzelfall hervor, es erwächst die Möglichkeit, eine Mannigfaltigkeit von Einzelfällen vergleichend zu überschauen und eventuell mit Evidenz zu urteilen: In allen Fällen sei das individuelle Moment ein anderes, aber 'in' jedem sei dieselbe Spezies realisiert; dieses Roth sei dasselbe wie jenes Rot – nämlich spezifisch betrachtet, sei es dieselbe Farbe – und doch wieder sei dieses von jenem verschieden – nämlich individuell betrachtet, sei es ein verschiedener gegenständlicher Einzelzug."

Strawson confers yet another title on type-(3) entities: "We can say, of a dead man, that *his kindness* and *his intelligence* are no more and that the world is the poorer for their loss. What we here speak of are non-substantial (though substance-dependent) particulars, *particularized qualities*."[107] – If we regard the concepts of kind-membership and inherence as primitive, we can define the concept of exemplification: x exemplifies y iff y is a property & for some z, z is a member of y & z inheres in x. Thus the blossom in Carmen's hair exemplifies redness just in case a particular instance of redness inheres in that blossom. This is the position Strawson arrived at in his last publication when he classified particularized qualities as *property-instances*.[108]

Now applying this ontological scheme to the field of intentional acts, we obtain the following result: the matter that is common to three datable acts of thinking that some roses are red, say, is a species three members of which inhere in those acts, just as the property of being red which is common to the blossom in Carmen's hair and her lips, say, is a species instances of which inhere in those brilliant items. Of course, mental acts are very unlike people, flowers and their parts. They belong rather with (substance-involving) *events* such as Socrates's suddenly turning pale and the gradual fading of Carmen's rose, which one could call dynamical individual moments, as opposed to the static ones we have been concerned with. Bolzano classifies acts of judging and acts of presenting as 'adherences' of thinking beings,[109] and he explains that there are higher-order adherences.[110] Hence particularized properties of mental acts-or-states are second-order adherences. So if propositions and notions are the species Husserl takes them to be, then the members of these species are adherences of adherences.

This Husserlian view is categorially economical in that it does not claim that there are kinds *as well* as notions and propositions, but rather that there are kinds some of which are notions and some of which are propositions. Furthermore, the Husserlian view allows us to define a concept Bolzano took to be indefinable, namely *x is the matter of y*, by invoking concepts that one needs in any case if one accepts the doctrines incapsulated in Aristotle's Ontological Square, namely the concepts of kind-membership and of inherence. The definition of *x is the matter of y* runs on lines parallel to those on which the definition of the concept of exemplification ran:

(MT-OF)
x is the matter of y if and only if:
x is a proposition or a notion & y is a mental act-or-state &

---

[107] Strawson (1974), p.131.
[108] Strawson (2006), cf. Strawson (1959), pp.168-169 with n. 1. Instead of '*x* inheres in *y*' Strawson says, '*x* is attributively tied to *y*.'
[109] Bolzano (1838), *passim*; Bolzano (WL), III, pp.10-12, 109, 112.
[110] Bolzano (1838), p.23.

for some z, z is a member of x & z inheres in y.

Now for Husserl, a kind K exists only if it is possible that something is a member of K, and he seems to understand the modality in the sense of 'conceptually possible.' So there is no such thing as the kind Round Square and the kind Regular Pentahedron, but there is such a thing as the kind Golden Mountain and the kind Perpetual Motion Machine.[111] Hence there is such a thing as the proposition x if and only if it is possible that a mental act occurs, or a mental state obtains, whose matter (*in specie*) is x. Can Bolzano accept this claim?[112]

The answer is: clearly not, as long as he sticks to his official doctrine according to which a proposition can be the complete matter only of an act of judgement or a state of belief. There is such a thing as the proposition that a triangle is not a triangle, but it is not possible, Bolzano rightly maintains, "that any thinking being seriously believes that a triangle is not a triangle."[113] Since there cannot be any such judgings and believings, it is *a fortiori* impossible that there are any particularized properties which inhere in them. So, by Husserl's lights, there is no species with which our self-evidently false proposition could be identified. But of course, if Bolzano were to give up his official doctrine (something he ought to do in any case, for reasons that are independent of the present issue) then this obstacle would disappear. He could then consistently concede to Husserl that for any p the proposition that p is a matter-species the members of which inhere *either* in judgings or believings that p *or* in acts of merely entertaining the thought that p.

Husserl is right when he remarks that there is "not the slightest hint" of his conception of the matter-of relation in Bolzano.[114] And yet, sometimes he takes himself to have discovered a trace of it in the *Wissenschaftslehre*. In one of his annotations to his copy of the *Wissenschaftslehre*, he remarks:

"That Bolzano means the judgement *in specie* is shown [below] by his approving quotation from Mehmel."[115]

When Bolzano tries to show that his conception of *Sätze an sich* is not a newcomer on the philosophical scene, he quotes, among others, a follower of Kant and Fichte in Erlangen: "[Gottlieb E. A.] Mehmel in his *Analytische Denklehre* [1803] says: 'The judgement, considered objectively, i.e., in abstraction from the mind whose

---

[111] Husserl (1903), p.159; Husserl (LU) II/1, p.343 / (LI) 2, p.79, (6) 450; cf. Künne (1983), ch. III, §1.

[112] Cf. the discussion of this issue in Textor (1996), pp.16-25.

[113] Bolzano (1935), 18.12.1834, p.84; cf. Bolzano (WL) I, p.176.

[114] Husserl (1903), p.157.

[115] Husserl's remark, referring to Bolzano (WL) I, p.85, is to be found at the bottom of p.77 of his copy. It is commented upon in Rollinger (1996):"Daß Bolzano das Urtheil *in specie* meint, zeigt das beifällige Zitat aus Mehmel."

action it is, is called 'proposition'."[116] Now abstraction is traditionally regarded as the way we gain cognitive access to a species when starting from particulars that belong to that species. So Husserl's claim in the margin next to the quotation from Mehmel is not implausible: "That would be the judgement *in specie*" (*Das wäre das Urteil* in specie). In his *Ideas pertaining to a Pure Phenomenology* of 1913, Husserl again refers to this quotation but he rightly emphasizes that what Bolzano was driving at was not the 'noetic' (or species-) conception of propositions but rather the 'noematic' conception.[117] As regards the 'approving quotation' Bolzano's approval cannot have been whole-hearted. That is shown by his unequivocal disapproval of Franz Exner's attempt at making sense of *Sätze an sich* by appealing to abstraction. His objection to Exner applies with equal force to Mehmel: our disregarding the fact that judgings are mental acts cannot result in their ceasing to *be* mental acts.[118]

Husserl himself gave up the species conception of propositions (and of notions) before 1913 when he published *Ideas*.[119] Note that what he came to regard as mistaken is not the view that a property is a species of individual moments, but only the *application* of this view to the matter of intentional acts-or-states. In a letter to his former doctoral student Roman Ingarden he leaves no doubt as to what he takes to be the main error in the *Logical Investigations*:

> "The mistake [in LI] mainly consisted in conceiving 'sense' and 'proposition' in the case of acts of judgement [...] as *species*. The independence of the existence of a proposition from the contingent act of judging and its agent does not by itself entail that what is ideally identical is a species."[120]

To be sure, one can consistently accept the independence claim without endorsing the species conception: Bolzano's theory is evidence for that. But of course, that doesn't explain why Husserl has come to regard the species conception as mistaken. In *Experience and Judgement*, based on various manuscripts Husserl wrote between 1910 and 1929 and posthumously published in 1938, he says:

> "To be sure, a proposition is general insofar as it points to infinitely many positing acts in which it is meant; but it is not general in the sense in which a species is general."[121]

---

[116]Bolzano (WL), I, p.85: "Das Urtheil objectiv, das ist, mit Abstraction von dem Geiste, dessen Handlung es ist, betrachtet, heißt ein Satz." Mehmel (1803), §217, again referred to in Bolzano (WL), I, 93, where Husserl underlines the quotation in his copy. (Fréchette's observation.) With an allusion to Mephisto's sarcastic remark on the *Collegium Logicum* Mehmel sent his book to Goethe (s. *Briefe an Goethe* - Regestausgabe, Weimar 1980 ff, No. 4/1019).

[117]Husserl (1913b), p.196 note; cf. also Husserl (1908), pp.33, 156 note.

[118]Bolzano (1935), 22.11.1834, pp.65-66; 18.12.1834, pp.82-84. Exner had received a copy of some parts of the manuscript of Bolzano (WL) in 1833. Since 1830 the book ms. travelled from one German publisher to the next, until it was finally published in Bavaria in 1837.

[119]His lectures in the summer term of 1908 document his change of mind: Husserl (1908).

[120]Husserl (BW), vol. 3, p.182: "Der Fehler lag vor allem in der Fassung des 'Sinnes' u. 'Satzes' bei Urtheilserlebnissen [...] als Species. Die Unabhängigkeit des Seins eines Satzes von dem zufälligen Urtheil u. Urtheilenden besagt noch nicht, daß das ideal-Identische ein specifisches ist." Cf. p.269.

[121]Husserl (1938), §64d: "Gewiß ist ein Satz insofern allgemein, als er auf eine unendliche Zahl

Husserl's first statement is rather odd: If an entity's being the intentional object of potentially infinitely many intentional acts were a good reason for calling it a general object, then the Tour Eiffel could also lay claim to this title. Husserl's second statement is just a flat denial of the earlier doctrine. So, where is the argument? Husserl continues:

> "Two acts of judgement that mean the same proposition mean *one and the same thing*: it is not the case that each of them by itself means a proposition-instance which is contained in it as a moment and that both only mean *similar* propositions [proposition-instances?] so that the one irreal proposition [that] 2 < 3 is only the species comprising all such particularities."[122]

This is rather opaque (hence hard to translate), and it seems to distort the earlier doctrine. I cannot make sense of Husserl's talk of '*x (ver)meint y*' in this and the preceding quotation, unless it means: *y* is the intentional object of *x*. But then the objection does not hit its target: The species conception did not at all imply that the matter-moment of an act of judging is the intentional object of that act. Its claim was that if there is something which is the intentional object of such an act then it is a state of affairs. Furthermore, the earlier doctrine did not *deny* that if two people judge that 2 < 3 then their acts have the very same matter: It was an attempt at explaining 'having the same matter' in terms of two other relations, membership and inherence. If Husserl does not have any other objection against his earlier view, and a more substantial objection, then I cannot help thinking that he would have done better better to stick to his old guns.

7 Are Questions Propositions?

To the best of my knowledge, the most extensive and most thorough discussion of a single contention in Bolzano's philosophy of logic that can be found in any of Husserl's books and articles published during his lifetime is contained in the last chapter of his *Logical Investigations*.[123] The topic of this discussion is a courageous if not outrageous Bolzanian contention which, at least on the face of it, flatly contradicts what most philosophers since Aristotle took for granted. *Questions*, Bolzano claims, *are a special kind of propositions and hence truth-evaluable*. Let me call this *Bolzano's Tenet*. In this section I shall reconstruct and evaluate both Bolzano's Tenet and Husserl's criticism thereof.

---

setzender Akte hinweist, in denen er eben vermeint ist; aber er ist nicht allgemein im Sinne der Gattungsallgemeinheit."

[122]"Zwei Akte des Urteilens, die denselben Satz meinen, meinen *identisch dasselbe*, und nicht meint jeder einmal für sich einen individuellen Satz, der als Moment in ihm enthalten wäre, und jeder nur einen *gleichen*, so daß der eine irreale Satz 2 < 3 nur das Gattungsallgemeine all solcher Vereinzelungen wäre."

[123]In 1920 Husserl emphasized that he had refrained from modifying the text of the 1st edition of LI only because in the meantime his views had changed too drastically (preface to the 2nd edition of Husserl (LU) II/2, p.vii / (LI) 2, p.179). I shall concentrate exclusively on his 1901 position, more precisely: on those aspects of that position which are relevant for an evaluation of Bolzano's thesis about questions.

What exactly is it that Bolzano maintains when he says that questions are a kind of propositions? Propositions are thinkables and sayables which can be singled out by that-clauses. They are truth-evaluable, hence, assuming bivalence as Bolzano does,[124] they are either true or false. Since propositions are neither mental nor linguistic, Bolzano's Tenet would be all too obviously false if by 'questions' he were to mean anything mental or linguistic. Now the term 'question' is multiply ambiguous, and for our inquiry it is most important not to get entangled in this ambiguity. We must distinguish

Questions1 : *mental acts of asking oneself a question*,

Questions2 : *illocutionary acts of asking a question*,

Questions3 : *interrogative sentences*, and

Questions4 : *askables*.

Wonderings, i.e., sense-1 questions, are voiced by sense-2 questions. Husserl occasionally labels the former 'internal questions (*innerliche Fragen*)' and the latter 'overt questions (*Anfragen*).' The second term (which in ordinary German has a far narrower application) is meant to register the fact that sense-2 questions are essentially addressed to someone. Sense-3 questions are linguistic vehicles of sense-2 questions; unsurprisingly Husserl calls them '*Fragesätze* (interrogative sentences).' Sense-4 questions, finally, are possible contents of sense-1 and of sense-2 questions. In Husserl's language, an askable is a '*Frageinhalt*,' and he identifies it with the 'meaning of the interrogative sentence (*Bedeutung des Fragesatzes*).'[125] (Actually, it can coincide with the conventional linguistic meaning of a sense-3 questions only if the latter are free of context-sensitive elements.) Askables are those thinkables and sayables which can be singled out by indirect sense-3 questions (for example, by the clauses in 'He asked *whether the conference had started*' or 'She asks *when the conference will end*'). So let us reformulate Bolzano's Tenet: *All askables propositions*.

Bolzano takes his Tenet to contradict Aristotle, and Husserl concurs.[126] But is the conflict really that straightforward? There can be no doubt that by 'λόγοσ αποφαντικός' Aristotle means a sequence of words which we would call a declarative sentence. Thus, when he maintains that only declarative sentences are sentences with a truth-value, he does not, strictly speaking, deny what Bolzano affirms. What Aristotle denies is that sense-3 questions are truth-evaluable. Bolzano translates Aristotle's term for declarative sentences by 'Satz,' and since he admittedly uses

---

[124] Bolzano (WL), II, p.7.
[125] Husserl (LU) II/2, pp.211-212 / (LI) 2, p.328.
[126] Aristotle, *De Interpretatione* 4, 16 b 33-17 a 7, quoted in part in Bolzano (WL) I, p.87 and vaguely referred to by Husserl in Husserl (LU) II/2, p.207 / (LI) 2, p.325.

this expression, most of the time, as short for '*Satz an sich*,' the distinction between interrogative sentences and askables gets momentarily blurred. (I shall return to this point at the end of this paper.)

In his exposition of the Tenet in the first volume of the *Wissenschaftslehre* (which Husserl exclusively draws upon in his discussion), Bolzano writes:

> "Of course, a [sense-4] question does not assert anything about the object it is concerned with; nevertheless it does assert something, namely that the questioner wants to be given some information about that object. That's why a [sense-4] question can also be true as well as false."[127]

In the second volume Bolzano sharpens this claim:

> "By a [sense-4] question I mean any proposition which asserts that the questioner wants the indication of a truth which has such-and-such a property."[128]

Let us not get confused by this confusing talk of 'asserting.' It does not mean any linguistic activity but alludes to the predicative structure of the proposition: 'Proposition *P* asserts something about something' means that according to *P* something holds of something.[129]

The constraining properties mentioned in the second quotation determine various *kinds* of askables.[130] Let us start with *yes/no* interrogatives. According to Bolzano's Tenet, what you say by uttering the interrogative

> (𝔍1) 'Is Vesuvius an active volcano?'

you could just as well say by uttering the following sentence

> (ℜ1) 'I want the indication of the true proposition in the pair {[Vesuvius is an active volcano],[Vesuvius isn't an active volcano]}'

Faute de mieux I shall call sentences of the ℜ-type rogatives, or when more precision is needed, canonical rogatives ('*rogare*' in Latin can be used in the sense of 'request').

Search interrogatives typically begin with interrogative pro-forms like 'who,' 'what,' 'when,' 'where,' 'why,' or 'how.' (Strangely enough, many English-speaking philosophers and linguists call such interrogatives wh-questions. This

---

[127] Bolzano (WL) I, p.88, quoted in Husserl (LU) II/2, p.207 / (LI) 2, p.325: "Eine Frage [...] sagt freilich über das, worüber sie fragt, nichts aus; darum sagt sie aber gleichwohl noch etwas aus: unser Verlangen nämlich, über den Gegenstand, wornach wir fragen, eine Belehrung zu erhalten. Sie kann eben deßhalb auch Beides, wahr und falsch seyn."

[128] Bolzano (WL) II, 71 f.: "Ich verstehe [...] unter einer Frage jeden beliebigen Satz, in welchem ausgesagt wird, daß man die Angabe einer Wahrheit verlange, die man durch eine gewisse Beschaffenheit, welche sie haben soll, näher bezeichnet hat." Cf. Bolzano (WL) II, pp.194, 292. Of course, 'Satz' here (as ever so often in Bolzano (WL)) is short for 'Satz an sich.'

[129] Bolzano (WL) I, pp.115, 118.

[130] Bolzano (WL) II, 72-76.

terminology is not to be recommended: It gets the spelling of 'how' wrong.[131]).
Bolzano distinguishes two kinds of search interrogatives. According to his Tenet,
instead of using

(𝔍2) 'Which object has the property of being an active volcano?'

you might equally well use the rogative

(𝔑2) 'I want the indication of a true proposition in which the property
of being an active volcano is ascribed to an object'

and what you say by uttering

(𝔍3) 'Which properties does Vesuvius have?'

you could also say by uttering the rogative

(𝔑3) 'I want the indication of several true proposition in which different properties are ascribed to Vesuvius'

It is due to Bolzano's prejudice in favour of the subject-predicate form that he takes search questions always to seek either subjects, as in (𝔍2), or predicates, as in (𝔍3), but let us put this aside here. Like many latter-day logicians Bolzano has no scruples to classify sentences as expressing sense-4 questions "even if they lack the shape which grammarians call the question-form."[132] Rogatives are meant to disclose the alleged fact that "the *real* subject [of a question] is the speaker."[133] Obviously Bolzano is thinking here of situations in which somebody puts a question to someone else.[134] The term 'indication (*Angabe*)' in his rogatives is tailor-made for such cases. But as propositions in general are possible contents both of judgings and assertings, so askables in particular are possible contents of mental acts as well as of speech-acts. Hence a slight modification is needed. One could obtain the required generality if one were to replace 'indication' by 'determination,' stipulating that this term is to cover answers given by addressees of sense-2 questions and answers found by the questioners of sense-1 questions. As for the latter case, Bolzano's contention would be that the real subject of what a thinker asks in a silent wondering is the thinker.

---

[131] The Stoics called yes/no questions 'ἐρωτήματα' and search interrogatives 'πύσματα.' In Frege (1918), p.62, the former are referred to as '*Satzfragen* (sentence questions),' the latter as '*Wortfragen* (word questions).' A common pair of labels in German is '*Entscheidungsfragen* (decision questions)' vs. '*Ergänzungsfragen* (completion questions).'

[132] Bolzano (WL), II, p.72: "obgleich sie gar nicht die Gestalt haben, welche die Grammatiker sonst eine Frageform nennen."

[133] Bolzano (WL), I, p.89: "das eigentliche Subject [...] ist der Sprechende."

[134] Bolzano (WL) II, p.72.

Sometimes Bolzano uses a different kind of rogatives. In the very section of the *Wissenschaftslehre* on which Husserl's discussion is based, sentences like our (ℑ1) and (ℑ2) get paraphrased as

(ℜ*1) 'I want to *know whether* Vesuvius is an active volcano'

(ℜ*2) 'I want to *know which* properties Vesuvius has.'

Bolzano points out that Antoine Destutt, Comte de Tracy also takes interrogatives to be replaceable by such rogatives: in his *Grammaire* Destutt claims that '*Avez-vous fini?*' has the same meaning as '*je desire* savoir si *vous avez fini.*'[135] In one respect Destutt and Bolzano come here rather close to the *epistemic imperative* approach advocated by Åqvist and Hintikka (and anticipated to a certain extent by Frege):[136] according to them the deep-structure of (ℑ1), for example, is the surface structure of the epistemic imperative

(E1) 'See to it that I know whether Vesuvius is an active Volcano!'

Now as compared with Destutt-Bolzano rogatives and Åqvist-Hintikka imperatives, *canonical* rogatives in the style of (ℜ1-3), which Bolzano favours in vol. II of the *Wissenschaftslehre*, have two advantages. Firstly, we can have the wish that is articulated by a canonical rogative 'even if,' as Bolzano puts it, "we very well *know* the truth we want to be specified (as happens, for example, when we ask in the manner of Socrates)."[137] Canonical rogatives do not call for a special treatment of examination questions (which are often not so very Socratic, I fear). Secondly, unlike Destutt-Bolzano rogatives and Åqvist-Hintikka imperatives, *canonical* rogatives do not suggest that at bottom all sense-3 questions are *in*direct sense-3 questions. This suggestion, which seems to get the priorities wrong, is also carried by the *performative* approach to questions favoured by David Lewis:[138] according to him, the deep-structure of (ℑ1) is the surface structure of the explicitly performative utterance

(P1) 'I hereby ask whether Vesuvius is an active volcano.'

Husserl's discussion in the *Logical Investigations* unfortunately oscillates between Destutt-Bolzano rogatives and performative renderings (which he does not recognize for what they are). Since Bolzano himself makes the point about 'asking in

---

[135] Bolzano (WL) I, p.88; Destutt de Tracy (1803), p.52. In this (non-canonical) form the view is to be found already in Bolzano's earliest notes on logic: Bolzano (1812), p.146, and it re-appears in Bolzano (WL) III, p.177.
[136] Åqvist (1965); Hintikka (1974), pp.104-110; cf. Frege (1918), p.62.
[137] Bolzano (WL), II, p.75: "selbst wenn uns die Wahrheit, nach der wir fragen (wie etwas beim Sokratisiren), recht wohl bekannt ist."
[138] Lewis (1983), pp.222-225.

the manner of Socrates,' I take it that his considered view is that which appeals to *canonical* rogatives in the style of (ℜ1-3). (That's why I called them 'canonical.') So, according to Bolzano's Tenet, what a speaker s says at time t by uttering an interrogative is the proposition which the corresponding canonical rogative expresses in the mouth of s at t, namely:

(*Rogation*) '[I want the indication of a truth of type X]$_{s,t}$.'

(where '[...]$_{s,t}$' is to be read as 'the proposition which is expressed by the bracketed sentence "..." with respect to speaker s at time t'). If we call such propositions *rogations*, we can formulate Bolzano's Tenet thus: *Askables are rogations*.

A rogation may be *false*. When does an utterance of an interrogative express a false rogation? Bolzano replies: whenever the speaker, at the time of her utterance, does not really have the desire that a truth of such-and-such a type be indicated to her. When does this happen? Bolzano answers: whenever she used the wrong words for voicing her desire or she only pretends to have that desire.[139] Now the first objection against Bolzano's Tenet in Husserl's *Logical Investigations* is directed at this point:

> "Truthfulness [veracity, sincerity] or untruthfulness, and in general appropriateness and inappropriateness are issues which can come up with respect to any kind of speech, but truth and falsity only with respect to asserting. Hence one can make several objections to somebody who makes an assertion: 'What you say, is false:' This is the *factual objection* . And 'You are not speaking sincerely' or 'You are not expressing yourself appropriately:' This is the objection of *insincerity* and *inappropriateness of speech*. When somebody poses a question, only the latter type of objections can be made. Perhaps he is pretending, or perhaps he is misusing certain words and saying something different from what he wants to say. But one would not make the factual objection, since the questioner is not making a factual claim."[140]

Surely it would be very odd to react to an utterance of an interrogative by remarking, 'What you say is false.'[141] Advocates of the *epistemic imperative* approach can accept this observation with equanimity. Adherents of the *performative* approach who follow Lewis (as against Austin) in declaring utterances of (*P*1) to be true don't have to be embarrassed by this observation either, for they can point out that the oddity of the comment 'What you say is false' results from the fact that (*P*1) utterances are self-verifying, hence true as a matter of course. Bolzano

---

[139] On the first case: Bolzano (WL) I, p.88, on the second: Bolzano (WL) II, pp.75/76.

[140] Husserl (LU) II/2, p.208 / (LI) 2, pp.325-326: "Von Wahrhaftigkeit oder Unwahrhaftigkeit, und überhaupt von Angemessenheit und Unangemessenheit kann bei jedem Ausdruck gleichmäßig die Rede sein. Von Wahrheit und Unwahrheit aber nur beim Aussagen. Dem Aussagenden kann man also Mehrfaches einwenden: Was du sagst, ist unwahr. – Dies ist die *sachliche Einrede*. Und: Du sprichst nicht wahrhaftig; oder auch: Du drückst dich unpassend aus. – Das ist der Einwurf der unwahrhaftigen und der inadäquaten Rede. Dem Fragenden kann man nur Einwände der letzteren Art machen. Er verstellt sich vielleicht oder gebraucht seine Worte unrichtig und sagt anderes, als er wirklich sagen will. Aber man wird ihm nicht die sachliche Einrede machen, da er eben keine Sache vertritt."

[141] Cf. also Husserl (LU) II/2, p.207 / (LI) 2, p.325.

could try to defuse the objection along the lines Husserl suggests at the end of the last chapter of *Logical Investigations*. If utterances of interrogatives express rogations they are in the same boat as many first-person present-tense ascriptions of experiences and mental acts: provided the speaker does not use his words improperly, "they *are* true or false," as Husserl puts it, "but truth coincides here with sincerity."[142] Saying 'That is false' in reaction to a question, Bolzano could reply to the objector, is as odd as (and no more odd than) saying this in response to an utterance of 'I am in pain.' Whether this move really extricates Bolzano's Tenet from the objection is debatable, but I shall not debate it here, since his Tenet has to face more formidable challenges.

If an utterance of an interrogative expresses a rogation, then, as Husserl points out, it abbreviates an occasion-sensitive expression:

"The circumstances of the utterance make it clear that it is the speaker who is asking a question. Hence the full import of the sentence does not coincide with its literal meaning, rather it is determined by the occasion, namely by the relation of the sentence to the person currently speaking."[143]

At this point an opponent of Bolzano's Tenet might object: "But this argument would equally apply to declarative sentences, so we would have to take the expression 'S is P' as abbreviating an occasion-sensitive expression 'I judge that S is P,' etc. in infinitum."[144] Husserl does not accept the alleged analogy:

"If we transform the sentence 'S is P' into the sentence 'I judge that S is P' [...] we obtain a meaning [...] which is not even equivalent to that of the original; for the plain sentence can be true while its subjectivized variant is false, and vice versa. But the situation is very different in the case at hand. Even if one refuses to speak here [in the case of questions] of truth and falsity, one will always find an assertoric form which 'essentially says the same thing' as the original interrogative form [...], e.g., 'Is S P?' = 'I want to know whether S is P' [...]. Hence in such sentential forms a relation to the speaker may after all be implied."[145]

---

[142] Husserl (LU) II/2, p.221 /L2, 334: "Sie sind zwar wahr oder falsch, aber Wahrheit fällt hier mit Wahrhaftigkeit zusammen." In Bolzano (WL) II, p.292 Bolzano classifies rogations as 'perceptual propositions (*Wahrnehmungssätze*),' and by 'perception' he means here introspection.

[143] Husserl (LU), II/2, p.210 / (LI), 2, p.327: "Die Umstände der Äußerung machen es ja ohne weiteres verständlich, daß der Redende selbst es ist, der da fragt. Also liegt die volle Bedeutung des Satzes nicht in dem, was er selbst nach seinem Wortlaute bedeutet, sondern ist durch die Gelegenheit, nämlich durch die Beziehung zur augenblicklich redenden Person bestimmt."

[144] Husserl (LU), II/2, p.211 / (LI), 2, p.327: "Das Argument würde doch nicht minder auf Aussagesätze passen; also müßten wir den Ausdruck 'S ist P' als gelegenheitliche Verkürzung für den neuen Ausdruck 'ich urteile, daß S P ist' interpretieren, und so in infinitum."

[145] Husserl (LU), II/2, p.213 / (LI) 2, pp.328-329: "Bei der Umwandlung des Satzes 'S ist P' in den Satz 'ich urteile, daß S P ist' [...] erhalten wir [...] Bedeutungen, [...] die den ursprünglichen nicht einmal äquivalent sind; denn der schlichte Satz kann wahr, der subjektivierte falsch sein, und umgekehrt. Ganz anders im Vergleichsfalle. Mag man in ihm von Wahr und Falsch zu reden ablehnen: man wir doch immer eines Aussage[form] finden, die 'wesentlich dasselbe besagt,' wie die ursprüngliche Frageform [...], z.B. 'Ist S P?' = 'ich wünsche [...] zu wissen, ob S P sei' [...]. Sollte in derartigen Satzformen also nicht doch eine Beziehung [...] zu dem Redenden impliziert sein."

Husserl's objection to the analogy is a counterpart to Lewis' objection to John Ross' proposal that declarative sentences abbreviate performatives: it gets truth-conditions wrong.[146] One may wonder whether these criticisms are sufficiently worked out as they stand. Let us replace Husserl's 'judge' by 'believe' and consider the following exchange: – Antonio: 'What is the weather like?' Belinda: 'I believe it is raining.' Carlota: 'No, you are wrong: it has stopped raining a while ago.' – If Husserl is right then *C* does not dissent from what *B* said. But doesn't *C* obviously do just that? Could *B* sensibly reply, 'Whether it is raining or not is entirely irrelevant to the question whether I was right in what I said. After all, I only made an autobiographical statement'?[147] The same kind of problem arises for Lewis' objection to Ross. Consider this snapshot from a court hearing: – Dolores: 'I hereby state that I never ever met that man.' Emilio: 'But here is a photograph taken last year which shows you in conversation with him.' – If Lewis is right, then *E* does not contradict *D*. But isn't that exactly what he does? If not, defendants and witnesses could easily escape every charge of perjury by beginning their assertions always with a performative prologue like 'I hereby state that,' thereby making them self-verifying.[148] So it is by no means obvious that an assertoric utterance of a sentence of the the form 'I believe (state) that *p*' does not, as Husserl puts it, "essentially say the same thing" as an assertoric utterance of the embedded sentence.

Husserl's main objection to Bolzano's Tenet is that it is descriptively wrong to take sense-1 questions, silent wonderings, to be predications about the questioner.[149] One can strengthen this criticism by pointing out that wonderings are conceptually less demanding than such predications: One cannot wonder whether it is raining without having the concept of rain, but one hardly needs the ability to think of oneself in the first-person mode in order to wonder whether it is raining. The same argument applies, *mutatis mutandis*, to sense-2 questions, illocutionary acts of asking a question. If somebody sincerely utters a *rogative* (and if he says what he wanted to say), then he ascribes to himself the property of having a certain wish. But if somebody sincerely utters an *interrogative* (and if she succeeds in saying what she wanted to say), nothing of the sort happens. She then *has* such a wish, for otherwise her utterance would not be sincere. But she does not ascribe to herself the property of having this wish. She might not yet have acquired the conceptual resources needed for self-ascriptions.

Let me add two further objections. Firstly, interrogatives and rogatives have different illocutionary act potentials. Somebody who seriously utters an interrogative poses a question. But somebody who seriously utters a rogative may not ask

---

[146]Lewis (1983), p.224.
[147]Cf. Wittgenstein (1967), Pt. II, sect. X, on "Moore's paradox."
[148]Cf. Lycan (2000), pp.181-186 on "Cohen's problem."
[149]Husserl (LU), II/2, p.218 / (LI) 2, p.332.

a question. He may rather *answer* a question, namely this one: 'Who wants the indication of a true proposition with such-and-such a feature?' (Saying 'Me, too' would be a bizarre reaction to the utterance of an interrogative, but since the utterance of a rogative may be an assertion about the speaker, as we just saw, the reaction 'Me, too' may very well be sensible.) Hence Bolzano's theory does not tell us what sense-2 questions are, in other words, what (the speech-act of) asking a question is. But then, Bolzano did not promise to contribute to the theory of illocutionary acts. Therefore I am not inclined to press this objection.

Secondly, if the utterance of an interrogative like

($\mathfrak{Z}$1) 'Is Vesuvius an active volcano?'

really were to express a rogation, then two different speakers, or the same speaker at two different times, would express different propositions with ($\mathfrak{Z}$1): after all, one of those rogations may be true whereas the other is false. (Recall the indices in the schema *Rogation* above.) But doesn't every competent speaker of English who utters ($\mathfrak{Z}$1) seriously pose the *same* question: *the* question whether Vesuvius is an active volcano? To be sure, the unwanted multiplication of questions could be avoided if one were to identify the proposition expressed by ($\mathfrak{Z}$1) in a certain context with the *conventional linguistic meaning* of the rogative ($\mathfrak{R}$1). But this would no longer be a Bolzanian proposition. For Bolzano, the proposition Carmen expresses by uttering a first-person sentence has got to be different from the proposition José expresses by uttering the same first-person sentence with the same conventional linguistic meaning: the first proposition is about Carmen, the second is about José, one of them may be true and the other false.[150] Basically the same objection applies to Lewis' approach. When you utter

(P1) 'I hereby ask whether Vesuvius is an active volcano'

you say something about yourself, whereas when *I* utter it I say something about myself. So we express different propositions by (*P*1). But we ask the same question when we both utter ($\mathfrak{Z}$1).

Husserl is right, then, in rejecting Bolzano's (and Lewis') Tenet that

(B) *All askables are propositions* (but not vice versa).

---

[150]Cf. Bolzano (WL) II, pp.77 f. on indexical sentences, and Künne (2005), in Künne (2008) ch. 5, §5. At least the multiplication which is due to the multitude of questioners would be avoided if one were to replace 'I want' rogatives by the impersonal formula 'It is desired (*es wird verlangt*) that a true proposition of type *X* be indicated.' Compare Bolzano's formulation of 'problem propositions (*Aufgabesätze*)' in Bolzano (WL) II, pp.72 and 68. But this proposal is implausible. A sincere utterance of the impersonal variant of a canonical rogative can coherently be continued by 'As for myself, I do not have the desire that a truth of type *X* be indicated.' But somebody who sincerely, and *in propria persona*, utters the corresponding interrogative cannot coherently go on like that.

But Bolzano can stick to a weaker contention:

(B*) *For each utterance of an interrogative there is a proposition (a rogation) which is true if and only if the questioner is sincere.*

This must be correct, for the pertinent proposition simply gives the sincerity condition of the utterance. The weak Bolzanian thesis (B*) applies, *mutatis mutandis*, to all utterances which have sincerity conditions, hence even to interjections like 'ouch.' This generality is intended by Bolzano.[151]

Husserl, on the other hand, seems to maintain that

(H) *No askable is a proposition* (and no proposition is an askable).

Although I sympathize with Bishop Butler's dictum that 'everything is what it is, and not another thing,' thesis (H) seems to me to be as false as Bolzano's strong Tenet. The truth of the matter, I think, is this:

(F) *Some askables are propositions, and all propositions are askables.*

If I ask myself, or others, whether Vesuvius is an active volcano, the content of this act is the same as the content of my judgement, and of my assertion, that Vesuvius is an active volcano. (The act-*matter* is the same, though of course not what Husserl calls the 'act-quality:' unlike judgings, sense-1 questions [wonderings] are not positing, and they have a special psychological mode. And equally obviously, the force of sense-2 questions [illocutionary acts of asking a question] differs widely from that of assertings.) Propositions can be singled out by that-clauses as well as by whether-clauses, for a declarative sentence in a given context expresses the very same proposition as the corresponding *yes/no* interrogative in the same context, and *vice versa*. Frege saw this.[152]

At the point we have now reached we can recognize that the following stance has a chance of being coherent: conceding that English yes/no interrogatives are not true or false (sc. in English, at a certain context) any more than any other interrogatives are, while maintaining that yes/no interrogatives, in contradistinction to search interrogatives, express propositions which are true or false (*simpliciter*). This is coherent if we take yes/no interrogatives to be an exception to the right-to-left half of a bridge-principle that is unexceptionable as regards *declarative* sentences: Sentence S is true in language L at context c if and only if what is expressed by S at c is true. This move would mitigate the tension between Aristotle's and Bolzano's views about questions, which Husserl emphasized at the outset of his discussion of Bolzano's Tenet.

---

[151] See Bolzano (WL) I, p.88.
[152] Frege (1918), p.62 & Frege (1919), pp.143-145. Cf. Künne (2009), ch.2, §3; ch. 3, §1.

Acknowledgements

This is the text of lectures I gave in Barcelona in September 2008. I did not remove the traces of the occasion for which it was written: may they serve as indicators of my gratitude for the opportunity to present this material at the *Institut d'Estudis Catalans*. Special thanks go to Ignacio Jané, Genoveva Martí, Manuel Martínez and Daniel Quesada for helpful questions and constructive criticism – and above all to Francesc Pereña Blasi, my best Catalan friend, who did much more for me than organize those lectures. A precursor of section 7 was read at the International Conference on *"Les Recherches logiques d'Edmund Husserl 1901-2001: Origines et Postérité de la Phénoménologie"*, held in Montréal in May 2001; the current text is a revised version of Künne (2008), ch. 6.

BIBLIOGRAPHY

Åqvist, L. (1965). *A new Approach the the Logical Theory of Interrogatives*, Filosofiska föreningen i Uppsala, Uppsala.

Beyer, C. (1996). *Von Bolzano zu Husserl. Eine Untersuchung über den ursprung der phänomenologischen Bedeutungslehre*, Springer, Dordrecht.

Bergman[n], [S.] H. (1909). *Das philosophische Werk Bernard Bolzanos, mit Benutzung ungedruckter Quellen kritisch untersucht*, Halle. Georg Olms, Nachdruck Hildesheim-N.Y., 1970.

Bergman[n], [S.] H. (1946). "Briefe Franz Brentanos an Hugo Bergmann", in *Philosophy and Phenomenological Research*, 7, pp.83-158.

Bolzano, B. (GA). *Bernard Bolzano Gesamtausgabe*, J. Berg et al. (eds). Reihe 1: Schriften; reihe 2A: Nachgelassene Schriften. Frommann, Bad Canstatt, Stuttgart, 1969.

Bolzano, B. (WL). *Dr. B. Bolzanos Wissenschaftslehre*, 4 Bande, Sulzbach, 1837. Leipzig 1929/1931, reprint Aalen 1970, 1981. in Bolzano (GA): 1,11-14, 1985, ff. English translation (partial): *Theory of Science*, edited by R. George, University of California Press, Berkeley and Los Angeles, 1972. English translation (partial): *Theory of Science*, edited by B. Terrell, D. Reidel, Dordrecht and Boston, 1973.

Bolzano, B. (1812 - circa). "Etwas aus der Logik", in Bolzano (GA), 2 A, p.5.

Bolzano, B. (1838). *Dr. Bolzano's Athanasia [...]. Zweite verbesserte Ausgabe*, Sulzbach. Reprint Minerva, Frankfurt/M, 1970.

Bolzano, B. (1843). "Aufsatz, worin eine von Hrn. Exner [...] angeregte logische Frage beantwortet wird", Prag. in Bolzano (GA), 1 18; translated in Bolzano (2004).

Bolzano, B. (1849). "Über die Eintheilung der schönen Künste", Prag. Reprint in *Untersuchungen zur Grundlegung der Ästhetik*, ed. by D. Gerhardus, Philo Verlagsges., Syndikat, Frankfurt/M., 1972.

Bolzano, B. (1851). *Drei philosophische Abhandlungen*, Leipzig. In Bolzano (GA): 2 A, 12/3.

Bolzano, B. (VZL). "Verbesserungen und Zusätze zur *Logik*", in Bolzano (GA), 2 A, 12/2.

Bolzano, B. (1935). *Der Briefwechsel B. Bolzano's mit F. Exner* edited by E. Witner, in *Spisy Bernarda Bolzano / Bernard Bolzano's Schriften* 4, Královská česká společnost nauk, Prag. Partially translated in Bolzano (2004).

Bolzano, B. (2004). *On the Mathematical Method, and Correspondence with Exner*, edited and translated by P. Rusnock & R. George, Rodopi, Amsterdam.

Destutt de Tracy, A. (1803). *Eléments d'idéologie*, volume 2: *Grammaire*, Paris.

Dummett, M. (1973). *Frege – Philosophy of Language*, Duckworth, London.

Frege, G. (1884). *Grundlagen der Arithmetik*, W. Köbner, Breslau.

Frege, G. (1892). "Über Begriff und Gegenstand", in: *Funktion, Begriff, Bedeutung*, ed. by G. Patzig, Vandenhoeck & Ruprecht, Göttingen, 2008.

Frege, G. (1918). "Der Gedanke", with original pagination in *Logische Untersuchungen*, ed. by G. Patzig, Vandenhoeck & Ruprecht, Göttingen, 2003; and in Künne (2009).

Frege, G. (1919). "Die Verneinung", with original pagination in *Logische Untersuchungen*, ed. by G. Patzig, Vandenhoeck & Ruprecht, Göttingen, 2003; and in Künne (2009).

Hintikka, J. (1974). "Questions about Questions" in M. Munitz & P. Unger (eds.), *Semantics and Philosophy*, New York University Press, New York.

Husserl, E. (GW). *Husserliana. Edmund Husserl, Gesammelte Werke*. van Breda H.L. and Ijselling, S. (eds.), The Hague, 1950, ff. Bernet, R. and Melle U. (eds.), Kluwer Academic Publishers, Dordrecht, 1988, ff.; Springer, New York, 2005, ff.

Husserl, E. (LU). *Logische Untersuchungen*, 3 vols, Halle, 1900-1901, 1913. vol. I (*Prolegomena*), II/1 and II/2 reprinted Tübingen 1980. in Husserl (GW): vol. XVIII (1975), XIX/1 and XIX/2 (1984).

Husserl, E. (LI). *Logical Investigations*. translated by J.N. Findlay, Routledge & Kegan Paul, London, 1970, 2001.

Husserl, E. (1894). *Intentionale Gegenstände*, ms. in Husserl (GW), vol.XXII.

Husserl, E. (1894b). "Intentionale Gegenstände", revised edition by Schumann, K., in *Brentano Studies*, 3; translated in Rollinger (1996), as Appendix 1.

Husserl, E. (1896). Review of Twardowski (1894), ms. in Husserl (GW), vol.XXII.

Husserl, E. (1903). Review of Palágy (1902), in Husserl (GW), translated in Mohanty (1977).

Husserl, E. (1903/04). "Bericht über deutsche Schriften zur Logik", in Husserl (GW), XXII.

Husserl, E. (1908). *Vorlesungen zur Bedeutungslehre*, in Husserl (GW), XXVI.

Husserl, E. (1910). Review of Marty (1908), in Husserl (GW), vol.XXII.

Husserl, E. (1913). "Entwurf einer 'Vorrede' zu den *Logischen Untersuchungen*", manuscript 1913, in *Tijdschrift voor Philosophie*, 1, 1939, pp.106-33; 319-39; in Husserl (GW), XX/1, 2002; translated by P.J. Bossert & C.H. Peters, Martinus Nijhoff Publishers, The Hague, 1975.

Husserl, E. (1913b). *Ideen zu einer reinen Phänomenologie und phänomenologischen Philosophie* Erstes Buch (1913). in Husserl (GW), III/1 (1976); translated by F. Kersten, Martinus Nijhoff Publishers, The Hague, 1982.

Husserl, E. (1938). *Erfahrung und Urteil*, Hamburg 1976. Translated by J.S. Churchill & K. Ameriks, Northwestern University Press, Evanston IL, 1973.

Husserl, E. (BW). *Briefwechsel*, vols. 1,3 and 7. In Husserl (GW): *Dokumente*. Kluwer Academic Publishers, Dordrecht, 1994.

Kerry, B. (1885-1891). "Über Anschauung und ihre psychische Verarbeitung", Acht Artikel, in *Viertelj. f. wiss. Philos.* 9 (1885): I; 10 (1886): II; 11 (1887): III-IV; 13 (1889): V-VI; 14 (1890): VII; 15 (1891): VIII.

Künne, W. (1983). *Abstrakte Gegenstände* Suhrkamp, Frankfurt am Main, 2nd Extended Edition 2008.

Künne, W. (1991). "The Nature of Acts: Moore on Husserl", in D. Bell & N. Cooper (eds.), *The Analytic Tradition*, Blackwell, Oxford.

Künne, W. (1997). " 'Die Ernte wird erscheinen [...]' Die geschichte der Bolzano-Rezeption [...]", in Ganthaler, H. & Neumaier, O. (eds.), *Bolzano und die österreichische Geistesgeschichte*, Academia-Verlag, St. Augustin. Revised and much enlarged in Künne (2008).

Künne, W. (1998). "Substanzen und Adhärenzen. Zur Ontologie in Bolzanos *Athanasia*", in *Philosophiegeschichte und logische Analyse*, 1, pp.233-250.

Künne, W. (2005). "Propositions in Bolzano and Frege", in Künne, Siebel, Textor (1997); in M.Beaney & E. Reck (eds.), *Gottlob Frege – Critical Assessments*, vol.1, Routledge, London; and in Künne (2008)

Künne, W. (2006). "Analyticity and Logical Truth. From Bolzano to Quine", in Textor (2006), revised in Künne (2008).

Künne, W. (2008). *Versuche über Bolzano / Essays on Bolzano*, Academia Verlag, St. Augustin.

Künne, W. (2009). *Die philosophische Logik Gottlob Freges. Ein Kommentar*, Klostermann Vittorio GmbH, Frankfurt.

Künne, W., Siebel, M., Textor, M. (1997) (eds.) *Bolzano and Analytic Philosophy* in *Grazer Philosophischen Studien*, vol, 53.

Lewis, D. (1983). "General Semantics", in *Philosophical Papers*, vol. 1, Oxford University Press,New York.

Lotze, R.H. (1874). *Logik*, Leipzig 1912. Translation edited by B. Bosanquet (1888), Garland, New York, 1980.

Lycan, W.G. (2000). *Philosophy of Language*, Routledge, London.

Marty, A. (1908). *Untersuchungen zur Grundlegung der allgemeinen Grammatik und Sprachphilosophie* Halle. Reprint Georg Olms, Hildsheim, New York, 1976.

Mehmel, G.E.A. (1803). *Versuch einer vollständigen analytischen Denklehre als Vorphilosophie und im Geiste der Philosophie*, Walther,Erlangen.

Meinong, A. von (1904). *Über Gegenstandstheorie*, in *Gesamtausgabe*, Bd. 2, Akademische Druck und Verlagsanstallt, Graz, 1971.

Mohanty, J. (1977). (ed.) *Readings in Husserl's 'Logical Investigations'*, Martinus Nijhoff Publishers, The Hague.

Montague, M. (2007). "Against Propositionalism", in *Noûs* 41 (2007), pp.503-518.

Palágy, M. (1902). *Der Streit der Psychologisten und Formalisten in der modernen Logik*, W. Engelmann, Leipzig.

Perry, J. (1994). "Intentionality (2)", in S. Guttenplan (ed.), *A Companion to the Philosophy of Mind*, Blackwell, Oxford.

Quine, W.v.O. (1960). *Word and Object*, MIT Press, Cambridge/MA.

Rollinger, R.D. (1996). *Husserl's Position in the School of Brentano*, Department of Philosophy, Utrecht. Springer, Dordrecht, 1999.

Russell, B. (1919). *Introduction to Mathematical Philosophy*, George Allen and Unwin, London.

Schnieder, B. (2002). *Substanz und Adhärenz*, Academia Verlag, St. Augustin.

Schnieder, B. (2006). "Particularised Attributes: an Austrian Tale" in Textor (2006).

Searle, J. (1983) *Intentionality. An Essay in the Philosophy of Mind*, Cambridge University Press, Cambridge.

Searle, J. (1994) "Intentionality (1)", in S. Guttenplan (ed.), *A Companion to the Philosophy of Mind*, Blackwell, Oxford.

Simons, P. (1992). *Philosophy and Logic in central Europe from Bolzano to Tarski*, Springer, Dordrecht.

Stepanians, M. (1998). *Frege und Husserl über Denken und Urteilen*, Mentis Verlag, Paderborn.

Strawson, P.F. (1959). *Individuals. An Essay in Descriptive Metaphysics*, Routledge Kegan & Paul, London.

Strawson, P.F. (1974). *Subject and Predicate in Logic and Grammar*, Methuen, London.

Strawson, P.F. (2006). "A Category of Particulars", in P.F. Strawson & A. Chakrabarti (eds.), *Universals, Concepts and Qualities*, Ashgate Publishing, Aldershot.

Tarski, A. (1936). "Über den Begriff der logischen Folgerung", translated in *Logic, Semantics and Metamathematics*, Clarendon Press, Oxford, 1956.

Textor, M. (1996). *Bolzanos Propositionalismus*, de Gruyter, Berlin - N.Y.

Textor, M. (2000). "Bolzano et Husserl sur analyticité", in *Les Études philosophiques*, 4.

Textor, M. (2006). editor, *The Austrian Contribution to Analytic Philosophy*, Routledge, London.

Twardowski, K. (1894). "Zur Lehre vom Inhalt und Gegenstand der Vorstellungen", München 1982; translated as "On the Content and Objects of Presentations", Martinus Nijhoff Publisher, The Hague, 1977.

Twardowski, K. (1926). "Selbstdarstellung", in *Grazer Philosophische Studien*, 39, 1991.

Wittgenstein, L. (1967). *Philosophische Untersuchungen*, in *Werkausgabe*, Bd. 1, Frankfurt am Main, 1984.

Tractarian Beginnings and Endings
Worlds, Values, Facts and Subjects

KEVIN MULLIGAN

ABSTRACT. This occasional paper from the proceedings of the Burkamp Club sets out part of the Austro-German context of the beginning and the end of Wittgenstein's *Tractatus*: Husserl's 1900-1901 account of the world as the totality of contingently obtaining states of affairs or facts; and Scheler's 1913-1916 account of persons as correlates but not parts of worlds, and of microcosms, selves, bodies, life, solipsism, God, value and happiness.

1 Introduction

Wittgenstein begins the *Philosophical Investigations*, after a quotation from Augustine, with the following idea:

> "The old view was based essentially on two assumptions, that are internally connected. It was believed that the functions of language could all be traced back to the naming function of words: *every word is a name for something, its meaning [Bedeutung]. [...] And it was thought that the sentence contains essentially an aggregate [Inbegriff] of names [Nennungen]*. And in accordance with this first assumption *the processes of language learning were accounted for as [a process of] learning to name objects*. Both claims are false; *the function of naming is only one of several functions of words* and the fact that language learning is not based only on acquisition of the naming function is being shown more and more by *systematic observation of children*. Matters are essentially more complicated than they seemed to the first simple theory; just how complicated they are cannot be somehow deductively inferred but must be grasped on the basis of *systematic observation of concrete cases of linguistic understanding*."[1]

This is, of course, not quite what Wittgenstein writes at the beginning of the *Investigations*. The passage just quoted is taken from p.107 of Karl Bühler's 1909 article "Über das Sprachverständnis vom Standpunkt der Normalpsychologie aus." Bühler there says of the view that the functions of language could all be traced back to the naming function of words that it is "most clearly formulated by Hobbes."[2] Wittgenstein's foil is a view to be found in a passage from Augustine and what he actually says about this view is:

---

[1] Bühler (1909), p.107; italics mine.
[2] Bühler (1909), p.107, cf. p.105.

"These words, it seems to me, give us a particular picture of the essence of human language. It is this: *the individual words in language name objects–sentences are combinations [Verbindungen] of such names [Benennungen]*. –In this picture of language we find the roots of the following idea: *Every word has a meaning [Bedeutung]*. The meaning is correlated with the word. It is the object for which the word stands.

Augustine does not speak of there being any *difference between kinds of word*. If you describe *the learning of language* in this way you are, I believe, thinking *primarily of nouns* like 'table,' 'chair,' 'bread,' *and of people's names*, and only secondarily of the names of certain actions and properties; and of the remaining kinds of word as something that will take care of itself."[3]

Elsewhere I have described the elements of Karl Bühler's philosophy of language, which finds its main expression in his remarkable *Sprachtheorie* of 1934, and investigated their relation to what Wittgenstein writes in the first twenty five sections of the *Investigations*.[4] The Austro-German context of Wittgenstein's ideas often throws light on the contents of his views both early and late. In what follows I look at the Austro-German context of the beginning of the *Tractatus* and of some passages near the end of the *Tractatus*.

2 Tractatus 1-2

The apodictic beginning of the *Tractatus* is:

1. Die Welt ist alles, was der Fall ist.

   The world is all that is the case.

1.1 Die Welt ist die Gesamtheit der Tatsachen, nicht der Dinge.

   The world is the totality of facts, not of things.

1.11 Die Welt ist durch die Tatsachen bestimmt und dadurch, dass es *alle* Tatsachen sind.

   The world is determined by the facts, and by their being *all* the facts.

---

[3] Wittgenstein (1968), §1, italics mine.
[4] Mulligan (1997).

A fact, we are then told, is the obtaining of states of affairs:

2. Was der Fall ist, die Tatsache, ist das Bestehen von Sachverhalten.

What is the case – a fact – is the obtaining of states of affairs.

In 1900 another Austrian philosopher, Husserl, had said that the world is a total unity:

> "Denn die Welt ist nichts anderes als die gesamte gegenständliche Einheit, welche dem idealen System aller Tatsachenwahrheit entspricht und von ihm unabtrennbar ist."[5]

> "For the world is nothing more than the unified objective totality corresponding to, and inseparable from, the ideal system of all factual truth."[6]

What is this totality a totality of? What is a factual truth? As we shall see, Husserl uses 'fact' and 'factual' in two different ways.

Every truth, Husserl thinks, is inseparable from an obtaining state of affairs. "The interconnection of things (Sachen)" and "the interconnection of truths" are "mutually inseparable" and

> "What holds of single truths or states of affairs, plainly also holds of interconnections of truths or of states of affairs."[7]

There are contingent and non-contingent truths and these correspond to states of affairs which obtain contingently and to states of affairs which obtain non-contingently respectively. Factual truths are contingent truths and are inseparable from contingent facts:

> "Facts [Tatsachen] are 'contingent': they might very well not have been the case, they might have been different."[8]

> "Every fact is [...] temporally determinate."[9]

Husserl's Humean use of 'fact' and 'factual' to refer to "the world of *matter of fact*."[10] is not the only way he uses the term 'fact.' He also occasionally uses the term 'fact' to refer not just to Humean matters of fact but to obtaining or objective states of affairs:

> "[...] der objektive Sachverhalt, die Tatsache [...]."[11]

> "[...] the objective state of affairs, the fact [...]."[12]

---

[5] Husserl (1975), §36(6), p.128.
[6] Husserl (1970), vol. I, §36(6), p.143.
[7] Husserl (1975), §62, p.230-231; Husserl (1970), vol. I, p.226.
[8] Husserl (1975), §37, p.129; Husserl (1970), vol.I, p.144.
[9] Husserl (1975), §36(3), p.126; Husserl (1970), vol.I, p.140.
[10] Husserl (1975), vol.I, §40, p.153.
[11] Husserl (1901), LU V, §33, p.479; cf. V §36, pp.490-491.
[12] Husserl (1901), LU V §33, cf. V §36 490-91.

A year later, another Austrian philosopher, Meinong, also says:

"Ein Objektiv, das besteht, wird auch als 'Thatsache' bezeichnet."[13]

"An objective, which obtains, is also called [a] 'fact'."

In order to understand this second use of 'fact' it is necessary to understand Husserl's account of states of affairs.[14] A "state of affairs is no thing (Ding)," he says[15]. States of affairs come in many varieties. They may be 'universal,' of the form *"the state of affairs that all A's are B"*[16], 'indeterminate, singular,' of the form *"the state of affairs that some A is B"*[17], 'general,' of the form *"the state of affairs the (a) A (in specie) is B"*[18]. They are positive or negative[19] or singular and 'primitive,'[20] relational or non-relational.[21]

Husserl's account of states of affairs, then, is an example of Austrian Baroque – there are many different "forms of states of affairs"[22] corresponding to the different logical forms of propositions. Wittgenstein's account, on the other hand, is an example of Austrian Romanesque. In particular, according to Husserl but not Wittgenstein, not all obtaining states of affairs obtain contingently, not all obtaining states of affairs are matters of fact.

The non-Humean use of 'fact' by Husserl and Meinong might be called the functorial use. A state of affairs obtains or does not obtain, holds or does not hold[23], is objective or is not objective[24]. Instances of

(1) The state of affairs that p obtains

are modifications or transformations of instances of the functorial construction

(2) It is the case that p.

(Compare the relation between 'The proposition that p is true' and 'It is true that p'). As we have seen, Husserl calls a state of affairs which is objective, which obtains, a fact. One modification or nominalisation of (1) allows us to refer, as

---

[13] Meinong (1977), p.69.
[14] Cf. Süssbauer (1995), Mulligan (1990).
[15] Husserl (1901), LU V §33, p.478.
[16] Husserl (1984), §23, p.167.
[17] Husserl (1984), §23 p.168.
[18] Husserl (1984), §23 p.168. In the previous quotations underlining is by Husserl, italics are mine.
[19] Husserl (1901), LU V §28, p.462.
[20] Husserl (1901), LU P §6, p.31.
[21] Husserl (1901), LU VI §48, pp.681,683.
[22] Husserl (1901), LU VI §51, p.688, §63, p.722.
[23] Husserl (1975), vol. I, §6, p.29.
[24] Husserl (1901), LU V, §33, p.479; cf. V §36, pp.490-491.

Husserl does in 1913, to "the obtaining [*Bestand*] of a state of affairs."[25] Yet another nominalisation allows us to refer, as does Wittgenstein, to "the obtaining [*Bestehen*] of states of affairs."[26]

Husserl and Wittgenstein attach great importance to the distinction between formal and non-formal (material) categories. Indeed this distinction and the uses to which they put it sharply distinguish their very different philosophies of logic and language from those of, say, Frege, Meinong and Russell.[27] According to Husserl, the category of state of affairs is a formal category:

> "In [...] Zusammenhang mit den [...] *Bedeutungskategorien*, stehen andere, zu ihnen korrelative Begriffe, wie Gegenstand, Sachverhalt, Einheit, Vielheit, Anzahl, Beziehung, Verknüpfung usf. Es sind die reinen oder *formalen gegenständlichen Kategorien.*"[28]

> "In [...] connection with [...] the *categories of meaning* [...] stand other correlative concepts such as Object, State of Affairs, Unity, Plurality, Number, Relation, Connection etc. These are the pure, the *formal object categories.*"[29]

In one of his lists of formal categories Husserl mentions the category of facts:

> "[T]he pure truths of logic are all the ideal laws which have their whole foundation in the 'sense,' the 'essence' or the 'content,' of the concepts of Truth, Proposition [Satz], Object, Property, Connection, Law, Fact [Tatsache] etc."[30]

'Fact' here probably refers not to the category of obtaining states of affairs but to the category of contingently obtaining states of affairs, to matters of fact.

Wittgenstein, too, asserts that the concept of fact is a formal concept, like the concepts of number and object.[31] And since a fact is the obtaining of a state of affairs we may ascribe to him the view that the concept of a state of affairs is a formal concept.[32]

As far as I can see, Husserl nearly always uses the term 'fact' to refer to a state of affairs which obtains contingently. The same is true of other phenomenologists who developed Husserl's account of states of affairs, for example, Reinach, but not of Meinong or his pupils.[33] The term 'fact' is also used by Bradley, Russell and McTaggart but they use the term to refer to the possession of qualities by things or to things which stand in relations. Thus in 1914 Russell writes:

---

[25] Husserl (1975), vol.I, §6, p.29.
[26] On Wittgenstein's idiom, cf. Dietrich (1973), pp.20-23.
[27] See also Husserl (1984), vol. III, §11, p.256 (Husserl (1970), vol. II, p.455). On Husserl and Wittgenstein on formal categories, cf. Mulligan (1993).
[28] Husserl (1975), §67, p.245.
[29] Husserl (1970), vol.I §67, p.237.
[30] Husserl (1975), §37, p.129; cf. Husserl (1970), vol.I, p.144.
[31] Wittgenstein (1974), 4.1272; from here on, decimals correspond to the standard enumeration of Tractarian sentences.
[32] Cf. Favrholdt (1964), p.122.
[33] Indeed Meinong notes that his use of 'fact' conflicts with the tendency to use this word of the objectives of empirical judgements, that is to say, Humean matters of fact. See Meinong (1977), p.69.

> "The existing world consists of many things with many qualities and relations [...] When I speak of a 'fact' I [...] mean that a certain thing has a certain quality, or that certain things have a certain relation."[34]

Thus the existing world, according to Russell, consists of facts. But these facts are not defined in terms of states of affairs.[35] They are British, not Austrian facts. And according to Husserl and Wittgenstein the world is identical with the totality of facts; no fact is a proper part of the world. One apparent difference between Husserl and Wittgenstein on contingent facts concerns the rôle played in Wittgenstein's theory by the claim that simple objects form the substance of the world (2.021).[36]

In *Ideas* (1913) Husserl seems to have had second thoughts about what the world is. He there seems to reject his earlier view, that the world is the unified totality of contingent facts, of facts which are matters of fact:

> "Die Welt ist der Gesamtinbegriff von Gegenständen möglicher Erfahrung und Erfahrungserkenntnis [...]."[37]

> "The world is the totality of objects than can be known through experience [...]."[38]

3 Tractatus 5.6-

The 5.6's of the *Tractatus* begin by telling us that the limits of my language mean the limits of my world and that logic pervades the world. Elsewhere I have explored the relation between these claims and Husserl's account in 1901 of "the absurd [counter-sensical, *widersinnige*] problem of the real meaning of the logical," the absurdity of the view that the course of the world could prescribe the "limits of the validity of the laws of logic."[39] In what follows I consider the Austro-German context of some of Wittgenstein's subsequent remarks about the metaphysical subject, the human soul, life, living bodies, solipsism, microcosms, God, value, happiness, the world and worlds, in particular a philosophy of all of these which is closer to Wittgenstein's views than is any other philosophy.

One type of subject referred to by Wittgenstein is the philosophical self or metaphysical subject:

> "The philosophical self [Ich] is [...] the metaphysical subject, the limit of the world, not a part of it."[40]

---

[34] Russell (1980), p.60.
[35] Cf. Correia and Mulligan (2007).
[36] But cf. Mulligan (2004), p.410, Note 14.
[37] Husserl (1928), §1, p.8.
[38] Cf. Husserl (1972), §1, p.46.
[39] Husserl (1984), II (VI, §65), p.728; Husserl (1970), II,p.830. Cf. Mulligan (1993).
[40] Wittgenstein (1974), 5.641.

In 1916 Max Scheler writes in his *Ethics* – presumably a good example of what Wittgenstein calls "gassing" about "the ethical"[41] – that a person is no part of *a* world:

> "The *person* is never a 'part' of a 'world' but always the *correlate* of a 'world'."[42]

Scheler's argument for this claim seems to be that "a person [...] is never an object [Gegenstand];" persons can be described but there is no such thing, he thinks, as knowledge by acquaintance of a person; whatever is in the world is an object.[43]

In the same year Wittgenstein noted that

> "The self [Ich] is no object."[44]

> "I objectively confront every object. But not the self [Ich]."[45]

Scheler and Wittgenstein distinguish between worlds and the world. Scheler raises the following question:

> "If to every 'person' there corresponds a 'world' and to every 'world' a 'person', we must ask [...] whether the 'idea' of *one identical real world* [...] has phenomenal fulfilment."[46]

According to Wittgenstein "the world [is] *my* world."[47]

Scheler calls the worlds which correspond to different persons *microcosms* and the one identical real world the *macrocosm*:

> "Let us call this idea of one identical real world the idea of the *macrocosm* [...]. If there is such a macrocosm [...a]ll microcosms, i.e. all individual 'personal worlds' are [...] parts of the macrocosm."[48]

Unlike Scheler Wittgenstein identifies the world and *his* world and says that he himself is his world. Like Scheler, Wittgenstein calls his world a microcosm:

> "I am my world (The microcosm)."[49]

> "It is true: man *is* the microcosm."[50]

---

[41] Wittgenstein (1969), pp.35-36.
[42] Scheler (1973), p.393; Scheler (1966), p.392.
[43] See Scheler (1966), p.386; Scheler (1973), p.387.
[44] Wittgenstein (1979), 7.8.16.
[45] Wittgenstein (1979), 11.8.16.
[46] Scheler (1973), p.396; Scheler (1966), p.395.
[47] Wittgenstein (1974), 5.62.
[48] Scheler (1973), p.396; Scheler (1966), p.395.
[49] Wittgenstein (1974), 5.63.
[50] Wittgenstein (1979), 12.10.16. The concept of a microcosm is employed by Schopenhauer and the Austrian philosopher Otto Weininger. Indeed Schopenhauer has much to say about most of the claims discussed in this section. On the parallels between Schopenhauer and Wittgenstein, see Lange (1992), Micheletti (1973) chs.3,4, Gabriel (1993), chs. 4-5. Scheler, too, takes up many Schopenhauerian ideas. In particular, he follows Schopenhauer in thinking that personal identity is a matter of the heart and the liver rather than of the head.

Although Scheler distinguishes between the world of a person and the world a central claim of his personalist, anti-solipsistic realism is that the one and only world is given to me – shows itself to me – in a way it cannot show itself to anyone else because it shows itself to me as the fact that my world is a part of the world:

> "the content [Gehalt] of the being of the world is, in every case, different for each person."[51]

Each such difference reveals the 'contingency' of the real world. This world cannot be the object of 'determinations' which employ "general concepts and propositions."[52] The fact that my world is part of the world cannot be said but shows itself:

> "the truth about the world [...] is, in a certain sense, a 'personal truth'."[53]

According to Wittgenstein, what solipsism means, that the world is my world, is correct but cannot be said. It shows itself.[54]

If each individual world corresponds to a person and if there is a macrocosm, what person corresponds to the macrocosm? Scheler's answer is

> "The personal counterpart of the macrocosm would be the idea of an infinite and perfect spiritual person."[55]

The idea of such a person is the idea of a person who, like every person, is no part of any world. If there is such a person, could she reveal herself? Scheler's answer runs as follows:

> "Any reality of 'God' has [...] its only possible foundation in a possible positive revelation [Offenbarung] of God in a positive person."[56]

Since no person is part of any world such a revelation could not take place in a world. In Wittgenstein's words,

> "God does not reveal [offenbart] himself *in* the world."[57]

What follows from the relations between a person or metaphysical subject, on the one hand, and worlds, on the other hand, for the place of value?

According to Scheler,

> "That which can be called *originally* 'good' and 'evil' [...] is the '*person*' [...]."[58]

---

[51] Scheler (1966), p.394; cf. Scheler (1973), p.395.
[52] Scheler (1966), p.393; cf. Scheler (1973), p.394.
[53] Scheler (1966), p.394; cf. Scheler (1973), p.395.
[54] Wittgenstein (1974), 5.62.
[55] Scheler (1966), p.387; cf. Scheler (1973), p.396.
[56] Scheler (1966), p.395; cf. Scheler (1973), p.396-7.
[57] Wittgenstein (1974), 6.432.
[58] Scheler (1973), p.28; Scheler (1966), p.49.

If no person is part of a world, is the ethical value of a person part of a world? Scheler's answer is affirmative.[59] Wittgenstein, however, says that "in [the world] there is no value."[60] The conclusion Wittgenstein draws from this, that "it is impossible for there to be propositions of ethics,"[61], is not drawn by Scheler. But Scheler does argue that one type of value cannot be described or attributed by ethics – individual values, the good or bad *for* a person.

One ethical proposition, according to Scheler, concerns the relation between ethical value and happiness:

> "That there is some connection between happiness [blissfulness, Glückseligkeit] and the positive ethical [sittlich] value of the person, between good behaviour and the positive feelings which accompany it, a connection that is more than a merely fortuitous and empirical association or the lack of one – this is the opinion of all who have thought seriously about this connection."[62]

The non-contingent connection, he thinks, is that "only the good person is happy," and "only the happy person acts in an ethically good way."[63] But it is wrong to think of such happiness (or unhappiness) as any sort of reward (punishment or reprisal):

> "Why do we need a so called 'reprisal' [Sanktion] here? It does not matter if a deed brings the agent [...] any number of states of unhappiness [...]. No reward [good which is a reward, Lohngut] that is supposed to function as a reward for what is ethically good can ever, by essential necessity, determine happiness as deep as the *happiness itself* out of which ethically good willing streams forth and which accompanies it. No punishment [evil which is a punishment, Strafübel] [...] can ever determine suffering as deep as the wretchedness out of which a bad deed flows or as deep as the feeling of displeasure that accompanies it."[64]

Wittgenstein distinguishes between reward and punishment in an ordinary and a non-ordinary sense. But it is not clear whether he thinks that the agreeableness of ordinary rewards is of the same kind as the agreeableness of non-ordinary rewards:

> "When an ethical [ethischen] law of the form, 'Thou shalt ...', is laid down, one's first thought is, 'And what if I do not do it?'. But it is clear that ethics has nothing to do with punishment and reward in the ordinary sense. So our question about the *consequences* of an action must be unimportant. – At least these consequences should not be events. For there must be something right about the question we posed. There must indeed be some sort of ethical reward and ethical punishment, but they must reside in the action itself.
>
> (And it is also clear also that the reward must be something pleasant and the punishment something unpleasant)."[65]

---

[59] Scheler (1966), p.392.
[60] Wittgenstein (1974), 6.41. Cf. "What is good and evil is essentially the self [das Ich], not the world," Wittgenstein (1979), 5.8.16.
[61] Wittgenstein (1974), 6.42.
[62] Scheler (1966), p.355; cf. Scheler (1973), p.354.
[63] Scheler (1966), p.359-360; cf. Scheler (1973), p.379.
[64] Scheler (1966), p.360; cf. Scheler (1973), p.359-60.
[65] Wittgenstein (1974), 6.422.

Scheler, as we have seen, thinks that the happiness of the good man is not any sort of reward and the unhappiness of the evil woman is not any sort of punishment. He does not, however, think that such happiness and unhappiness lie in certain actions themselves. They are rather the sources of such actions and accompany these.

The non-contingent link between goodness and the heart has consequences for the world of a person or man. Bliss (Seligkeit) and despair (Verzweiflung), Scheler says, are the correlates of a person's ethical value or disvalue and "fill the whole of [...] our world."[66] As Wittgenstein puts it, "the world of the happy man [Glücklichen] differs from that of the unhappy man."[67] The context of Wittgenstein's claim and 6.422 suggest that the happy man is one whose will is good, the unhappy man one whose will is evil. Scheler says that it is a mistake to identify bliss and happiness, despair and unhappiness. Bliss and despair lie deeper than (un)happiness.[68]

Wittgenstein asserts that "if good or evil willing changes the world, it can only change the limits of the world," the world "must so to speak wax (zunehmen) or wane (abnehmen) as a whole" (6.43). Scheler thinks that there is something which makes of "the entire moral world" a "great whole" which "rises (steigt) and falls (fällt) as a whole" – the principle of solidarity or shared responsibility. Scheler illustrates this point by imagining a "world court (Weltgericht)" and "its highest judge".[69]

Wittgenstein's metaphysical subject or philosophical self is not man, the body, the human body, the human soul of which psychology treats.[70] Nor is it the thinking, presenting subject, since this does not exist.[71] Scheler's person is not identical with any man, body, soul or self. But there are, he thinks, selves. Unlike persons, selves are objects. In self-knowledge, self-love, self-hate and self-control a self is an object for a person. "The self is itself just an object among other objects."[72] It is not simple, it is no conglomerate but rather a unity.[73] The view that empirical selves are complex unities which are part of the world was very popular amongst Brentano's heirs. It is to be found in (early) Husserl, Stumpf and Witasek. Thus Husserl says in 1900 that "the self belongs to the world".[74]

According to the Scandinavian interpretation of the *Tractatus*, Wittgenstein there accepts, in addition to the metaphysical subject, what is often called an 'em-

---

[66] Scheler (1966), p.345.
[67] Wittgenstein (1974), 6.43.
[68] Scheler (1966), pp.109,126.
[69] Scheler (1966), p.523; on the rôle of the notion of a last judgment (ein jüngstes Gericht) in Wittgenstein and other Austrian thinkers cf. Smith (1978).
[70] Wittgenstein (1974), 5.641
[71] Wittgenstein (1974), 5.631.
[72] Scheler (1966), p.375; cf. Scheler (1973), p.375.
[73] Scheler (1966), p.414-5.
[74] Husserl (1975), §36, p.128.

pirical self.'⁷⁵ Thus Favrholdt argues that the human soul Wittgenstein refers to at 5.641 is just such an empirical self and that it is composite. It contains thinkings and thoughts, that is, those facts which are pictures. Although the metaphysical subject neither thinks nor has ideas nor believes, the human soul does think (5.631, 5.542), have ideas (5.631) and believes (5.542).⁷⁶ I shall assume in what follows that Favrholdt is right about this.

A living body (Leib) is an object like any other object. What is the relation between a person, a metaphysical subject and an empirical self, on the one hand, and living bodies, on the other hand? Wittgenstein says

> "Were I to write a book, 'The world as I found it [*wie ich sie vorfand*]', I should have to include a report on my body [Leib], and should have to say which parts were subordinate to my will, and which were not, etc., this being a method of isolating the subject or rather of showing that in an important sense there is no subject; for it alone could *not* be mentioned in that book."⁷⁷

Scheler reports Avenarius' assertion

> "that there is a simple finding [*Vorfinden*] (which neither presupposes nor contains an ego [...]) and [...] that this datum contains nothing more than a lived body [Leib] and its environment [Umwelt]."⁷⁸

And agrees with Avenarius that perception of a living body does not require the assumption of a self.⁷⁹ The self which Wittgenstein could not mention in his book is what he calls the metaphysical subject. When Scheler says that perception of a living body does not require the existence of a self, the self he has in mind is something which can be perceived and is no person.

Scheler illustrates his claims by considering the perception of a cube, an example discussed also by Wittgenstein.⁸⁰ My perception of a cube, Scheler says, involves no awareness of me. For this a new act is required, the object of which is a self and what is given to this self of the cube.⁸¹ Nor does perception of a cube involve any awareness of the existence and place of a bodily organism. For such awareness a new act is required:

> "[T]he fact that I see by way of some activation of my eyes, rather than my ears, lies neither in the intuition of the function of seeing nor in that of the seen thing."⁸²

---

⁷⁵ Stenius (1960), pp.220-221; Hintikka (1958), Favrholdt (1964). Cf. also Dietrich (1973), pp.170-180.
⁷⁶ Favrholdt (1964) p.96ff., 111f., 147f..
⁷⁷ Wittgenstein (1974), 5.631; my italics.
⁷⁸ Scheler (1966), p.402; Scheler (1973), p.404; emphasis mine.
⁷⁹ Scheler (1973), p.406; Scheler (1966), p.404; but contrast Scheler (1973), p.378, Scheler (1966), p.377.
⁸⁰ Wittgenstein (1974), 5.5423.
⁸¹ Scheler (1973), p.56; Scheler (1966), p.75.
⁸² Scheler (1973), p.57; Scheler (1966), p.76.

Wittgenstein says that "nothing in the visual field allows one to infer that it is seen by an eye."[83] And he rules out any awareness of a metaphysical self to whom a cube, for example, could be perceptually given.

'Solipsism' may refer to the doctrine that there is more than one metaphysical subject or person. But Scheler and Wittgenstein[84] also use the term to indicate an opposition, apparent or real, to realism. Scheler argues that the assumption that the world and the way it is given depend on what he calls a self "leads necessarily to solipsism."[85] But every act of knowing involves an immediate knowledge of the essential independence of things, both oneself and what is perceived of the external world, with respect to the execution of this act. So such solipsism is self-evidently absurd (widersinnig).[86] Thus 'solipsism' here is opposed to realism about selves and about the external world.

According to Favrholdt, Wittgenstein's empirical self or human soul is composite and contains beliefs, thinkings, thoughts and presentations. The metaphysical self does not believe, think or enjoy presentations. But it does, it seems, will. For the world is independent of my will.[87] In the *Notebooks* we are told that the willing subject exists.[88] Another possibility is that the will, *"in so far as it is the subject of ethical attributes"* is outside the world, unlike the *"will as a phenomenon,"* something which is of interest only to psychology.[89] Does the metaphysical subject do anything else? Favrholdt argues that

> "in a certain sense, [the metaphysical subject] *has* knowledge, because to compare propositions with reality and decide whether they are true or false is to receive knowledge."[90]

> *"The ontological picture theory presupposes the existence of a subject which does not think but takes in the logic of the world by an act of seeing."*[91]

This seeing is one type of knowledge, knowledge of internal relations. I shall assume that Favrholdt is right about this.

According to Scheler in 1916 a person wills, perceives, intuits, judges and has ideas or presentations and only a person, not a self, performs these 'spiritual' acts. Selves, on the other hand, see, hear, attend, and only in a self do these 'psycholog-

---

[83] Wittgenstein (1974), 5.633.
[84] Wittgenstein (1974), 5.64.
[85] Scheler (1973), p.378; Scheler (1966), p.378. Wittgenstein says in the *Notebooks* that the subject which is no part of the world "is a presupposition of its existence" (Wittgenstein (1979), 2.8.16). Scheler rejects the view that the world depends on what he calls a self and asserts that persons, who are not parts of any world, and worlds are mutually interdependent.
[86] Scheler (1973), p.379; Scheler (1966), p.378-9)
[87] Wittgenstein (1974), 6.373.
[88] Wittgenstein (1979), 5.8.16.
[89] Wittgenstein (1974), 6.423, cf. 5.631; Favrholdt (1964), pp.98-99.
[90] Favrholdt (1964), p.102.
[91] Favrholdt (1964), p.167.

ical' functions occur.[92] But before and after 1916 Scheler sets out a quite different distribution of roles for persons and selves. He demotes judging from the category of spiritual acts to that of psychological functions and asserts that the presentations or ideas (*Vorstellungen*) 'of the psychologist' are not acts but functions. Thinking, belief and assertion too, he says, are merely psychological functions. On this view, then, only a person wills, perceives and intuits, that is, enjoys knowledge by intuition, and only selves judge, believe, think and have presentations (in one sense of this term).[93] Indeed a self is a unity of thinkings, judgings, and presentings as is the empirical self of the *Tractatus*, "the human soul, with which psychology deals."[94]

A person, according to Scheler, is simply a unity, a unity of spiritual acts such as willing, perceiving and intuiting.[95] Is the metaphysical subject of the *Tractatus* a unity? If it wills and knows, what is the relation between it and such willing and knowing? It is often argued that the metaphysical subject of the *Tractatus* must be simple because it is a limit of the world, a point.[96] On some views of ordinary boundaries these certainly can be complex. Perhaps the same is true of some Tractarian boundaries.[97] After all, some of the boundaries of the world can, perhaps, be altered by good or evil willing.[98]

4 Conclusion

What does Wittgenstein mean when he says that certain things show themselves and cannot be said? What exactly is it that is supposed to show itself? It is tempting to answer the first question by saying that showing is a close relative of what

---

[92] Scheler (1966), pp.387-390.
[93] Scheler (1955), pp.234-6, 219, 248, 230.
[94] Wittgenstein (1974), 5.641.
[95] Scheler (1966), pp.389,382.
[96] Wittgenstein (1974), 5.632, 5.64; cf. Favrholdt (1964), pp.96,148.
[97] In order to evaluate this suggestion it would be necessary to understand the relation between time, the metaphysical subject and its willing and knowing. Scheler's view is a suggestive object of comparison. Scheler's persons and their acts (like the persons and acts of his personalist contemporary, McTaggart) are atemporal, see Scheler (1966), pp.385-387. A person is an atemporal unity and the correlate but no part of its world. Wittgenstein seems to think that it is possible to live atemporally by living in the present, Wittgenstein (1974), 6.4311. On the relation between this and related claims and the views of Weininger, cf. Schulte (2004). He also says that the world and life are one, Wittgenstein (1974), 5.621. Scheler, too, is happy to talk of life in this *lebensphilosophische* way (Scheler (1912), p.363) but does not identify any world with life. The correlate of a living body (*Leib*) is a milieu (*Umwelt*), of a person a world. In 1923 Scheler mentions a 'world-view' in which "the world 'as a whole' is given as one *universal organism*, permeated by 'one' life," see Scheler (1973a), p.92, Scheler (1954), p.81. Scheler's philosophy of the living body and of Life leads him to describe the view that "the mechanical view of nature gives a picture of absolutely real nature" as one that is based on an 'illusion' (Scheler (1966), p.420). Wittgenstein's account of Life is one – but not the most important – source of his view that "the entire modern world-view is based on the illusion that the so called laws of nature are the explanations of natural phenomena" (6.371).
[98] Wittgenstein (1974), 6.43.

the phenomenologists called intuition, and to answer the second question by saying that what shows itself is just what, according to the phenomenologists, can be intuited. The latter comprises all non-contingent connections, analytic and synthetic, within and between states of affairs and their elements, which hold in virtue of the essence of these elements or of the concepts they fall under. (A fairly typical example is provided by the penultimate quotation from Husserl in section 2 above). By grasping these essences in intuition, categorial and non-categorial, we see the connections. Among these states of affairs are *Wertverhalte*, ethical and aesthetic, but also all logical laws, for logical laws are merely one type of formal ontological law. Intuition is of essences and of *Wesensverhalte*. Then the mistake of the phenomenologists, if it is a mistake, would be to attempt to say what they see, the result of which is mere 'gassing.'[99]

A philosopher who succumbed to this double temptation seems to be Gustav Bergmann. "[W]hat was once intuited [erschaut] now shows itself" is his comment in 1938 on Wittgenstein.[100] In order to determine whether the temptations should be resisted it is important to ask: To whom, if to anyone, does what shows itself show itself? Who, for example, enjoys an "intuition of the world sub specie aeterni?"[101] Who intuits? As we have seen, according to Scheler, persons intuit.

---

[99] Scheler's radical view is that phenomenologists do not or should not try to say what they mean: "a priori contents can only be shown [aufgewiesen]," they can neither be 'proved' nor 'deduced,' see Scheler (1966), p.69. As he says in a text published in 1933, "The meaning of the words used by the phenomenologist and the units of meaning of the sentences he writes or speaks do not 'say' what he means [...] Words and sentences are here invitations [...] to see [...] what the phenomenologist means," Scheler (1957) p.465, cf. p.391.

[100] Bergmann (1988), p.173.

[101] Wittgenstein (1974), 6.45. Wittgenstein says (a) that the feeling or '*Gefühl*' of the world as a limited whole (6.45) is mystical and (b) identifies the mystical and that the world is (6.44). ad (a) - If feelings require a cognitive underpinning and if the object of an emotion is the object of its cognitive underpinning, the most likely candidate for the cognitive underpinning of the feeling of the world as a limited whole is the '*Anschauung*' of the world as a limited whole, that is to say, sub specie aeterni (6.45). Part of the immediate context of this line of thought is the account given by yet another Austrian philosopher, Heinrich Gomperz, of the role of grasping the world as a whole in his philosophy of inner freedom and mysticism (atheistic, pantheistic and theistic): "If a man at a particular moment grasps the world as a whole as a good [...], then he is happy at that moment," Gomperz (1904), 7-8, cf. 22, 197. In the *Notebooks* Wittgenstein says that "the good life is the world seen sub specie aeternitatis" (7.10.16). One of the features of mysticism, according to Gomperz, is the movement from the realist to the idealist, even solipsistic attitude, Gomperz (1904), 296, 310. ad (b) – In 1917 Scheler says that the fact that something is is the object of philosophical wonder (*Verwunderung*, Scheler (1917), 65, cf. 96) and that insight into this fact is "presupposed by the constitution of the sense of the word 'doubt about something'," Scheler (1917), 65. According to Wittgenstein, that the world is is the mystical and the mystical shows itself (6.522) and doubt presupposes that something can be said (6.51). The proximity of these remarks suggests that it is *because* that the world is shows itself that scepticism about whether the world is is nonsensical (*unsinnig*, 6.51; Husserl calls sceptical theories 'absurd' (*widersinnig*), cf. Husserl (1975), §32 p.120). In a lecture in 1929 Wittgenstein refers to an experience he describes by saying "I wonder at the existence of the world," Wittgenstein (1993), p.40, cf. 43. In the same lecture Wittgenstein refers to "the experience of absolute safety," an experience which has been "described by

(Husserl's transcendental ego plays a similar role). Does what shows itself show itself to the metaphysical subject of the *Tractatus*? Wittgenstein nowhere says that this is the case.

Acknowledgments

Work on this paper was supported by the Swiss FNS project on properties and relations and the Swiss NCCR project on values and emotions. Thanks to Ingvar Johansson and Barry Smith.

BIBLIOGRAPHY

Bergmann, G. (1988). "Erinnerungen an den Wiener Kreis. Brief an Otto Neurath", in F. Stadler, editor, *Vertriebene Vernunft. Emigration und Exil österreichischer Wissenschaftler*, pp.171–180, Jugend und Volk, Vienna, Munich. English Translation: "Memories of the Vienna Circle. Letter to Otto Neurath (1938)", F.Stadler (ed.), *Scientific Philosophy: Origins and Developments*, Vienna Circle Institue Yearbook, I, 1993, pp.193-208.

Bühler, K. (1909). "Über das Sprachverständnis vom Standpunkt der Normalpsycologie aus", in *Bericht über den dritten Kongress für experimentelle Psychologie*, pp.94–130. Leipzig.

Bühler, K. (1934). *Sprachtheorie. Die Darstellungsfunktion der Sprache*. Fischer, Jena. Reprint Fischer, Stuttgart 1965. English translation by D.Goodwin, *Theory of Language. The representational Function of Language*, Benjamins, Amsterdam, 1990.

Correia, F. and Mulligan, K. (2007). *Facts*, in E.N. Zalta, editor, *Stanford Encyclopedia of Philosophy (Winter 2007 Edition)*. The Metaphysics Research Lab Center for the Study of Language and Information, Stanford University, 2007. http://plato.stanford.edu/entries/facts/.

Dietrich, R.A. (1973). *Sprache und Wirklichkeit in Wittgensteins Tractatus*, Max Niemeyer Verlag, Tübingen.

Favrholdt, D. (1964). *An Interpretation and Critique of Wittgenstein's Tractatus*, Munksgaard, Copenhagen.

Gabriel, G. (1993). *Grundprobleme der Erkenntnistheorie. Von Descartes zu Wittgenstein*, Ferdinand Schöningh, Paderborn.

Gomperz, H. (1915 (1904)). *Die Lebensauffassung der griechischen Philosophen und das Ideal der inneren Freiheit*, Eugen Diederichs, Jena.

Hintikka, J. (1958). "On Wittgenstein's 'Solipsism'", *Mind*, 265, pp.88-91

Husserl, E. (1928). *Ideen zu einer reinen Phänomenologie und phänomenologischen Philosophie*, volume I. Max Niemeyer, Halle.

---

saying we feel safe in the hands of God," Wittgenstein (1993), p.42. Scheler is one philosopher tempted by such descriptions: "Thus it is in God alone that the individual person may know himself to be safe or judged," Scheler (1966), p.550. Gomperz, on the other hand, describes the safety of the atheist, see Gomperz (1904), p.17, cf. p.297, and Anzengruber that of the pantheist.

Husserl, E. (1970). *Logical Investigations*, volume I,II. Routledge & Kegan Paul, London. English translation of Husserl (1975), Husserl (1984), Husserl (1984a).

Husserl, E. (1972). *Ideas. General Introduction to pure phenomenology*. Collier Books, New York.

Husserl, E. (1975). *Logische Untersuchungen, Vol. I*, Husserliana XVIII. Nijhoff, The Hague.

Husserl, E. (1984). *Logische Untersuchungen, Vol. II, Part 1*. Husserliana XIX/1. Nijhoff, The Hague.

Husserl, E. (1984a). *Logische Untersuchungen, Vol. II, Part 2*. Husserliana XIX/2. Nijhoff, The Hague.

Lange, E.M. (1992). *Wittgenstein und Schopenhauer*. Junghans-Verlag, Cuxhaven.

Meinong, A. (1977). "Über Annahmen", In *Gesamtausgabe, vol. IV*. Akademische Druck- u. Verlagsanstalt, Graz.

Micheletti, M. (1973). *Lo Schopenhauerismo di Ludwig Wittgenstein*. Editrice La Garangola, Padua.

Mulligan, K. (1990). "Husserl on States of Affairs in the Logical Investigations", *Epistemologia*, special number on "Logica e Ontologia", XII, pp.207âÄŚ234.

Mulligan, K. (1993). "Proposizione, stato di cose ed altri concetti formali nel pensiero di Wittgenstein e Husserl", in *L'uomo, un segno*, A.G. Gargani, editor, fascicolo speciale "Wittgenstein contemporaneo", pp.41–66. Marietti, Genova.

Mulligan, K. (1997). "The Essence of Language: Wittgenstein's Builders and Bühler's Bricks", *Revue de Métaphysique et Morale*, 2, pp.193–216, 1997.

Mulligan, K. (2004). "Essence and Modality. The Quintessence of Husserl's Theory", in M. Siebel & M. Textor (eds.), *Semantik und Ontologie. Beiträge zur philosophischen Forschung*, pp.387–418, Ontos Verlag, Frankfurt.

Russell, B. (1980). "Logic as the Essence of Philosophy", In *Our Knowledge of the External World*, George Allen & Unwin, London.

Scheler, M. (1912). "Über Ressentiment und moralisches Werturteil. Ein Beitrag zur Pathopsychologie der Kultur", *Zeitschrift für Pathopsychologie*, I, pp.268-368.

Scheler, M. (1917). "Vom Wesen der Philosophie", *Summa*, I/1, 40-70.

Scheler, M. (1954). *The Nature of Sympathy*. English translation of Scheler (1973a), Routledge & Kegan Paul, London.

Scheler, M. (1955). "Vom Umsturz der Werte", in *Gesammelte Werke, vol. III*, Francke Verlag, Berne.

Scheler, M. (1957). *Schriften aus dem Nachlass*, in *Gesammelte Werke, vol. X*, Francke Verlag, Berne.

Scheler, M. (1966). "Der Formalismus in der Ethik und die materiale Wertethik. Neuer Versuch der Grundlegung eines Ethischen Personalismus", in *Gesammelte Werke, Vol. II (1913-1916)*,

Francke Verlag, Berne.

Scheler, M. (1973). *Formalism in Ethics and Non-Formal Ethics of Values*, Northwestern University Press, Evanston. English translation of Scheler (1966).

Scheler, M. (1973a). *Wesen und Formen der Sympathie, Gesammelte Werke*, Vol. VII, Francke Verlag, Berne.

Schulte, J. (2004). "Wittgenstein and Weininger. Time, Life, World", in D.G. Stern, editor, *Wittgenstein reads Weininger*, pp.112–137, Cambridge University Press, Cambridge.

Smith, B. (1978). "Law and Eschatology in Wittgenstein's Early Thought", *Inquiry*, 21, pp.425-41.

Stenius, E. (1960). *Wittgenstein's 'Tractatus', a critical exposition of its main lines of thought*, Blackwell, Oxford.

Süssbauer, A. (1995). *Intentionalität, Sachverhalt, Noema. Eine Studie zu Husserl*, Alber, Freiburg.

Wittgenstein, L. (1968). *Philosophical Investigations*. Blackwell, Oxford.

Wittgenstein, L. (1969). *Briefe an Ludwig von Ficker*, (ed.) G. H. v. Wright, vol.I, Brenner-Studien.

Wittgenstein, L. (1974). *Tractatus Logico-Philosophicus*. Routledge & Kegan Paul, London.

Wittgenstein, L. (1979). *Notebooks 1914-1916*. University of Chicago Press, Chicago.

Wittgenstein, L. (1993). "A lecture on Ethics", in J. Klagge & A. Nordmann (eds.), *Philosophical Occasions 1912-1951*, Hackett Publishing Company, Indianopolis & Cambridge.

Gentzen's Original Proof of the Consistency of Arithmetic Revisited

JAN VON PLATO

ABSTRACT. Gentzen's first published proof of the consistency of arithmetic through transfinite induction was preceded by another proof based on different ideas. Criticisms by Bernays and Gödel led Gentzen to change it. The manuscript with the original proof had been typeset before the changes, and was saved in the form of galley proofs.

A discussion of the circumstances of Gentzen's proof is presented, followed by a proof of the consistency of arithmetic based on Gentzen's original proof. The latter can be simplified considerably, into what should be the most accessible of his altogether four different proofs of the consistency of arithmetic.

1 Gentzen's four proofs

Gödel made public his first incompleteness theorem at a meeting in Königsberg in September 1930. Johann von Neumann was present and immediately grasped the result, then arrived within a couple of months at the second incompleteness theorem about the unprovability of the consistency of arithmetic. This is shown by von Neumann's letter to Gödel, from 20 November 1930.[1] Von Neumann also conjectured that the consistency of mathematics would be unprovable in some absolute sense. However, his plans to publish came to nothing when it turned out that also Gödel had arrived at the second incompleteness theorem and included it in the manuscript of his famous paper of 1931. That paper was submitted some days before the arrival of von Neumann's letter.

A couple of years later, after the Gödel-Gentzen translations from Peano arithmetic to Heyting Arithmetic in 1932–33, it became clear that the consistency of arithmetic is provable constructively, even if not finitistically. Thus, von Neumann's conjecture turned out to be too pessimistic. Gentzen succeeded in finding a proof of consistency of arithmetic towards the end of 1934. Criticisms by Bernays and Gödel made him change the proof into one that uses the now famous method of transfinite induction up to the ordinal ε_0. As is indicated in the published paper of 1936, the most crucial part of the proof was changed early in 1936. Whatever

---

[1] See Gödel (1986–2003), vol. IV.

was left of Gentzen's manuscripts and other papers of this period after his death in 1945, they seem to have been destroyed in 1964. The original proof of consistency of arithmetic was submitted to *Mathematische Annalen*, the leading journal of mathematics at that time, in August 1935, and was preserved in the form of galley proofs. An English translation of the parts differing from the 1936 version was published in Manfred Szabo's edition of Gentzen's *Collected Papers* in 1969.[2] The German original was published in 1974 with the title "Der erste Widerspruchsfreiheitsbeweis für die klassische Zahlentheorie" and with an introduction written by Bernays.[3]

As explained in Bernays' paper "On the original Gentzen consistency proof for number theory" of 1970,[4] the criticism was that the proof "implicitly included an application of the fan theorem." In 1970, he held his own earlier criticisms doubtful. Kreisel, in fact, reports a result by which the addition of the fan theorem to Peano arithmetic would not make its consistency provable.[5]

Gentzen's original proof is the easiest one to understand of his altogether four different proofs of the consistency of arithmetic. It uses a system of natural deduction for classical logic, and the basic idea can be described as follows: Assume that a false formula C is derivable from some assumptions Γ. Think of an adversary, a 'falsifier,' who tries to create a 'worst possible' situation in which C fails. Say, if C is $\forall xA$, a falsifying instance $A[n/x]$ is chosen. Your task is to show that in this situation, also some formula in Γ can be falsified. Then, if C is derivable from Γ and no formula in Γ can be falsified, neither can C. In this terminology, the consistency of arithmetic follows from the availability of a 'winning strategy' in the game against the 'falsifier' opponent.

In Gentzen's second proof, of 1936, transfinite induction replaces the earlier argument. The presentation is not optimal because only parts of the article were rewritten.

In 1938, Gentzen published a third proof, also based on transfinite induction, but with a different logical calculus, namely a version of the multisuccedent sequent calculus **LK** of Gentzen's thesis.[6]

Finally, the fourth proof that was published in 1943 also uses transfinite induction to ε_0. Gentzen shows that transfinite induction up to any ordinal strictly less than ε_0 is equivalent to ordinary induction in Peano arithmetic, but that the transfinite induction principle for ε_0 itself is unprovable in Peano arithmetic, even if it can be expressed in it. This result establishes incompleteness directly, without Gödel's method of coding the provability predicate. Also, consistency is an immediate

---

[2] Gentzen (1969).
[3] Gentzen (1974).
[4] Bernays (1970).
[5] Kreisel (1987), p.174.
[6] Gentzen (1934-35).

consequence, because everything is provable in an inconsistent system. These results of Gentzen's last paper clarified the situation created by Gödel's theorems, especially the second one.

We present a proof of the consistency of arithmetic based on the idea of Gentzen's original 1935 proof. It makes explicit the 'game-theoretic' nature of that proof and uses a system of natural deduction that is a slight modification of the one used by Gentzen. Contrary to Gentzen, the logic is intuitionistic because the rule of indirect proof that leads to classical logic is not needed.

2 A system of natural deduction for arithmetic

We shall use a minimal system of arithmetic, obtained by taking only the connectives $\&, \neg$, the false formula $\bot$, and the universal quantifier. The rule of indirect proof is admissible, i.e., its conclusion derivable whenever its premiss is derivable, when $\vee$ and $\exists$ are left out of the language of Heyting arithmetic. There is thus for each formula derivable in Peano arithmetic a classically equivalent formula derivable in the system used in this paper. The system of natural deduction, called **NI**, is divided into the standard patterns of introduction and elimination rules:

$$\dfrac{A \quad B}{A \& B} \, \&I \qquad \dfrac{\overset{1}{(A^m)}}{\vdots} \qquad \dfrac{A[y/x]}{\forall x A} \, \forall I$$
$$\dfrac{\bot}{\neg A} \, \neg I, 1$$

Table 1. The introduction rules of natural deduction.

Rule $\forall I$ has the standard variable restriction: the *eigenvariable* y must not be free in any assumptions $A[y/x]$ depends on. In the rules, any numbers $m \geqslant 0$ of assumptions can be discharged. If $m = 0$, there is a *vacuous* discharge, and if $m > 1$, there is a *multiple* discharge. Otherwise a discharge is *simple*. Each instance of a rule must use a fresh discharge label.

Next to the introduction rules we have elimination rules:

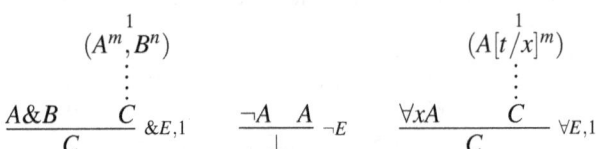

Table 2. The elimination rules of natural deduction.

As before, any multiplicities $m, n \geq 0$ of assumptions can be discharged. In each rule, the formula with the connective or quantifier is the *major premiss* of that elimination rule, and the other premisses are *minor*. Rules $\&E$ and $\forall E$ are different from those of Gentzen; the latter follow by setting, in turn, $C \equiv A$ or $C \equiv B$ in $\&E$, and $C \equiv A[t/x]$ in $\forall E$, and by leaving unwritten the superfluous minor premisses:[7]

$$\frac{A\&B}{A} \&E_1 \qquad \frac{A\&B}{B} \&E_2 \qquad \frac{\forall xA}{A[t/x]} \forall E$$

Table 3. Gentzen's elimination rules.

General elimination rules lead to a remarkable simplification in the definition of derivations in *normal form*. Such normal form was Gentzen's first aim in his way to proving the *subformula property* of derivations in natural deduction. This property says that all formulas in a normal derivation are subformulas of open assumptions or of the conclusion, and it is the central tool in the structural analysis of proofs.

DEFINITION 2.1 A derivation in **NI** is *normal* if all major premisses of elimination rules are assumptions.

The standard procedure for transforming a derivation into normal form, suggested through a couple of examples in Gentzen (1934-35) and worked out in detail in Prawitz (1965), consists of the removal of consecutive introduction-elimination pairs. Such pairs are known as *detour convertibilities*, or *'Umwege'* in Gentzen's terminology. Prawitz considered also *permutation convertibilities*, instances of $\vee E$ or $\exists E$ that conclude a major premiss of an E-rule.

DEFINITION 2.2 An E-rule with a major premiss derived by an I-rule in a derivation is a *detour convertibility*.

---

[7] The more general form of conjunction elimination was first introduced by Schroeder-Heister (1984). A systematic study of such general elimination rules is contained in von Plato (2001).

A detour convertibility on $A\&B$ is a part of derivation of the form, with the converted derivation part on the right:

$$
\begin{array}{c}
\vdots \quad \vdots \\
\dfrac{A \quad B}{A\&B}\,\&I \quad \overset{1}{(A^m,B^n)} \\
\quad \vdots \\
\dfrac{\quad C}{C}\,\&E,1 \\
\vdots
\end{array}
\qquad \rightsquigarrow \qquad
\begin{array}{c}
\vdots \quad \vdots \quad \vdots \quad \vdots \\
A,\; \overset{m\times}{\ldots},A \quad B,\; \overset{n\times}{\ldots},B \\
\vdots \\
C \\
\vdots
\end{array}
$$

The assumptions on which A depends are multiplied m times, and similarly for B.

DEFINITION 2.3 An E-rule with a major premiss derived by an E-rule in a derivation is a *permutation convertibility*.

The novelty of general elimination rules is that permutation conversions apply to all cases in which a major premiss of an E-rule has been derived. There is a good number of permutation convertibilities, all similar, and we show only one of them here.

A permutation convertibility on major premiss $C\&D$ derived by $\&E$ on $A\&B$ and its conversion are

$$
\begin{array}{c}
\overset{1}{(A^m,B^n)} \\
\vdots \quad \vdots \quad \overset{2}{(C^k,D^l)} \\
\dfrac{A\&B \quad C\&D}{C\&D}\,\&E,1 \quad \vdots \\
\dfrac{\quad E}{E}\,\&E,2 \\
\vdots
\end{array}
\qquad \rightsquigarrow \qquad
\begin{array}{c}
\overset{1}{(A^m,B^n)} \quad \overset{2}{(C^k,D^l)} \\
\vdots \quad \vdots \quad \vdots \\
\quad \dfrac{C\&D \quad E}{E}\,\&E,2 \\
\dfrac{A\&B \quad\quad E}{E}\,\&E,1 \\
\vdots
\end{array}
$$

Observe, in particular, how the *height* of derivation of the major premiss $C\&D$ in the lower E-rule is diminished by one in the permutation (as measured by the steps of inference from the assumptions closed by the rule).

The main result about natural deduction is that all derivations can be converted to normal form. Normal derivations obey the subformula property that is, however, somewhat complicated to prove when general elimination rules are used. We shall therefore prove it later, in connection with a modification of the notation for derivation trees into what is called 'sequent calculus style,' and move instead ahead into arithmetic proper.

Our system of Heyting arithmetic, **HA** for short, consists of a language with one primitive predicate, the equality $a = b$ between numbers, the rules **NI**, a number of arithmetical axioms, and the rule of induction. The axioms include that equality is reflexive, symmetric and transitive. Further, functions can be defined by recursion, and equals can be substituted in expressions. Finally, there are the two axioms of infinity for the successor function:

1. **Reflexivity:** $a = a$,

2. **Symmetry:** $a = b \supset b = a$,

3. **Transitivity:** $a = b \ \& \ b = c \supset a = c$.

4. **Replacement for successor:** $a = b \supset s(a) = s(b)$.

Functions can be defined by recursion, with corresponding replacement axioms, as exemplified by the sum function:

5. **Recursion for sum:** $\quad x + 0 = 0, \quad x + s(y) = s(x + y)$.

6. **Replacement for sum:** $a = b \supset a + c = b + c \ \& \ c + a = c + b$.

7. **Infinity1:** $\neg s(a) = 0$.

8. **Infinity2:** $s(a) = s(b) \supset a = b$.

We note that the last axiom is provable in **HA** through the recursive definition of a predecessor function and its replacement axiom. To the above axioms is added a rule of induction:

$$\begin{array}{ccc} & \overset{1}{(A[y/x]^m)}, \Delta & \overset{1}{(A[t/x]^n)}, \Theta \\ \Gamma & \vdots & \vdots \\ \vdots & & \\ A[0/x] & A[sy/x] & C \\ \hline & C & \end{array} \ \textit{Ind},1$$

Table 4. The rule of induction.

In the rule, A is any formula, called the *induction formula*, and y is the eigenvariable of the rule, not free in Γ, Δ, Θ, or C. In the derivation of the minor premiss C, the term t can be arbitrarily chosen. If t is a number instead of a variable, we have an instance of *numerical induction*. The rule of induction has no major premiss, but it has an arbitrary consequence and therefore behaves like an elimination rule. Gentzen's original work used a special case of the above induction rule:

$$\frac{A[0/x] \quad A[sy/x]}{A[t/x]} \, Ind,1$$

with discharged assumption $(A[y/x]^m)^1, \Delta$ above.

Table 5. Gentzen's rule of induction.

The merit of the more general form of rule is that our simple definition of a normal derivation works also in arithmetic: If the conclusion C of *Ind* is a major premiss of an elimination, the elimination can be permuted to above the instance of *Ind* and we have therefore:

THEOREM 2.4 (NORMALIZATION) Derivations in **NI**+*Ind* convert to normal form.

The subformula property of normal derivations is lost, because of the induction rule in which track of induction formulas is lost.

We shall now state, somewhat bluntly, that the arithmetical axioms do not cause any real problems in a proof of the consistency of arithmetic, but all problems stem from the rule of induction. Therefore we concentrate our attention on the system **NI**+*Ind*.

In Gentzen's work, the logic was classical and the *rule of indirect proof* explicit:

$$\frac{\bot}{A} \, Raa,1$$

with discharged assumption $(\neg A^m)^1$ above.

Table 6. The rule of indirect proof.

(Here *Raa* stands for 'reductio ad absurdum.') By standard results as in, say, Troelstra and van Dalen (1988), this rule is admissible for equations. It then follows that it is admissible for arbitrary formulas in the calculus **NI**+*Ind*. For these reasons, there is no need to consider rule *Raa* or other rules equivalent to it, a fact that Gentzen mentions but does not make use of.[8]

---

[8] See paragraph 12.4 of the original proof in Gentzen (1974).

3 Natural deduction in sequent calculus style

Following Gentzen, it will be useful to introduce a notational variant of natural deduction, called 'natural deduction in sequent calculus style.' This just means that the open assumptions are shown locally on each line of a derivation, with the notation $\Gamma \to C$ in which C is a conclusion and Γ a *list* of those open assumptions C depends on. Expressions of the form $\Gamma \to C$ are called *sequents*, a word the use of which as a noun was begun by Kleene.[9] It corresponds to Gentzen's German '*Sequenz*' that just means sequence. Thus, sequent calculus stands for a 'calculus with lists of assumptions.' We consider two lists of assumptions to be equal if they differ only in order. The left, *antecedent* parts of sequents form *multisets* of formulas in which only the multiplicity of each formula is counted. If Γ and Δ are two such multisets, then Γ, Δ is their multiset union.

As a particular case of a sequent, we have $A \to A$ which corresponds to making the assumption A in the standard notation for natural deduction derivations. A sequent of the form $A \to A$ is called an *initial sequent*. Derivations begin with such sequents. The rules for the rest of the calculus are as follows:

$$\frac{\Gamma \to A \quad \Delta \to B}{\Gamma, \Delta \to A\&B} \&I \qquad \frac{\Gamma \to A\&B \quad A^m, B^n, \Delta \to C}{\Gamma, \Delta \to C} \&E$$

$$\frac{A^m, \Gamma \to \bot}{\Gamma \to \neg A} \neg I \qquad \frac{\Gamma \to \neg A \quad \Delta \to A}{\Gamma, \Delta \to \bot} \neg E$$

$$\frac{\Gamma \to A[y/x]}{\Gamma \to \forall x A} \forall I \qquad \frac{\Gamma \to \forall x A \quad A[t/x]^m, \Delta \to C}{\Gamma, \Delta \to C} \forall E$$

$$\frac{\Gamma \to A[0/x] \quad A[y/x]^m, \Delta \to A[sy/x] \quad A[t/x]^n, \Theta \to C}{\Gamma, \Delta, \Theta \to C} Ind$$

Table 7. Natural deduction and induction rule in sequent calculus style.

In the rules, $m, n \geqslant 0$ indicate the number of copies of formulas used from the assumption parts of premises. Rule $\forall I$ has the variable restriction that the *eigenvariable y* must not occur free in the conclusion of the rule, and the same for rule *Ind*.

We say that the sequent $\Gamma \to C$ is derivable if there is a derivation with the rules such that the last line is $\Gamma \to C$. It is called the *endsequent* of the derivation. A useful way to read a sequent $\Gamma \to C$ is to say that 'Γ gives C.' The natural reading 'C is derivable from Γ' does not go well together with saying that the sequent $\Gamma \to C$ is derivable.

The notion of normal derivation is as before, with major premises (leftmost premisses) of elimination rules assumptions as in definition 2.1. Normal instances of the E-rules have the forms

---

[9] Kleene (1952), p.441.

$$\frac{A\&B \to A\&B \quad A^m, B^n, \Delta \to C}{A\&B, \Delta \to C} \&E \qquad \frac{\neg A \to \neg A \quad \Delta \to A}{\neg A, \Delta \to \bot} \neg E$$

$$\frac{\forall x A \to \forall x A \quad A[t/x]^m, \Delta \to C}{\forall x A, \Delta \to C} \forall E$$

Table 8. Normal instances of E-rules in sequent calculus style.

The major premises do not carry any information that would not be visible otherwise from the conclusions of these rules. If they are left unwritten, the above rules look exactly like the rules of sequent calculus, and more specifically, what is called 'sequent calculus in natural deduction style' in Negri and von Plato (2001). Introduction rules operate on the right, succedent parts of sequents, and elimination rules on the left, antecedent part. (Well, negation rules operate on both parts.) One virtue of our calculus of natural deduction in sequent calculus style is that the subformula property of normal derivations is immediate: We observe that for each logical rule in a normal derivation, all formulas in the premises are subformulas of the conclusion, with the convention for rule $\neg I$ that $\bot$ is a subformula of $\neg A$. This would not be the case if Gentzen's E-rules were used.

It is straightforward to define translations between derivations in the above calculus and natural deduction with derivations written in the usual way, with assumptions as leaves in a derivation tree: Given a derivation in **NI**+*Ind*, start from the leaves, then work down according to the last rule used always collecting the open assumptions into an antecedent. It is well defined what multiset of open assumptions each formula in a derivation determines. Translation in the other direction starts from the endsequent of the derivation in sequent calculus style.[10]

Rules $\neg I$ and $\neg E$ for negation are special cases of those for implication when the negation of a formula A is defined as $A \supset \bot$. As mentioned, the rule of indirect proof reduces to atomic formulas in intuitionistic natural deduction free of disjunction and existence, so in particular for the calculus of Table 7. The rule of indirect proof is written in sequent calculus style as

$$\frac{\neg A^m, \Gamma \to \bot}{\Gamma \to A} Raa$$

We have so far not mentioned the rule of natural deduction that usually has to be added when the false formula $\bot$ is put to use, namely the rule of falsity elimination. Written in sequent calculus style, it has the form

$$\frac{\Gamma \to \bot}{\Gamma \to A} \bot E$$

---

[10] See Negri and von Plato (2001), ch. 8, for such translations.

We see that this rule is an instance of *Raa* with $m = 0$. It follows that the calculus of Table 7 is equivalent to the $\&, \neg, \forall$-fragment of Peano arithmetic. (This observation can be made because rule $\bot E$ is not used in the proof of admissibility of *Raa*.)

4 The falsification game

Given a sequent $\Gamma \to C$, we define, in Gentzen's words, a *reduction procedure* for it. Reduction is effected by suitable moves in a 'falsification game' in which first certain '**F**-moves' are taken by the 'falsifier,' followed by '**Y**-moves' that you take.

The falsifier's moves:

FVar. *The sequent $\Gamma \to C$ has free variables. The falsifier chooses at will numbers to instantiate these until there are no free variables left.*

F&. *The sequent is $\Gamma \to A\&B$ and the falsifier chooses at will either $\Gamma \to A$ or $\Gamma \to B$.*

F¬. *The sequent is $\Gamma \to \neg A$ and the reduced sequent is $A, \Gamma \to 0 = 1$.*

F∀. *The sequent is $\Gamma \to \forall x A$ and the falsifier chooses at will some instance $\Gamma \to A[n/x]$.*

Order of precedence: *Among the **F**-moves, move **FVar** comes before the other **F**-moves.*

Think of the falsifier as one who is omniscient and knows how to end up with the worst possible case for you, here, a false equation as a conclusion. Each step the falsifier takes simplifies the succedent of the sequent to be reduced until an equation $m = n$ remains. If the equation is true, the falsifier lost (and you won). Otherwise, when no **F**-move is applicable and $m = n$ is false, it is your turn to show that some of the assumptions must be false, too. To do this, you have the following moves:

Your moves:

Y&. *The sequent is $A\&B, \Gamma \to m = n$ with $m = n$ false. The reduced sequent is $A, B, \Gamma \to m = n$.*

Y¬. *The sequent is $\neg A, \Gamma \to m = n$ with $m = n$ false. The reduced sequent is $\neg A, \Gamma \to A$.*

Y∀. *The sequent is $\forall x A, \Gamma \to m = n$ with $m = n$ false. The reduced sequent is $A[k/x], \forall x A, \Gamma \to m = n$ in which you choose k.*

Order of precedence: **F**-*moves come always before* **Y**-*moves.*

Your aim in the game is to produce a false formula in the antecedent part of a sequent, whenever the falsifier has produced one in the succedent. Note that if a negation at left is reduced, the falsifier is in turn, unless it was a negation of an equality. Thus, your overall aim is to construct a winning strategy in the falsification game, to meet whatever choices the falsifier may make.

DEFINITION 4.1

(i) A sequent is *irreducible* if no reduction move applies to it.

(ii) A sequent $\Gamma \to m = n$ is in *endform* if either $m = n$ is true or there is some false equality in Γ.

(iii) A sequent $\Gamma \to C$ is *unfalsifiable* if for each choice of **F**-moves there are **Y**-moves such that $\Gamma \to C$ reduces to endform.

The aim of Gentzen's consistency proof is to show that all derivable sequents are unfalsifiable. It follows that the sequent $\to 0 = 1$ is not derivable, because it is irreducible but not in endform: no atom in the antecedent is false, because there are none.

5 The consistency proof

We show first that initial sequents reduce to endform, then that if the premises of the rules of **NI**+*Ind* reduce to endform, also their conclusions reduce to endform.

LEMMA 5.1 Initial sequents $A \to A$ reduce to endform.

Proof. The proof is by induction on the length of A. Assume **FVar**-moves to have been taken, so that there are no free variables. There are four cases:

1. A is an equality $m = n$, and we have $m = n \to m = n$. By the decidability of numerical equality, if $m = n$ is true, $A \to A$ is in endform, and the same if $m = n$ is false.

2. *A* is *B&C*. Then *B&C* → *B&C* reduces by **F&** to *B&C* → *B* or to *B&C* → *C*.

In the first case, consider the first time when the reduction of *B&C* → *B* by arbitrary **F**-moves gives a sequent of the form *B&C*, Γ → *m* = *n*, i.e., the first time for a **Y**-move. The sequent *B* → *B* is reducible by the inductive hypothesis, so the same sequence of **F**-moves as for *B&C* → *B* gives the reducible sequent *B*, Γ → *m* = *n*. Application of **Y&** to *B&C*, Γ → *m* = *n* gives *B*, *C*, Γ → *m* = *n*. When formula *C* in the antecedent is left intact, the sequent reduces exactly as *B*, Γ → *m* = *n*.

If *B&C* → *B&C* is reduced by **F&** to *B&C* → *C*, the proof is as above, with *C* in place of *B*.

3. *A* is ∀*xB*. Then ∀*xB* → ∀*xB* reduces by **F∀** to ∀*xB* → *B*[*k*/*x*] for some *k* that instantiates ∀*xB* and as in **2**, a series of **F**-moves gives a sequent of the form ∀*xB*, Γ → *m* = *n*. By the inductive hypothesis, *B*[*k*/*x*] → *B*[*k*/*x*] reduces to the reducible sequent *B*[*k*/*x*], Γ → *m* = *n*. Now apply **Y∀** to ∀*xB*, Γ → *m* = *n* to get *B*[*k*/*x*], ∀*xB*, Γ → *m* = *n*. Leaving intact ∀*xB* in the antecedent, this sequent reduces to endform.

4. *A* is ¬*B*. Then ¬*B* → ¬*B* reduces by **F¬** to ¬*B*, *B* → 0 = 1. By the inductive hypothesis, *B* → *B* reduces to endform. Application of **Y¬** to ¬*B*, *B* → 0 = 1 gives ¬*B*, *B* → *B* and, leaving intact ¬*B*, the sequent ¬*B*, *B* → *B* reduces to endform like *B* → *B*. QED.

Given two derivations in natural deduction, one concluding *A* and the other having *A* as an open assumption, they can be put together through a **composition**. We have the derivations

$$
\begin{array}{cc}
\Gamma & A, \Delta \\
\vdots & \vdots \\
A & C
\end{array}
$$

These derivations are composed together by substituting the assumption of *A* in the second derivation by the derivation of *A* in the first:

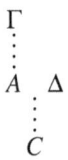

The only condition is that the possible discharge labels and eigenvariables of quantifier rules have been chosen distinct, which can always be effected by renaming them. In the notation of natural deduction in sequent calculus style, the composition of derivations has the form

Gentzen's Original Proof

$$\frac{\Gamma \to A \quad A, \Delta \to C}{\Gamma, \Delta \to C} \; Comp$$

We show that rule *Comp* is admissible:

LEMMA 5.2 (ADMISSIBILITY OF COMPOSITION OF DERIVATIONS.) *If the sequents $\Gamma \to A$ and $A, \Delta \to C$ are derivable in* **NI**+*Ind and possible labels and eigenvariables distinct, also the sequent $\Gamma, \Delta \to C$ obtained by composition is derivable.*

Proof. The proof is by induction on the last rule in the derivation of the second premiss of *Comp* that we may assume to be normal. If the premiss is an initial sequent, the conclusion of *Comp* is identical to the first premiss and the step of *Comp* can be removed. For the last rule, there are several cases:

1. The last rule is an *I*-rule. If it is &*I*, then $C \equiv D\&E$ and the part of derivation is, with A in, say, the first premiss of &*I* and $\Delta \equiv \Delta', \Delta''$:

$$\frac{\Gamma \to A \quad \dfrac{A, \Delta' \to D \quad \Delta'' \to E}{A, \Delta \to D\&E} \; \&I}{\Gamma, \Delta \to D\&E} \; Comp$$

Rule *Comp* is permuted up as follows:

$$\frac{\dfrac{\Gamma \to A \quad A, \Delta' \to D}{\Gamma, \Delta \to D} \; Comp \quad \Delta'' \to E}{\Gamma, \Delta \to D\&E} \; \&I$$

If the rule is $\forall I$ or $\neg I$, the permutation works similarly.

2. The last rule is an *E*-rule. If it is &*E*, then $A \equiv D\&E$ and the part of derivation is:

$$\frac{\Gamma \to D\&E \quad \dfrac{D\&E \to D\&E \quad D, E, \Delta \to C}{D\&E, \Delta \to C} \; \&E}{\Gamma, \Delta \to C} \; Comp$$

The instance of *Comp* is permuted to the first premiss of &*E*. By normality, it is an initial sequent, and the instance of *Comp* becomes removed. The transformed derivation becomes a non-normal instance of &*E*:

$$\frac{\Gamma \to D\&E \quad D, E, \Delta \to C}{\Gamma, \Delta \to C} \; \&E$$

If the rule is $\forall E$ or $\neg E$, the permutation works similarly.

3. The last rule is *Ind*. Formula A is in the antecedent of one of the premisses of *Ind* and *Comp* is permuted up to that premiss, as in **1**.

By the permutations **1–3**, *Comp* reaches initial sequents and becomes removed. QED.

If several compositions have been applied, they can be eliminated by eliminating first uppermost ones.

LEMMA 5.3 (CLOSURE OF REDUCTION TO ENDFORM UNDER COMPOSITION.) If the sequents $\Gamma \to A$ and $A, \Delta \to C$ are reducible to endform and possible labels and eigenvariables distinct, their composition into $\Gamma, \Delta \to C$ is reducible to endform.

Proof. The proof is by induction on the length of the composition formula A. We can assume possible free variables to have been removed by **FVar**.

1. $A \equiv m = n$. Then the first premiss of *Comp* reduces to $\Gamma^* \to 0 = 1$, or $\Gamma^* \to m = n$ if move **F¬** was never applied. Assume **F**-moves to have been applied to the conclusion $\Gamma, \Delta \to C$ until $\Gamma, \Delta, \Delta^* \to k = l$ is produced, in which $k = l$ can be assumed false and Δ^* consists of those formulas, possibly none, that applications of **F¬** have brought to the antecedent. Leaving Δ, Δ^* intact, the sequence of **Y**-moves that reduces $\Gamma \to m = n$ to the endform $\Gamma^* \to 0 = 1$ (or $m = n$), reduces $\Gamma, \Delta, \Delta^* \to k = l$ to an endform.

We note that if $\Gamma \to m = n$ is reducible and $m = n$ false, the equation $0 = 1$ can replace $m = n$: Compose $\Gamma \to m = n$ with the sequent in endform $m = n \to 0 = 1$ to get $\Gamma \to 0 = 1$.

2. $A \equiv A\&B$. The composition is

$$\frac{\Gamma \to A\&B \quad A\&B, \Delta \to C}{\Gamma, \Delta \to C} \; Comp$$

By assumption, $\Gamma \to A\&B$ is reducible, so both of $\Gamma \to A$ and $\Gamma \to B$ are. Consider the second premiss $A\&B, \Delta \to C$. Either there is no application of **Y&** to $A\&B$ in its reduction and $A\&B$ can be removed. Then $\Delta \to C$ is reducible, and therefore also $\Gamma, \Delta \to C$. Else **Y&** is applied at some stage to a reducible sequent $A\&B, \Delta^* \to 0 = 1$ with the reducible sequent $A, B, \Delta^* \to 0 = 1$ as result. We now apply *Comp* twice:

$$\frac{\Gamma \to B \quad \dfrac{\Gamma \to A \quad A, B, \Delta^* \to C}{B, \Gamma, \Delta^* \to 0 = 1} \; Comp}{\Gamma, \Gamma, \Delta^* \to 0 = 1} \; Comp$$

By the inductive hypothesis, *Comp* applied to shorter formulas maintains reducibility, so $\Gamma, \Gamma, \Delta^* \to 0 = 1$ is reducible. The reduction of $\Gamma, \Delta \to C$ by the arbitrarily chosen **F**-moves that reduce $A\&B, \Delta \to C$ to $A\&B, \Delta^* \to 0 = 1$, gives the sequent $\Gamma, \Delta^* \to 0 = 1$ that is reducible to endform by the same **Y**-moves as $\Gamma, \Gamma, \Delta^* \to 0 = 1$.

3. $A \equiv \forall xAB$. The composition is

$$\frac{\Gamma \to \forall xA \quad \forall xA, \Delta \to C}{\Gamma, \Delta \to C} \; Comp$$

By assumption, $\Gamma \to \forall xA$ is reducible, so $\Gamma \to A[n/x]$ is reducible for any choice of n. As in **2**, either there is no application of **Y**$\forall$ to $\forall xA$ in the reduction of the second premiss and $\forall xA$ can be removed. Then $\Delta \to C$ is reducible, and therefore also $\Gamma, \Delta \to C$. Else **Y**$\forall$ is applied at some stage to a reducible sequent $\forall xA, \Delta^* \to 0 = 1$, with the reducible sequent $A[k/x], \forall xA, \Delta^* \to 0 = 1$ as result. With the instance k also in the first premiss, application of *Comp* to the shorter formula $A[k/x]$ gives

$$\frac{\Gamma \to A[k/x] \quad A[k/x], \forall xA, \Delta^* \to 0 = 1}{\forall xA, \Gamma, \Delta^* \to 0 = 1} \; Comp$$

The conclusion is reducible by the inductive hypothesis. This same sequent is obtained from $\Gamma, \Delta \to C$ by the arbitrary moves that were applied to $\forall xA, \Delta \to C$, and therefore $\Gamma, \Delta \to C$ is reducible.

4. $A \equiv \neg A$. The composition is

$$\frac{\Gamma \to \neg A \quad \neg A, \Delta \to C}{\Gamma, \Delta \to C} \; Comp$$

In the reduction of the second premiss of *Comp*, if **Y**$\neg$ is never applied to $\neg A$, it can be deleted and what remains, the sequent $\Delta \to C$, is reducible. Then also $\Gamma, \Delta \to C$ is reducible. Otherwise there is a reducible sequent $\neg A, \Delta^* \to 0 = 1$, to which in turn **Y**$\neg$ is applied to give the reducible sequent $\neg A, \Delta^* \to A$.

The first premiss of *Comp* reduces by **F**$\neg$ to $A, \Gamma \to 0 = 1$. Application of *Comp* to the shorter formula A gives

$$\frac{\neg A, \Delta^* \to A \quad A, \Gamma \to 0 = 1}{\neg A, \Gamma, \Delta^* \to 0 = 1} \; Comp$$

The conclusion is reducible by the inductive hypothesis.

As above, if in the reduction of $\neg A, \Gamma, \Delta^* \to 0 = 1$ move **Y**$\neg$ is never applied to $\neg A$, it can be deleted and the remaining sequent $\Gamma, \Delta^* \to 0 = 1$ is reducible. This is the sequent produced from $\Gamma, \Delta \to C$ by the arbitrary initial **F**-moves that gave $\neg A, \Delta^* \to 0 = 1$, so $\Gamma, \Delta \to C$ is reducible.

If instead **Y**$\neg$ is applied, there is a reducible sequent $\neg A, \Gamma^* \to 0 = 1$ such that the instance of move **Y**$\neg$ gives the reducible sequent $\neg A, \Gamma^* \to A$. Composition with $A, \Gamma \to 0 = 1$ gives $\neg A, \Gamma, \Gamma^* \to 0 = 1$ that is reducible. Therefore, continuing this analysis, at some stage the formula $\neg A$ in the antecedent of the result of composition must remain unreduced and can be deleted. The resulting sequent is then reducible. QED.

THEOREM 5.4 *If the sequent $\Gamma \to C$ is derivable, it is unfalsifiable.*

Proof. The proof is by induction on the height of a derivation that we can assume to be normal. If $\Gamma \to C$ is an initial sequent, it is unfalsifiable by lemma 5.1. Otherwise consider the last rule of the derivation and show that if the premisses reduce to endform, also the conclusion reduces. The cases are six logical rules and *Ind*:

1. The last rule is &*I*. The sequents $\Gamma \to A$ and $\Delta \to B$ are assumed reducible. Rule **F&** applied to the conclusion $\Gamma, \Delta \to A\&B$ gives $\Gamma, \Delta \to A$ or $\Gamma, \Delta \to B$ so the conclusion is reducible.

2. The last rule is $\forall I$. The conclusion is $\Gamma \to \forall x A$, and it reduces by **F$\forall$** to $\Gamma \to A[m/x]$. The premiss $\Gamma \to A[y/x]$ is by assumption reducible, with y the eigenvariable. Rule **FVar** produces a sequent $\Gamma \to A[n/x]$ that is reducible for any choice of n, in particular, the choice m. Therefore the conclusion of $\forall I$ is reducible.

3. The last rule is $\neg I$. Reduction rule **F$\neg$** applied to $\Gamma \to \neg A$ produces the premiss $\neg A, \Gamma \to 0 = 1$ that is by assumption reducible.

4. The last rule is &*E*. The premisses are $A\&B \to A\&B$ and $A, B, \Delta \to C$. In the reduction, the conclusion $A\&B, \Delta \to C$ reduces to some $A\&B, \Delta^* \to C^*$ by **F**-moves, until it is turn for a **Y**-move. Move **Y&** gives the sequent $A, B, \Delta^* \to C^*$. Its reducibility follows: The sequent $A, B, \Delta \to C$ is reducible by assumption, so in particular, with the same **F**-moves that were applied to the conclusion, a reducible sequent $A, B, \Delta^* \to C^*$ is produced.

5. The last rule is $\forall E$. The second premiss is $A[t/x], \Gamma \to C$. **F**-moves applied to the conclusion $\forall x A, \Gamma \to C$ produce the sequent $\forall x A, \Gamma^* \to C^*$ and **Y$\forall$** gives $A[t/x], \Gamma^* \to C^*$. Since the premiss is reducible, the same **F**-moves as for the conclusion give $A[t/x], \Gamma^* \to C^*$ that must be reducible.

6. The last rule is $\neg E$. **Y$\neg$** applied to the conclusion $\neg A, \Delta \to 0 = 1$ followed by **F**-moves gives some $\neg A, \Delta^* \to A^*$. The same moves applied to the second premiss $\Delta \to A$ give the reducible sequent $\Delta^* \to A^*$. Therefore, leaving $\neg A$ intact, also $\neg A, \Delta^* \to A^*$ is reducible.

7. The last rule is *Ind*. The premisses are $\Gamma \to A[0/x], A[y/x], \Delta \to A[sy/x]$, and $A[t/x], \Theta \to C$. We may assume $t \equiv n$. In the second premiss, any application of rule **FVar** gives a reducible sequent, so that $A[m/x], \Delta \to A[sm/x]$ is reducible for any m. An n-fold composition of $\Gamma \to A[0/x]$ with $A[0/x], \Delta \to A[s0/x], \ldots, A[n-$

$1/x], \Delta \to A[n/x]$, each derivable, gives the sequent $\Gamma, \Delta^n \to A[n/x]$ that is derivable by lemma 5.2 and reducible by lemma 5.3. One more composition with $A[n/x], \Theta \to C$ gives the sequent $\Gamma, \Delta^n, \Theta \to C$ that is likewise derivable and reducible. Therefore also $\Gamma, \Delta, \Theta \to C$ is reducible. QED.

Closure of reducibility under composition is needed with rule *Ind*.

COROLLARY 5.5 *The system* **NI**+*Ind is consistent.*

Proof. The sequent $\to 0 = 1$ is irreducible but not in endform, therefore it is not derivable. QED.

As noted, our system is classically equivalent to Peano arithmetic.

6 Comparison with Gentzen's proof

The above proof of the reducibility of derivable sequents contains several simplifications as compared to Gentzen's original proof. His E-rules and rules for negation were

$$\frac{\Gamma \to A\&B}{\Gamma \to A}\&E_1 \qquad \frac{\Gamma \to A\&B}{\Gamma \to B}\&E_2 \qquad \frac{\Gamma \to \forall xA}{\Gamma \to A[t/x]}\forall E$$

$$\frac{\Gamma \to \neg\neg A}{\Gamma \to A}DN \qquad \frac{A,\Gamma \to B \quad A,\Delta \to \neg B}{\Gamma,\Delta \to \neg A}Wid$$

Table 9. Gentzen's original E-rules and negation rules.

Wid stands for the German *'Widerlegung,'* or refutation of the assumption A that leads to contradictory conlusions. The rule is substantially the same as our $\neg I$ in Table 7. Gentzen's calculus is classical, but the rule of double negation can be left out because it has the same effect as the rule of indirect proof that is admissible. We have not determined how difficult it would be to obtain such an admissibility result for Gentzen's calculus. As noted, Gentzen knew that classical logic is not necessary.

Gentzen's rules $\&E_1, \&E_2$, and $\forall E$ are 'special elimination rules,' obtainable from our rules by making the second premiss in E-rules an initial sequent, with two possibilities for &:

$$\frac{\Gamma \to A\&B \quad A \to A}{\Gamma \to A}\&E_1 \qquad \frac{\Gamma \to A\&B \quad B \to B}{\Gamma \to B}\&E_2$$

$$\frac{\Gamma \to \forall xA \quad A[t/x] \to A[t/x]}{\Gamma \to A[t/x]}\forall E$$

Table 10. Special elimination rules.

Gentzen's rules come out by leaving the second premiss unwritten. Now the simple notion of normality, by which the first premiss in E-rules is an initial sequent, is lost. Normality was used above in the proof of theorem 5.4 and of lemma 5.2, closure under composition of derivable sequents. Gentzen does not deal with the latter issue, but a proof of closure under composition for his rules would not be easy, contrary to our lemma 5.2.

The reducibility of derivable sequents is, in Gentzen's words, "the most essential part of the proof of consistency." In our concise proof of theorem 5.4, the rule of composition had to be used only in the case of rule *Ind*. With Gentzen's rules, it has to be used with rules *DN*, *Wid*, and *Ind*.

7 Modification of the original proof

Gentzen's original consistency proof was arguably the most important result in logic and foundations after Gödel's incompleteness theorems. It aroused immediate interest, especially through Bernays who stayed at the Institute for Advanced Study in Princeton in 1935–36. He had sailed in September 1935 over the Atlantic, in the company of Gödel and with Gentzen's manuscript in his briefcase. Correspondence between Bernays and Gentzen documents to some extent the objections to Gentzen's first proof. Gödel made a suggestion to change the proof, of which we only know, in Gentzen's words, that "the modification suggested by Gödel was known to me, but it is in reality not usable from the finitary point of view, due to its impredicative character."[11] At that time, Gentzen took his proof to be finitary, even if not representable in arithmetic. He writes in the original version that the methods used in the proof are undoubtable (*'unbedenklich'*) or "at least essentially more secure than the doubtable forms of inference of pure number theory." The reason is that his methods are "finitary in a specific sense," as defined in his paper.[12]

After the criticism of Bernays, Gödel, and possibly also von Neumann, Gentzen changed his original proof. The reason was not that there was objectively something to criticize: as mentioned, it can be shown that the 'fan theorem' Bernays thought to be the illicit moment behind Gentzen's proof, would not make the consistency of Peano arithmetic provable. Before any criticisms and before the first proof, Gentzen had conceived a different argument for the termination of the reduction procedure, one that uses transfinite induction. A specific notation is created to this purpose in the published version of 1936; however, in his two later proofs, of 1938 and 1943, Gentzen used the standard notation for ordinal numbers.

The 1936 paper suffered modifications in two long sections, which created in-

---

[11] See Menzler-Trott (2001), p.58, and also Kreisel (1987), pp.173–174.

[12] "That this is the case shall be considered more closely in Section V," at p.98 of the German version. In the printed paper, the passage was changed into: "To what extent this is the case [...]."

congruities with earlier parts that remained unchanged. At the same time, Gentzen took the occasion to simplify the proof through a modification of the system of natural deduction. Last, the general procedure for the reduction of sequents was not used at all, but a new one for the reduction of derivations.

In the modified second proof of the consistency of arithmetic, the logical rules of natural deduction are simplified by admitting the following sequents among the initial sequents:

$$A\&B \to A \quad A\&B \to B \quad A,B \to A\&B \quad \forall xA \to A[t/x]$$

$$A, \neg A \to 0 = 1 \quad \neg\neg A \to A$$

Of the logical rules only $\&I$, $\forall I$ and $\neg I$ remain. All the other ones follow from the added initial sequents through Gentzen's *chain rule*:

$$\frac{\Gamma_1 \to A_1 \quad \ldots \quad \Gamma_n \to A_n \quad A_1,\ldots A_n, \Delta \to C}{\Gamma_1,\ldots \Gamma_n, \Delta \to C} \; Chn$$

In terms of standard natural deduction, the effect of the chain rule is that the open assumptions $A_1,\ldots A_n$ of a derivation are substituted through their derivations in a *simultaneous substitution*.

In Bernays (1970), a proof of consistency is sketched that uses the idea of Gentzen's unpublished proof, but the logical calculus used is not the one of the original proof. It is, instead, the above modification taken from the printed proof of 1936. The reader has some difficulty in gathering these details from Bernays' presentation.

The main novelty of the printed, second proof is the assignment of ordinal numbers to derivations and the proof that each reduction step on derivations diminishes its ordinal. However, Gentzen was not satisfied with the way the proof came to be presented. His style, as such, creates difficulties with its long, verbose arguments the details of which have to be laboriously worked out by the reader. Added to this come the incongruities created by substantial changes that were not worked through for the whole paper. Thus, in 1938, Gentzen returned to the topic and published a third proof of the consistency of arithmetic. Its aim was, in his own words, "to detail out the basic ideas and to make each single step of the proof as understandable as possible." As mentioned, this proof used the classical sequent calculus of Gentzen's doctoral thesis, and made explicit use of the assignment of transfinite ordinal numbers to derivations. This methodology proved itself in Gentzen's last paper of 1943, the culmination of the studies on incompleteness and consistency of arithmetic.

BIBLIOGRAPHY

Bernays, P. (1970). "On the original Gentzen consistency proof for number theory", in J. Myhill et al., eds, *Intuitionism and Proof Theory*, pp.409–417, North-Holland, Amsterdam.

Gentzen, G. (1933). "Über das Verhältnis zwischen intuitionistischer und klassischer Arithmetik", submitted for publication but withdrawn, first published in *Archiv für mathematische Logik*, vol. 16 (1974), pp.119-132. English tr. in Gentzen (1969).

Gentzen, G. (1934-35). "Untersuchungen über das logische Schliessen", *Mathematische Zeitschrift*, vol. 39, pp.176-210 and 405-431.

Gentzen, G. (1936). "Die Widerspruchsfreiheit der reinen Zahlentheorie", *Mathematische Annalen*, vol. 112, pp.493–565.

Gentzen, G. (1938). "Neue Fassung des Widerspruchsfreiheitsbeweises für die reine Zahlentheorie", *Forschungen zur Logik und zur Grundlegung der exakten Wissenschaften*, vol. 4, pp.19–44.

Gentzen, G. (1943). "Beweisbarkeit und Unbeweisbarkeit der Anfangsfällen der transfiniten Induktion in der reinen Zahlentheorie", *Mathematische Annalen*, vol. 120, pp.140–161.

Gentzen, G. (1969). *The Collected Papers of Gerhard Gentzen*, ed. M. Szabo, North-Holland, Amsterdam.

Gentzen, G. (1974). "Der erste Widerspruchsfreiheitsbeweis für die klassische Zahlentheorie", *Archiv für mathematische Logik*, vol. 16, pp.97–118.

Gödel, K. (1933). *Zur intuitionistischen Arithmetik und Zahlentheorie*, as reprinted in Gödel (1986–2003), pp.286–295.

Gödel, K. (1986–2003). *Collected Works*, vols. I–V, ed. S. Feferman et al. Oxford University Press, Oxford.

Kleene, S. (1952). *Introduction to Metamathematics*, North-Holland, Amsterdam.

Kreisel, G. (1987). "Gödel's excursions into intuitionistic logic", in P. Weingartner and L. Schmetterer, eds, *Gödel Remembered*, pp.65–186, Bibliopolis, Naples.

Menzler-Trott, E. (2001). *Gentzens Problem: Mathematische Logik im nationalsozialistischen Deutschland*, Birkhauser, Basel. English version to be published by the American Mathematical Society.

Negri, S. and von Plato, J. (2001). *Structural Proof Theory*, Cambridge University Press, Cambridge.

von Plato, J. (2001). "Natural deduction with general elimination rules", *Archive for Mathematical Logic*, vol. 40 (2001), pp.541–567.

Prawitz, D. (1965). *Natural Deduction: A Proof-Theoretical Study*. Almqvist & Wicksell, Stockholm.

Schroeder-Heister, P. (1984). "A natural extension of natural deduction", *The Journal of Symbolic Logic*, vol. 49, pp.1284–1300.

A.S. Troelstra and D. van Dalen (1988). *Constructivism in Mathematics: An Introduction*, volume 1. North-Holland, Amsterdam.

PART II

FROM PHILOSOPHY TO LOGIC

Judgements, Contents and their Representations
CATARINA DUTILH NOVAES

ABSTRACT. In his defense of the epistemological view of logic, G. Sundholm often points out that the real objects of logical investigation should be judgements and inferences, as opposed to contents and consequences. Sundholm recognizes Frege as a kindred spirit in this sense, and deplores the exile of judgements and inferences from logic after Frege. In this paper, I argue that Frege's notational choices are in fact at odds with his conceptual assumption that judgements are the primitive elements of logic, and thus may have at least partially been responsible for the exile deplored by Sundholm. Moreover, I argue that the difficulties related to the formulation of a purely judgement-based logical notation constitute a real challenge for the epistemological view of logic, and may even suggest that logic is indeed after all about contents and consequences rather than about judgements and inferences.

1 Introduction

Among the many lessons that my thesis supervisor Göran Sundholm taught me, one of the most important ones is that logic is essentially about *inferences*, i.e., moves from knowledge of the premises to knowledge of the conclusion, and thus not about *consequences*, i.e., abstract relations between contents.[1] For this reason, Sundholm maintains that the key elements in logic must be judgements, not contents or propositions. Frege is a great source of inspiration for his views of logic, and he has indeed advocated 'logical atavism:'[2] we should turn to Frege for inspiration, not to our immediate predecessors, i.e., those belonging to the meta-mathematical tradition consolidated after the 1930's. According to Sundholm, logic was fundamentally misled after Frege, in no small measure due to the reduction of judgements to contents and of inferences to consequences.

In this paper, I suggest that this fact may partially be Frege's own fault, and this for two reasons related to his choice of notation in the *Begriffsschrift*. Firstly (and

---

[1] Whether I have come to agree completely with his distinctive view of logic is a different matter (which shall be further addressed in this paper); But his epistemological approach to logic has always been stimulating and thought-provoking. Indeed, I consider it to have been an essential element of my formative years as his PhD student.

[2] Sundholm (2001).

less importantly), his notation to denote assertion, the famous judgement stroke, is cumbersome and extremely inconvenient for type-setting. Secondly (and more importantly), while Frege deems judgements to be the basic elements of logic, the notation for judgements is not a primitive element of his notational system: It is formed by the application of the functor judgement stroke '|' to a capital letter preceded by the horizontal dash representing judgeable content '$-A$,' yielding '$\vdash A$.' This notational choice suggests that contents, not judgements, are the primary entities for logic.

I first survey Sundholm's conception of logic (which he views as essentially Fregean) as an *organon* for the production of new knowledge, where judgements and inferences occupy the central stage. I then discuss the tension that emerges from Frege's emphasis on judgement and his notational choices. Thirdly, I briefly examine what happened to the notions of inference and judgement in the developments after Frege, and conjecture that his notational choices may have played a role in the exile of judgement from logic. Fourthly, I contrast Frege's representation of judgements as the combination of content and judgement stroke with medieval representations of judgements and contents. Judgements – *propositiones*, assertions – were also the (sentential) primitive elements for medieval logicians, and many of them (such as Abelard and Buridan) recognized the so-called Frege Point in some sense or another. However, in their logical language – the regimented form of Latin that was medieval academic Latin – judgements receive a more basic syntactic representation than contents. I shall focus on Abelard, who arguably 'discovered' propositionality more than seven centuries before Frege, and thus shares a significant portion of common ground with Frege. In conclusion, these considerations shall allow me to reflect on the importance of notational devices mirroring the conceptual framework underlying a logical system, and to question what the almost unanimous attribution of a canonical form to contents instead of to judgements may mean for the status of the epistemological conception of logic.

2 Sundholm on logic: judgement, inference and knowledge

As is well known, Sundholm belongs to the 'constructivist school' in logic, the school having its roots in the Dutch intuitionistic movement, which was then further developed in particular by Swedish logicians such as Dag Prawitz and Per Martin-Löf, but also by Michael Dummett. Roughly put, this school is characterized by an emphasis on the epistemological import of logic and on the grounds for the knowledge obtained (not exclusively, but primarily) through logic. The correctness of a judgement (for example, that a certain inference is valid) must be solidly grounded: One must actually present that in virtue of which the judgement is correct, its ground. In the case of an inference, one must present the *proof* leading from premises to conclusion; More generally, one must display either the procedure – the appropriate act – or the traces thereof – the construction – that has allowed the agent to produce the piece of knowledge expressed in a judgement.

Sundholm has advocated his epistemological view of logic in several of his writings. His main motto may be said to be: "Logic is an epistemological tool for obtaining new knowledge from known premises."[3] In particular, Sundholm views Frege as a kindred spirit, and interprets Frege's much-debated contention that a valid inference must have true – in fact, *known* – premises in the light of this epistemological vision of logic.[4] If the premises are not known, then the conclusion will not be properly grounded, and thus no new knowledge will be obtained by means of the passage from premises to conclusion, even if the passage itself is warranted:

> "Nothing at all can be inferred from false premises. A mere thought, that has not been recognized as true, cannot be a premise. Mere hypotheses cannot be premises."[5]

Sundholm deplores the almost total absence of this epistemological component in most of 20th century work in logic, in particular under the influence of Hilbert and Tarski, but with precedents already in Bolzano and even in the medieval notion of *consequentia*.[6] He locates the ultimate 'fall' in the period immediately after Frege, in particular with Whitehead and Russell's *Principia Mathematica*, which was then further consolidated both in Hilbert's formalistic program and in Tarski's model-theoretic tradition. Indeed, in the formalistic program, there is an outright conflation of judgements and contents.[7] Moreover, Tarski's wide-ranging influence was also particularly nefast: Tarski made the supreme *faux-pas* of reducing the validity of an inference to the holding of a consequence. There is, however, a fundamental difference between inferences and consequences: (valid) inferences fulfill the epistemological role of producing new knowledge, but consequences simply cannot do that. Therefore, if consequences become the chief object of logic, then logic can no longer have the epistemological import that Sundholm and Frege assign to it. As Sundholm puts it, logic in the 19th and 20th century gradually abandoned its status of 'epistemo-logic' and became increasingly 'onto-logic.'[8]

Let us thus take a closer look at the crucial differences between inference and

---

[3] Sundholm (1998), p.31.

[4] Sundholm (2002), p.572.

[5] Letter to Jourdain, Frege (1976), p.H8. English translation from Sundholm (1998). As is well-known – and aptly discussed by Bell in his (1979), III –, this view poses serious problems concerning the possibility of proofs by absurdity, as in such proofs the initial premise is not seriously asserted but rather posited precisely so that a contradiction can be derived from it, thus proving its falsity.

[6] Sundholm (1998).

[7] See Sundholm (1988). In the proof-theoretic tradition stemming from the work of Gentzen the significance of inferences for logic is still preserved, even though Gentzen's sequent calculus is an uninterpreted calculus. See Sundholm (2002).

[8] Sundholm (1988).

consequence in order to understand why consequences cannot produce new knowledge. Here is how Sundholm defines these two concepts:[9]

- A consequence is a relation between propositions that may hold;
- An inference is an act of passage from judgement(s) to judgement that may be valid.

The two most visible differences are thus that (i) a consequence is an abstract *relation*, floating in a Platonic realm somewhere, whereas an inference is an *act* performed by an agent; (ii) the relata of a consequence are two sets (the set of premises and the set of conclusion(s)) of *propositions*, whereas the relata of an inference are two sets of *judgements*.[10] Judgements are themselves acts performed by an agent, so an inference is an act acting upon acts in such a way that it "effects a passage from known judgements to a novel judgement that becomes known in virtue of the inference in question."[11] Again, this conception of inference is essentially the one held by Frege, which Sundholm fully endorses:

> "An inference [...] is an act of judgement that is drawn according to logical laws from judgements previously made. Each premise is a certain proposition which has been recognized as true, and also in the conclusion-judgement a certain proposition is recognized as true."[12]

Now, given that producing new knowledge is obviously an act conducted by an agent, the mere fact that an inference is an act and a consequence is not (it is a relation, independent of any agent's doings) immediately implies that consequences can obviously not produce anything, and thus a fortiori cannot produce knowledge: A product is by definition the outcome of an act of producing. But naturally, the proponents of the reduction of (valid) inferences to consequences (that hold) do not maintain that inferences and consequences are entirely (conceptually) equivalent; Rather, they maintain that each valid inference corresponds to a consequence that holds, *mutatis mutandis*. In other words, given a certain inference, in order to evaluate its validity it is sufficient to convert it appropriately into a consequence (premises and conclusion become propositions; the act of inferring becomes a relation) and then evaluate whether the consequence holds. With this reduction, logic is then free to focus on the study of consequences, as presumably knowledge of the validity of inferences would simply follow from it. The underlying assumption is that of extensional equivalence between consequences that hold and, *mutatis mutandis*, valid inferences.

---

[9] Sundholm (1998), p.30.

[10] This characterization of consequences is essentially that of Bolzano. In Tarski (2002), for example, consequences do not have this Platonic ontological status, but this dissimilarity is immaterial for the present purposes.

[11] Sundholm (1998), p.27.

[12] Frege (1906), p.387; english translation from Sundholm (1998)

What are the necessary and sufficient criteria for a putative consequence to hold? This is of course one of the main debates in recent philosophy of logic. After decades of Tarskian, model-theoretic homogeny, the debate was ignited anew with the publication of Etchemendy's *The Concept of Logical Consequence*.[13] I have argued elsewhere,[14] and following the medieval notion of *consequentia*, that the fundamental aspect of the notion of consequence is the incompatibility of the truth of the premises with the falsity of the conclusion, i.e., what Etchemendy calls the representational notion of (logical) consequence, rather than the property of substitutivity with preservation of validity of a consequence's non-logical terms, i.e., what Etchemendy calls the interpretational notion of (logical) consequence. But this dispute is immaterial for the present purposes, as in both cases a consequence may hold independent of an agent's (grounded) assent to it. That is to say: A given consequence C may hold even if no agent has been through the process of proving that it indeed holds, or in any case not the particular agent performing a (putative) inference I whose validity is presumed to be grounded on the fact that the corresponding consequence C does hold.

That consequences thus described are not adequate tools for the production of new knowledge is made patent by the phenomenon of 'lucky knowledge.' The point is that, if knowledge is indeed justified true belief,[15] then a consequence fails to produce new knowledge because, even if the belief thereby produced happens to be a true one by sheer chance, it is still not properly justified. 'Lucky knowledge' is thus *un*justified true belief, and hence it is not knowledge properly speaking (just as a false friend is not a friend properly speaking). Suppose I am asked to state the number of windows in the Empire State Building; I haven't got a clue, but nevertheless I utter '1.347,' just as I could have uttered any other number (but somehow 'something' tells me this is the right number). Now, suppose that coincidentally the number of windows in the Empire State Building is indeed 1.347; Well, it turns out that I have made a judgement that happens to be true, even though I had no grounds for it. Clearly, it would be very strange to count this judgement of mine as real knowledge[16] – even though it was (it expressed) a true belief – precisely because it was not justified. The issue is now that if valid inferences are defined as those corresponding to a consequence that holds, many of those purported inferences will be 'lucky inferences:' If all it takes for a consequence to hold (and thus

---

[13]Etchemendy (1990).

[14]Dutilh Novaes (2005).

[15]I am here assuming that, even if being a justified true belief does not count as a sufficient condition for knowledge, as suggested by the Gettier cases, it still counts at least as a necessary condition for knowledge.

[16]Notice that this does not even count as a Gettier case, as I have no grounds whatsoever for my assertion, not even the 'wrong' grounds of the typical Gettier cases. Indeed, the view that valid inferences can be reduced to consequences that hold makes no effort at all to account for the justificatory aspect of knowledge.

for the corresponding inference to be valid) is the impossibility of its antecedent(s) being true while its consequent is false, then an agent need not be in possession of the grounds for the legitimate move from premises to conclusions in order for her judgement of the validity of an inference to be correct. It will be sufficient that the corresponding consequence holds in some abstract, objective way, even if she has no knowledge of this fact.

An example may help illustrate the point. We now know that Fermat's Last Theorem (FLT) is a consequence of the basic axioms of number theory: Because we have a proof of FLT, we know that it is impossible for the basic axioms of number theory to be the case while FLT is not the case. Now, before Wiles' proof of the theorem, this consequence (basic axioms of number theory $\Rightarrow$ FLT) obviously already held, even though we did not know it to be the case then. But if all it takes for an inference to be valid is that it corresponds to a consequence that holds, then somebody claiming to effect the inference from the basic axioms of number theory to the truth of FLT *prior* to the discovery of the proof of FLT would, according to this criterion, be effecting a valid inference and thus producing new knowledge. But this is of course absurd: Our true belief in FLT only became justified once a *proof* of FLT was found.[17] Before that, such a purported 'inference' was merely a 'lucky inference.' Indeed, this very proof is precisely the (valid) inference from the basic axioms of number theory to FLT, and only thereby do we really acquire new knowledge, i.e., new *justified* true belief.

Hence, as Sundholm has been arguing relentlessly, if logic is indeed essentially concerned with the production of new knowledge, its basic elements can neither be consequences nor inferences whose validity depend solely on the holding of the corresponding consequences. A stricter criterion for the validity of inferences is required in order to avoid the undesired consequences of 'lucky knowledge,' one that emphasizes the actual epistemic process of arriving at the conclusion assuming knowledge of the premises, e.g., (as described by Frege) judgements 'drawn according to logical laws' (see passage above). Now, this epistemic process is entirely constituted of acts of judgements: An agent first judges the premises to be true, and then she judges the move from the premises to the conclusion to be a valid, grounded inference (and thus in fact undertakes this move); She is thereby entitled to real knowledge of the conclusion, i.e., she has grounds to assent to the conclusion.

Moreover, if inferences thus construed are viewed as the primary elements of logic, consequences can in turn be approached on the basis of inferences. Once a putative inference is established as a valid inference (by means of a proof), then we know that it corresponds to a valid consequence, as clearly it is impossible for

---

[17]Interestingly, before the proof was found, FLT was widely but merely intuitively thought to be true indeed. But obviously, at that point it was also widely thought that the (true) belief in FLT was not justified.

the contents of the judgements corresponding to the premises to be the case while the content of the judgement corresponding to the conclusion is not the case.[18] As for the (presumed) consequences that hold 'out there,' but whose corresponding inferences have not (yet) been established as valid, their role in and for logic would be completely idle, as they cannot be involved in the production of new knowledge. So one need not deny the existence of consequences that do not (yet) correspond to valid inferences (implying thus that the class of valid inferences is strictly contained in the class of consequences that hold,*mutatis mutandis*); But they are simply of no interest to logic construed as an epistemological tool.

Of course, it may still be argued that logic is *not* essentially concerned with the production of new knowledge, in which case a completely different picture of what logic is about must be provided. But if one subscribes to this view of logic, then clearly one ought to take judgements, not contents/propositions, as its basic elements. This is indeed Frege's explicit position; Implicitly, however, given his notational choices in the *Begriffsschrift*, he seems to take contents as primitive elements, thus hindering the dissemination of his own vision of logic, as I shall argue now.

3 Frege on judgement, and his notational 'slip'

In this section, I argue that, while from the point of view of Frege's philosophies of mind and language, a thought (content, proposition) may be more basic than a judgement, for logic (as Frege understands it) judgements must be treated as basic elements (as argued in the previous section), but that this fact is not properly represented in his notational choices in the *Begriffsschrift*.

Frege presented his views on content, truth and judgement in several of his writings, most notably in *Über Sinn und Bedeutung* and in *Der Gedanke*. These views are at the crossroad between epistemology, philosophy of mind and philosophy of language, and have thus a series of important implications that will be of no concern for us here. For our purposes, a characterization of the gist of his notion of judgement should suffice.[19] Here are two passages where Frege comes very close to providing a full-fledged definition of the concept:

"A judgement for me is not the mere comprehension of a thought, but the acknowledgment of its truth."[20]

---

[18] Assuming, of course, the soundness of the deductive system being used.

[19] Notice however that a thorough characterization of Frege's notion of judgement would require significantly more effort. Indeed, Bell (1979) is a book-length study on the topic, where Bell offers a critical analysis of Frege's positions, in particular concerning assertion and the judgement stroke (Part III). Bell outlines many important issues and tension that emerge from Frege's treatment of this notion, some of which echo the main objection to Frege's notation being presented here.

[20] Frege (1892), p.34, note 7. This translation replaces the term 'admission' in M. Black's translation by 'acknowledgement,' for the German 'Anerkennung,' as suggested by Pagin (2001).

"Judging can be regarded as advancing from a thought to its truth value. Naturally, this cannot be a definition. Judging is something quite of its own kind and incomparable."[21]

The idea is the following: Just as a term has a *Sinn* and a *Bedeutung* (although not all term have a *Bedeutung*, of course), so does a sentence. The *Sinn* of a sentence is the thought (*Gedanke*) it expresses, and its *Bedeutung* is its truth-value, i.e., either the True or the False (one of Frege's most controversial views).[22] Now, a judgement is, on the mental level, the acknowledgement of the truth of a thought, which is otherwise merely entertained; On the linguistic level, a judgement is the assertion (the expression of the belief) that the thought expressed by the sentence actually obtains. Thus seen, a judgement seems indeed to be a derivative entity, formed by the combination of a thought with the 'acknowledgement of its truth:' Judging is the act from a thought to a truth-value, and hence its product, the judgement, has a thought as one of its ingredients. That a thought may be seen as a more primitive entity also transpires from the observation that it can occur without the accompanying judgement, as is the case when one merely entertains a thought, but a judgement cannot occur without the accompanying thought.

This idea is, of course, what underlies Geach's formulation of the so-called Frege point (what we here refer to as 'judgement' is in Geach's terminology 'assertion'):

"A thought may have just the same content whether you assent to its truth or not; a proposition may occur in discourse now asserted, now unasserted, and yet be recognizably the same proposition [...] I shall call this point about assertion the Frege point."[23]

The general idea – present in several other influential theories such as Searle's speech-act theory, among others – is that any speech-act is a combination of a particular force with a given content, and that speech-acts with different forces may nevertheless share the same content. For example, the following different speech-acts

'John, put beer in the fridge' (a command)

'John puts beer in the fridge' (an assertion)

---

[21] Frege (1892), p.35. Pagin (2001), fn.3: "I have departed from Max Black by using the gerundives 'judging' and 'advancing' for the German 'Urteilen' and 'Fortschreiten', instead of 'judgements' and 'advances'. I have also preferred 'of its own kind' to 'peculiar' for the German 'einzigartiges'. Finally, Black has the indefinite 'a truth value' rather than the reflective 'its truth value', where the German is 'seinem Warheitswert'."

[22] As pointed out by Bell ((1979), p.86), the view that a sentence is the name of a truth-value is somewhat at odds with Frege's insistence that the complete judgement stroke ⊢ can only be appended to expressions that are propositional in character. But what emerges from *Über Sinn und Bedeutung* is that, concerning the sense/reference mechanism, there is no fundamental difference between single terms and sentences.

[23] Geach (1965), p.449.

'Does John put beer in the fridge?' (a question)

are presumed to have different forces but to share the same content, namely the content usually expressed by the that-clause *'that John puts beer in the fridge.'* Both Frege and Geach in fact focus on assertion, just as we here focus on judgement, thus disregarding other kinds of speech-acts;[24] but one of the strengths of the force-content distinction is that it allows for a neat, unified treatment of all kinds of speech-acts.

This account of speech-acts, and of judgements/assertions in particular, has not gone unchallenged,[25] but it is still widely held to be correct. In particular, it emerges in many of Frege's writings, in particular *The Thought*, and again, its most significant implication for the present purposes is the idea that a judgement is a compound entity (thus not a primitive entity), composed of a content (proposition, thought) and the judging force. But important though it is, my purpose here is not to offer a critical analysis of the force-content distinction as such, and at any rate not regarding philosophy of language and philosophy of mind.

As for logic, however, I have argued in the previous section that contents should not be the primitive elements of logic if one is to hold an epistemological view of the purpose of logic. In fact, contents seem to have a very small role to play within logic thus seen; But if this is so, then for the sake of theoretical parsimony, contents should not receive a specific form of representation in a logical language, or in any case this form of representation should not be a primitive element of the syntax. But, this is not what one encounters in Frege's *Begriffsschrift*.

In §2 of the *Begriffsschrift* Frege introduces the symbols to be employed to represent judgement. He writes:

"A judgement will always be expressed by means of the sign

$$\vdash$$

which stands to the left of the sign, or the combination of signs, indicating the content of the judgement."[26]

A few syntactic features of the use of $\vdash$ are already quite significant: $\vdash$ can neither be iterated nor occur embedded in a given expression, i.e., there is at most one occurrence of $\vdash$ per formula, all the way to the left. Hence, it is not possible to express something like: 'I assert that I assert p to be the case' or 'If I assert p to be the case, then I assert q to be the case.'

---

[24] For the present purposes, there is no significant dissimilarity between assertions and judgements, except for the fact that judgements may be understood either as the purely mental operation or as its verbal manifestation, whereas assertions correspond exclusively to the latter. This fact was noticed by Frege himself, see Frege (1918), p.62.
[25] Wittgenstein in the *Philosophical Investigations* is a prominent example. See Mastop (unpub.) for further references.
[26] Frege (1879), p.11.

The vertical stroke is supposed to represent the act of judging the content to be the case, and accordingly is known as the 'judgement stroke;' The horizontal stroke accounts for the fact that the content represented by a sign or signs is a combination of ideas and thus not a simple idea. A simple idea such as 'house' is not apt to be judged as either true or false; Only a combination of two or more ideas – say, 'house' plus 'large' – is judgeable. The horizontal stroke is thus the 'content stroke.' It is patent that a representation of a judgement with this notation mirrors its threefold structure (as described above), involving (i) ideas; (ii) the combination thereof, yielding a thought; (iii) the advancing from a thought to a truth-value.

There is no point in disputing the fact that Frege really held a threefold view of the structure of a judgement: The issue raised here is whether this structure must be fully represented for *the purposes of logical reasoning*. For Frege, the objects on which logic operates in order to draw inferences (the quintessential logical operation) are judgements, as is patent in the discussion on inference in §6 of the *Begriffsschrift*, and as discussed in the previous section. Accordingly, he proposes to represent the inference from a conditional judgement and the judgement of its antecedent to the judgement of its consequent as:[27]

$$\frac{\vdash B \to A \\ \vdash B}{\vdash A}$$

Of course, the important thing to bear in mind is the fundamental difference between the occurrence of a content in a categorical judgement and its occurrence in a hypothetical judgement of the form 'If A, then B.' In the latter, neither the content expressed by the antecedent nor the content expressed by the consequent is asserted/judged to be the case; In both cases, one could say that these two thoughts are merely comprehended or entertained, and that the assertive force is exclusively borne by the logical connective in question. What is asserted is the relation between the two contents, not the contents themselves.

At the same time, this distinction also seems to entail that no occurrence of *modus ponens* (just to take a particularly familiar example) can ever be valid, as it would always constitute a fallacy of equivocation. If *modus ponens* is correctly expressed by the following formulation:

If p then q
But p
Thus q

---

[27] In the following I represent a conditional in the usual way here, instead of in Frege's two dimensional way, purely for typographical convenience (which only goes on to prove that Frege's notation is indeed not extremely practical).

then it would seem that both propositional variables p and q are used equivocally: in the first premise, they are used to indicate the unasserted corresponding contents, while in the second premise and conclusion they are used to indicate the assertion of these contents. The matter is not simply of notation: If the objects that the propositional variables stand for are not the same (assertive vs. unassertive contents), then one may go as far as wondering whether the argument actually goes through; As is well-known from syllogistic, the terms of the premises and of the conclusion of an argument must stand for the same things (respectively) in order for the argument to be valid. So the dilemma seems to be that either the things that the propositional variables stand for really are the same – in which case the second premise and the conclusion are redundant, that is if in the conditional premise the propositional variables already have assertive force – or they are not the same, in which case the argument will not go through on pain of equivocation. This is a real dilemma, but Frege seems to provide a solution to it that is as good as any in that it outlines the *difference* as well as the *sameness* of a categorical and a hypothetical occurrence of a content: On the one hand they are the same, as they share the same content, but on the other hand they are not the same, as they do not share the same force. Ultimately, this is an issue of propositional identity, as Frege is well aware of (*Begriffsschrift*, §3).[28]

The point is though that, in order to emphasize the difference/sameness relation between the expressions of asserted and of unasserted content, one may either choose to take the asserted expression as the starting point on which a given operation is performed in order to obtain the unasserted expression, or one may choose to reverse the order of priority. Frege opted for the latter, and at least at first sight his choice seems quite felicitous: By 'extracting' the assertive force from a judgement and representing it separately, the validity of *modus ponens* is guaranteed by the fact that the propositional variables do stand for the same things (respectively) in premises and conclusion, namely for contents. At the same time, it is clear that a different force is attached to them in their hypothetical occurrences in the first premise and in their categorical occurrences in the second premise and conclusion.

Now, in many (but not all) natural languages, the lack of assertive force in the embedded sentences of conditionals is represented by the use of a different verbal form, e.g., the subjunctive verbal moods in German, Latin, Portuguese, Italian and Spanish, to mention but a few examples. While we seem to do fine without such a distinction in our daily use of English,[29] where subjunctive moods are hardly ever used (except for the occasional occurrence of phrases such as '*If I were you...*'), Frege could argue that, for logical and scientific purposes, it is of the utmost importance that this distinction be represented (cf. his critique of the

---

[28] This point is raised and discussed in Bell (1979), III.3.

[29] But English features other devices to indicate a force other than the assertive force. For example, it features subject-verb inversion to signal the interrogative force.

imperfections of the *'Sprache des Lebens'* in the Preface of the *Begriffsschrift*). In Frege's native German the use of non-indicative moods in non-assertive contexts (also in indirect speech) is very common, and the connection between Frege's judgement-stroke and the expressions of asserted and unasserted content in the 'language of life' with different verbal moods has not gone unnoticed.[30]

However, the crucial point is that, in an artificial notation, a choice is to be made as to what is to count as the *canonical form* of the expression of a content: Should we take the assertive expression of a content as the canonical form, and accordingly the non-assertive expression of it as a derivative form, or vice-versa?[31] It is clear that in natural languages the assertive expression of a content, i.e., the expression of a judgement, is usually taken to be the canonical form, corresponding to the use of indicative verbal moods in such cases. By contrast, non-assertive expressions of content are (often, though not always) rendered in a derivative form, for example by means of subjunctive verbal moods, nominalizations or that-clauses, which are clearly derivative/non-canonical with respect to 'plain' sentences with indicative verbs.

Frege, though, opted for the reverse solution: Prompted perhaps by the notion of judgement as a compound of content and force (even though, of course, the *Begriffsschrift* antedates *Über Sinn und Bedeutung* by more than a decade), he took the expression of unasserted thoughts to be more primitive or canonical. The expression of a judgement is obtained by means of the application of the functor 'judgement stroke' to the expression of an unasserted thought; The expression of a judgement is thus non-primitive and derivative in Frege's notation.

While it may be motivated by a more general account of the 'phenomenology' of judgements, this notational choice is in tension with Frege's own insistence on the primacy of judgements for logic, insofar as judgements alone can advance knowledge. Notice that I certainly do not wish to imply that this tension is an outright contradiction or anything of the kind. All I am suggesting is that a different notational choice, one where the expression of judgements is assigned the role of canonical form and the expression of unasserted content is derived from the canonical form, might have been more in tune with the key role played by judgements, but not by contents, in Frege's vision of logic. A non-canonical form could then be used whenever a non-assertive expression of content (e.g., embedded sentences in conditionals) is at stake.[32]

---

[30] See Bell (1979), III.4 and Dummett (1993).

[31] Here I assume that assertions/judgements are the chief speech-acts for logic. But the point could be expanded so as to include other kinds of speech-acts. The general question is whether the canonical form should correspond to the whole speech-act or to the content separated from the force.

[32] The challenge is though whether one can still obtain an operational logic with this notational choice, i.e., one that deals adequately with the modus ponens dilemma just discussed.

4 The aftermath

Frege's notational choice seems to me to have backfired: While he explicitly wanted judgements to be at the central stage, the fact that contents are represented as more basic than judgements may have suggested that one could simply do away with the 'cumbersome' strokes for judgement and content. Indeed, to quite a few people they seem to have appeared as unnecessary philosophical subtleties with no serious import for logic: Why not treat the logical relations between contents directly, instead of taking the detour via judgements and inferences? Isn't the judgement stroke redundant at best, but possibly even meaningless?[33]

In fact, Frege's notational choice concerning judgements seems to have been infelicitous for (at least) two reasons. Firstly, the bi-dimensional, non-linear representation of hypothetical judgements, in fact the notation of the *Begriffsschrift* as a whole, is hopelessly cumbersome and a type-setting nightmare. So the *Begriffsschrift* notation was virtually bound to die with Frege, especially after the appearance of the *Principia* notation, which is undoubtedly simpler, but whose simplicity in fact hides conceptual impoverishment. Secondly, as judgements are represented as derivative entities, one is led to wonder whether the whole judgement apparatus is really required for logical purposes. The generations following Frege seem by and large to have concluded that judgements are essentially superfluous for real logical analysis: gradually, judgements became almost entirely neglected. In effect, in 20th century logic judgements and inferences became 'endangered species,' as the title of Sundholm's inaugural lecture in Leiden suggests.[34] Rather than focusing on how an agent can attain new knowledge (justified judgements) on the basis of knowledge already available, a large portion of logicians became concerned exclusively with the logical relations between contents, and in particular with the relation of consequence (in the model-theoretic tradition). The metamathematical formalist tradition moved even further away from the pivotal role of judgement and inference for logic: In this tradition, well-formed formulas are no longer meaningful objects, and the goal is simply to study their properties and mutual relations, given a certain formalism. The formal languages typically used within the formalist tradition simply lack the resources to differentiate judgements from contents. Here is Sundholm's description of this phenomenon:

> "Instead of propositions, formulas are used as formalistic substitutes. These well-formed formulas serve furthermore as end-formulas of the formal proof-trees, and are thus also the formalistic counterparts of judgements. From a formalistic point of view, there is no distinction between judgements and their contents. The sign ' ⊢', which was used as an assertion-indicator by Frege, Russell and Heyting, is in the metamathematical tradition a mathematical predicate of provability, which is to be applied to well-formed formulas."[35]

---

[33] Wittgenstein, *Tractatus*, prop. 4.442: "Frege's 'judgement stroke' ⊢ is quite meaningless."
[34] Sundholm (1988).
[35] Sundholm (1988), p.13. My translation from the original Dutch text.

A few words on Whitehead and Russell's *Principia Mathematica* are in order here, as it seems to represent the beginning of the 'fall.' The two authors saw themselves as continuing Frege's logicist program, in particular by salvaging Frege's logic from paradox; But they made the conscious choice of not using Frege's difficult notation. Rather, in terms of notation they acknowledge the foremost influence of Peano (*Principia*, p.4). Presumably, however, they would have been committed to maintaining the most important conceptual features underlying Frege's logical system, in particular the emphasis on judgements and inference.

Whitehead and Russell do maintain the sign of assertion as one of the (primitive) symbols in their logical language. But significantly, it is introduced (in p.8) only *after* the introduction of the symbols standing for propositional contents (propositional letters), of the functional operations on contents (negation, disjunction etc.), and of truth-values. Remember that, in the *Begriffsschrift*, the notion of judgement is introduced in §2, and only the general idea of using letters as symbols of generality precedes it (in §1). Moreover, an embryonic state of the judgement-content conflation can be perceived in the remarks on inference immediately following the introduction of the sign of assertion,[36] more specifically on the inference from $\vdash p$ and $\vdash p \supset q$ to $\vdash q$:

> "...for the sake of drawing attention to the inference which is being made, we shall write instead
>
> '$\vdash p \supset \vdash q$'
>
> which is to be considered as a mere abbreviation of the threefold statement
>
> '$\vdash p$' and '$\vdash p \supset q$' and '$\vdash q$'."

In this passage, a *propositional* operation, denoted by '$\supset$,' is suddenly applied to judgements; granted, Whitehead and Russell view it as a 'mere abbreviation,' but even as an abbreviation it seems to indicate a categorical mistake, namely that of taking judgements and contents interchangeably.

The rest of the story of the exile of inference and judgement from logic has been told by Sundholm at different places,[37] so we need not rehearse its details once again. The point that I wish to stress, however, is the role that Frege himself may have played in this development: By opting to represent contents, not judgements, in a canonical/primitive form, he may have induced the idea that contents, not judgements, are the primitive elements for logic. The first signs of the content/judgement conflation can already be perceived in Whitehead and Russell's *Principia Mathematica*, as we have seen; From then on, the predominance of content/consequence over judgement/inference only became accentuated – with the exception of the few dissident voices who still insisted on the epistemological role

---

[36] Whitehead and Russell (1910-1913), p.9.
[37] Sundholm (1988), Sundholm (1998), Sundholm (2002).

of logic, i.e., precisely the intuitionist/constructivist tradition to which Sundholm belongs. In other words, perhaps the 'logical atavism' advocated by Sundholm should have a slight revisionist component: Frege's particular notational choices concerning judgements and contents might be in need of revision.

Of course, I am not claiming that Frege's notational choices are the sole responsible for the exile of inference and judgement from logic; Obviously, this development is significantly more intricate, and certainly many other causes contributed for its unfolding. It seems to me that Frege's notation is not even a major force behind the 'ontological turn' in logic: but one cannot help but wonder what the outcome would have been, had Frege been more faithful to his own conceptual commitments in his notational choices. Would judgement and inference have stood at least a fairer chance? And could he have made a different choice, given his conceptual commitments? That he could (at least in theory) have made a different choice will be argued for in the next section.

But before we move on to the next section, there is one observation I would still like to make. In attributing priority to judgements over contents in Frege's logical system, I am essentially relying in what he explicitly *says*, for example in the passages quoted in section 1 above. However, it may be argued that something else emerges from what he actually *does*, i.e., from the actual functioning of his logical machinery: It is conceivable that, in practice, the priority of contents over judgements is an integral, organic part of his logical system, *malgré* Frege himself. In this case, his notational choice would in fact be firmly grounded on a real albeit tacit aspect of the system and thus no coincidence at all. If this is so, then Frege's responsibility in the subsequent expulsion of judgements from the realm of logic would go well beyond notational choices: By switching the focus entirely onto contents, later logicians might simply have been developing a veiled but crucial aspect of Frege's own logic. Now, this issue cannot be adjudicated to a full extent at this point, but for now I leave it as a suggestion: It is at least theoretically possible that Frege did make the right notational choice after all, and that his conceptual insistence on judgements was in fact misguided and ultimately at odds with the actual machinery of his logical system. It would appear in any case that this (hypothetical) primacy of contents would not simply follow from the recognition of the force-content distinction (as I will argue below), but rather from other technical and structural aspects of Frege's logic, which would have to be looked into in more detail at a different occasion.

5 Medieval representations of judgement and content

Frege's representation of judgements as composed of contents and of assertive force may at first sight seem simply to follow from his 'discovery' of propositionality (as captured in Geach's 'Frege point.') Once one recognizes that new propositional contents may be formed by means of propositional operations (typically: negation, conjunction, disjunction etc.) taking contents as arguments and yielding different contents as values, then it may seem that one has no choice but to represent contents as primitive elements of a logical system. To be sure, these operations do not take judgements as their arguments, nor do they yield new judgements: *Inferences* are the operations that take judgements onto other judgements (and they are themselves judgements too). Of course, sometimes there is symmetry between propositional operations and inferences: From contents A and B one may form the content $A\&B$, just as from the judgements that A is the case and that B is the case one may (correctly) arrive at the judgement that A and B are the case. But this obviously does not hold irrespectively: From any two contents A and B one may form the content $A \to B$, but the judgements that A is the case and that B is the case are in no way sufficient warrant for the judgement that $A \to B$ is the case.

Nonetheless, in this section I will argue that Frege's theoretical assumptions and 'discoveries' did not necessarily force upon him the particular notational choice that he eventually made, namely that of taking the representation of contents, not of judgements, as canonical/primitive. In order to motivate this claim, I will present a counterexample, i.e., an author who held views strikingly similar to Frege's views, but who nevertheless made the reverse notational choice: the 12th century logician Peter Abelard.[38]

There are remarkable points of resemblance between the positions held by Frege (discussed so far) and the views held by Abelard. In fact, Christopher J. Martin credits Abelard for many of Frege's most important 'discoveries,' and claims that Abelard was an even greater logician than Frege.[39] Of course, Frege made these discoveries independently, as he most certainly had no knowledge of Abelard's ideas (in fact, most of Abelard's legacy simple got lost already in the generations immediately following him). But Abelard's logical starting point (what he had inherited from the previous traditions) was in fact extremely meager, while Frege could rely on much more elaborate material as his starting point, so in this sense Abelard's accomplishment is more impressive.

For our purposes, there are three important points of similarity among Frege and Abelard:

- Both hold that a sentence (typically) expresses a content: a *dictum* for Abelard, a thought for Frege;

- Both recognize the force-content distinction and that different speech-acts

---

[38]Technically, the choice in question was not really 'made' by Abelard, as he was mostly following conventions and usages already established. But the point to be made here is simply that propositionality and force-content distinction are perfectly compatible with a notational choice taking the expression of assertions as the canonical form.

[39]Personal conversation.

(with different forces) can nevertheless express the same content;[40]

- Both identify propositional operations; They are thus both 'inventors' of propositionality.[41]

In short, most of the points discussed so far that may have motivated Frege's decision to represent contents instead of judgements as primitive entities in his logical notation are to be found in Abelard as well. Nevertheless, Abelard did *not* take the representation of contents, i.e., *dicta*, as the canonical linguistic form. Granted, Abelard is often rather careless in his use of language; In particular, he frequently uses important terminology inconsistently.[42] This is of course a major flaw in a logician; In effect, a considerable amount of charity is needed when interpreting the linguistic expression of his ideas. But when charity is used, a solid and sophisticated logical system emerges.[43]

Before we inspect Abelard's linguistic devices for representing judgements and contents, a few words on his famous *dicta* (contents) are required. A *dictum* is what an assertion says to be the case; in Abelard's formulation, "that which is said by a proposition [assertion]."[44] Thus, a *dictum* is very much like a Fregean *Sinn/Gedanke*, insofar as a *Sinn* is what is expressed by a locution. But unlike Frege, Abelard resists any kind of reification of *dicta*: Even though a sentence says something, one is however not entitled to conclude that there is some *thing* that is said by a sentence. *Dicta* have thus a delicate, if not to say problematic, ontological status.[45]

For our purposes, what matters is how contents (i.e., *dicta*) and judgements are represented. If the points of similarity between Frege and Abelard raised above would indeed necessitate taking the representation of contents as canonical and the representation of judgements as derivative, then one would expect Abelard to follow a similar notational convention. But this is not what one finds in his writings. Abelard represents what we here refer to as judgements (in his terminology, *enuntiatio* or *propositio*) by canonical sentences of the nominative-subject/ indicative-verb form; Here, he is simply following common usage, so this is no

---

[40] For Abelard, see Abelard (LI *De in.*) 3.05.13 and 3.01.101.

[41] For Abelard, see Martin (2004), section II.1.

[42] See Martin (2004), p.166.

[43] Another point of similarity relating Frege to Abelard is that Abelard's logical system was also proven to be ultimately untenable, just as Frege's system was shown to be ultimately paradoxical. Alberic of Paris seems to have put forward a killing argument against Abelard's logic, showing that it allows for the derivation of "If Socrates is human and Socrates is not an animal, then it is not the case that Socrates is human and Socrates is not an animal" – a very embarrassing conclusion indeed. In C.J. Martin's terms: "Abelard's various intuitions about the propositional connectives cannot be reconciled," Martin (2004), p.191. But these embarrassments do not affect the present analysis.

[44] Abelard (LI *De in.*), 3.04.26.

[45] See King (2004), section IV.4.

great feat. More significant, though, are the devices he uses to refer to *dicta*; Indeed, *dicta* are represented by means of constructions *derived* from the canonical form of sentences described above. The two typical constructions to refer to *dicta* are the accusative-subject/ infinitive-verb construction and the that(*quod*)-clause construction. Both are particularly conspicuous in two passages where he discusses the notion of truth,[46] namely in (Abelard (LI *De in.*)) 3.01.100 and 3.04.26. The formulations used by Abelard are *"Verum est Socratem sedere"* and *"Verum est Socratem esse hominem"*, and the nominalized clauses are seen as indicating *dicta* (even though, properly speaking, a *dictum* is not a thing).

One might expect that similar constructions would be used for the occurrence of unasserted contents in conditionals. Abelard does recognize that, in a true conditional such as "If Socrates is a pearl, then Socrates is a stone,"[47] the whole is (correctly) asserted while neither the antecedent nor the consequent is asserted (*proponitur*)[48] – if they were asserted, then the asserted conditional would be false, as both are false. But the conditional is true, which plainly shows that the two embedded sentences are not asserted even though the whole is. Perhaps disappointingly, the two embedded sentences are represented in the canonical nominative-indicative form, i.e., the same form that is typically used to represent asserted contents: *"Si Socrates est margarita, Socrates est lapis."* But of course, in a conditional, the two relata are not assertions but contents. In Abelard's defense, one might say that this was simply the linguistic convention to formulate such conditionals at the time; But ideally, in a language where the judgement-content distinction is consistently represented, the embedded sentences in a conditional should be formulated with the appropriate convention for expressing contents.

Be that as it may, and even though Abelard is not entirely consistent in his representations of judgements (assertions) and contents (*dicta*), it seems to me that a lesson can still be learned from the common points between Frege and Abelard and their divergence concerning the choice of the canonical form of representing judgements or contents. As we have seen, Frege chooses contents to be represented canonically, i.e., as primitive elements of the syntax, and the representation of judgements is obtained from the representation of contents by means of the application of a functor, the judgement stroke. Abelard, by contrast, represents judgements by means of a canonical form; Contents are in turn represented by constructions obtained from this canonical form, namely the accusative-plus-infinitive construction or that-clauses. This approach suggests that contents are obtained from judgements, whereas Frege's approach suggests that judgements are

---

[46] For Abelard, assertions (*propositiones*) are not the actual bearers of truth and falsity, properly speaking. Rather, these bearers are the *dicta*, which are true when they describe things as they are (*ita est in re*) and false otherwise. Assertions are only derivatively true or false, i.e., when they 'say' a *dictum* that is true or false.

[47] At the time, it was believed that pearls were stones.

[48] Abelard (TI), 90.

obtained from contents. The upshot is thus that Frege's views on the force-content distinction and on propositionality do not seem to force a particular choice for the representation of contents and judgements. In any case, taking only these elements into account,[49] it seems that Frege could just as well have taken the representation of judgements as a primitive element of his syntax, instead of that of contents; This would in any case have been more attuned to the primacy he explicitly attributes to judgements in logic. More generally: had he done so, judgements and inferences might have had a better future in subsequent logical developments.

6 Conclusion

Generally, I take it to be important that a given logical notation emulate for as much as possible the conceptual structure underlying the logical system it represents. There is however a palpable tension in Frege's notational choices for representing judgements and contents and his epistemological view of logic. The comparison with Abelard was intended to suggest that, at least from a purely conceptual point of view, propositionality and the force-content distinction are not incompatible with a notation where the expression of judgements, not of contents, takes a canonical form.

Naturally, there are several fundamental differences between Abelard and Frege and their respective logical systems. The most important of them seems to be the fact that Frege explicitly introduces a formal language, with artificially created symbols, whereas Abelard expresses his ideas in the academic Latin of the 12th century. In fact, academic Latin was to become much more regimented in the subsequent centuries (in particular in the 14th century onwards), and with Abelard these conventions were still in their germinal state. Related to this is the fact that Frege's logic allows for a much higher level of abstraction. But precisely because Frege stipulated a new logical language, he could in principle simply have given a canonical form to the representation of judgements. So either this choice was somehow arbitrary, or there are hidden elements in his logic forcing the particular choice he actually made.

The challenge is of course to design a logical system/notation that has judgements represented canonically, as primitive elements, but which still functions appropriately for the purposes of logical reasoning and manipulation. Such a system would then be truly faithful to the tenets of the epistemological view of logic. But it would, for example, have to offer a satisfactory solution to the *modus ponens* dilemma discussed above, and that seems *prima facie* quite a challenge. Suppose for instance that in such a language judgements are represented by capital letters

---

[49] As suggested above, there might be other aspects of Frege's logical system that do necessitate this particular choice. The comparison with Abelard seems in any case to indicate that one would have to look further for these aspects, as propositionality and the force-content distinction alone seem not to be sufficient to force the choice of a canonical representation of contents instead of judgements.

($A, B, C, \ldots$) and that contents are represented derivatively, i.e., their representations are obtained by means of an operation on judgements, for example: $\underline{A}, \underline{B}, \underline{C}$ (the underlying representing the content-forming operation starting from judgements). In such a notation, some of the usual rules for the logical connectives might be formulated as:

$$\frac{\underline{A} \to \underline{B} \quad \underline{A}}{\underline{B}} \qquad \frac{\underline{A \& B}}{\underline{A}} \qquad \frac{\underline{A}}{\underline{A \vee B}}$$

Now, isn't this a little bit odd? How can A be applied to $\underline{A} \to \underline{B}$ in order to obtain B, if neither A nor B (but only the corresponding contents) actually occur in the first premise? Could this be a categorical mistake? From this perspective, Frege's solution appears as rather ingenious indeed.

In modern systems that do take judgements seriously, such as Martin-Löf's Constructive Type-Theory (CTT, which is, as we know, an important source of inspiration for Sundholm), judgements are syntactically construed on the basis of the expressions of contents (propositions). A typical judgement in CTT is $a:A$, which is the assertion to the effect that the object that a stands for (a proof, a construction) is of *type A*, which may mean for example that it is a proof of the proposition (content) that A stands for. It seems to me that this sort of type-assertion is indeed a 'pure' assertion, not a manipulation of contents in disguise. However, (overtly inspired by Frege's judgement stroke) Martin-Löf also represents an assertion of content A as A *true*.[50] Thus, while his analysis of the concepts of judgement and content are extremely lucid and sophisticated, it seems to me that Martin-Löf does perpetuate at least some of Frege's notational choices, and thus still does not deliver a notational system that does full justice to the intuition that judgements are the basic elements for logic (even though he is undoubtedly the best contender in this respect).

But what if the endeavor of designing a notation where judgements are truly primary proves to fail or to be exceedingly difficult? What if a logical system/notation where judgements are represented canonically and contents are represented derivatively turns out to be significantly less manageable (if at all manageable) than systems/notations where the reverse occurs? If this turns out to be the case, this outcome may suggest two conclusions concerning the epistemological vision of logic:

1. A radical conclusion: The vision of logic as an epistemological tool is fundamentally misguided. Logic is not about judgements and inferences; It is about contents and consequences.

2. A moderate conclusion: The emphasis on judgements and inferences belongs to the *philosophy* of logic, to what we think we are doing when doing

---

[50] See for example Martin-Löf (1996).

logic, but not to logic itself, i.e., to the general principles and operations characterizing the enterprise as such.

In the spirit of conclusion 1), it might be added that Frege's system is a somewhat awkward hybrid being, combining propositionality insights with an old-fashioned conception of the purpose of logic, which are nevertheless in tension with each other. In particular, his choice to represent contents instead of judgements canonically would belong to the insightful and successful part of his program. No wonder that in subsequent developments judgements and inferences became endangered species: They simply were not in their natural habitat in the first place, and thus had not much of a chance of survival.

Conclusion 1) would undoubtedly enrage Sundholm, Frege and the partisans of the epistemological view of logic in general. Still, it seems to me that they owe us an explanation for the fact that, even in the different logical systems that emphasize the role of judgements and inferences mentioned in this paper (Frege's in particular but also CTT), contents – not judgements – are represented canonically. Aren't contents then the most basic logical entities after all?

Conclusion 2) does not challenge the propriety of the epistemological view of logic as a whole, but relegates the importance of judgements and inferences to the realm of what we do with, and think about, logic. On this view, we may choose to give logic the epistemological import that Sundholm and Frege attribute to it, but we need not do so, as it is not an inherent feature of logic as such. Logic can be applied to the production of new knowledge, but in truth it essentially concerns relations of incompatibility and truth-preservation between contents (and ultimately between the states-of-affairs that contents correspond to). When we intend logic to be used as an epistemological tool, then we are indeed concerned with the assertion of these contents, with judgements, but otherwise logic does just fine without them.

But of course, these are conditional conclusions: *If* it is impossible, too difficult, or just considerably more difficult to work with a notation having judgements as primitive elements, this *might* indicate that the epistemological view of logic is 'less natural' after all. I of course do not exclude the possibility of a successful judgement-based notation, but I must say that I am unconvinced by those that have been proposed so far. Moreover, issues such as those underlying the *modus ponens* dilemma discussed above seem to me to be real challenges for such a judgement-based notation. Thus, although I am conceptually very sympathetic to the epistemological view of logic, I am somehow discouraged by the difficulties devising a notation faithful to its basic tenets, which makes me wonder whether logic is not after all essentially about contents and the states-of-affair that they represent.

Either way, it seems to me that the canonical form of representation attributed to contents in logical systems that nevertheless emphasize the role of judgements, such as Frege's, is something in want of an explanation. It is in any case something to be pondered on, and which may give us further insights in the debate opposing the epistemological and the ontological views of logic.

BIBLIOGRAPHY

Abelard (TI). "Tractatus de intellectibus". In P. Morin, editor, *Abélard: Des intellectionis*, Vrin, Paris, 1994.

Abelard (LI *De in.*). *De in. Glossae super Periermeneias (Commentary on Aristotle's De Interpretatione)*, Corpus Christianorum continuation medievalis, Brepols, Turnhout, forthcoming.

Bell, D. (1979). *Frege's Theory of Judgement*. Oxford University Press, Oxford.

Dummett, M. (1993). "Mood, Force and Convention", in M. Dummett, *The Seas of Language*, pp.202-203, Oxford University Press, Oxford.

Dutilh Novaes, C. (2005). "In search of the intuitive notion of logical consequence", in L. Behounek and M. Bilkova, editors, *LOGICA Yearbook 2004*, pp.109–123, Filosofia Publisher, Institute of Philosophy, Czech Academy of Sciences, Prague.

Etchemendy, J. (1990). *The Concept of Logical Consequence*. Harvard University Press, Cambridge.

Frege, G (1879). *Begriffsschrift*, in J. van Heijenoort, editor, *From Frege to Gödel – A source book in Mathematical Logic*, pp.1–82. Harvard University Press, 1977. Third printing.

Frege, G. (1892). "Über Sinn und Bedeutung", *Zeitschrift für Philosophie und philosophische Kritik*, (100), pp.25–50. Translated as "Sense and Reference" by M.Black, *The Philosophical Review*, 57 (1948), pp.207-2301.

Frege, G. (1906). "Über die Grundlagen der Geometrie", *Jahresberichte der Deutschen Mathematiker-Vereinigung*, 15, pp.377–403.

Frege, G. (1918). "Der Gedanke. Eine logische Untersuchung", *Beiträge zur deutschen Idealismus*, 2. Translated in P. Geach and M. Black, editors, *Translations of the Philosophical Writings by Gottlob Frege*, pp.58–77. Blackwell, 1952.

Frege, G. (1976). *Wissenschaftliche Briefwechslung*. Felix Meiner, Hamburg.

Geach, P. (1965). "Assertion", *Philosophical Review*, 74, pp.449–465.

King, P. (2004). *Metaphysics*, in J. Brower and K. Guilfoy, editors, *The Cambridge Companion to Abelard*, pp.65–125, Cambridge University Press, Cambridge.

Martin, C.J. (2004). *Logic*, In J. Brower and K. Guilfoy, editors, *The Cambridge Companion to Abelard*, pp.200–222, Cambridge University Press, Cambridge.

Martin-Löf, P. (1996). "On the meanings of the logical constants and the justification of the logical laws", *Nordic Journal of Philosophical Logic*, 1(1), pp.11–60.

Mastop, R. (unpub.). *Doing away with the sense-force distinction*, Manuscript.

Pagin, P. (2001). "Frege on truth and judgement", *Organon*, F VIII, pp.1–13.

Sundholm, B.G. (1988). *Oordeel en gevolgtrekking*, Inaugural address, Leiden University.

Sundholm, B.G. (1998). "Inference versus consequence", in *LOGICA Yearbook 1997*, pp.26–35, Filosofia Publisher, Institute of Philosophy, Czech Academy of Sciences, Prague.

Sundholm, B.G. (2001). "A plea for logical atavism", in *LOGICA Yearbook 2000*, pp.151–162, Filosofia Publisher, Institute of Philosophy, Czech Academy of Sciences, Prague.

Sundholm, B.G. (2002). *A century of inference: 1837-1936*, in P. Gärdernfors, J. Wolenski, and K. Kijania-Placek, editors, *The Scope of Logic, Methodology and Philosophy of Science*, Volume II of the 11th International Congress of Logic, Methodology and Philosophy of Science, Cracow, August 1999, pp.565–580, Kluwer Academic Publisher, Dordrecht.

Tarski, A. (2002). "On the concept of following logically", *History and Philosophy of Logic*, 23, pp.155–196.

Whitehead, A.N. and Russell, B. (1910-1913). *Principia Mathematica*. Cambridge University Press, Cambridge. Second edition edition, 1910, 1912, 1913. Abridged as *Principia Mathematica*, Cambridge University Press, Cambridge, 1962.

Epistemic Modalities
GIUSEPPE PRIMIERO

ABSTRACT. I present an analysis of the notion of epistemic modalities, based on an appropriate interpretation of two basic constructivist issues: verification and epistemic agency. Starting from an historical analysis of conditions for judgments, I analyze first the reading of necessity with respect to apodictic judgements, and then that of possibility with respect to hypothetical judgement. The analyis results in a formal treatment of rules for judgemental modal operators, whose aim is to preserve epistemic states corresponding to verified and unverified assumptions in contexts. In the conclusion, further tracks of research are indicated for designing a semantic framework and defining multi-agents systems.

1 Introduction

Since Brouwer's dissertation *Over de grondslagen der Wiskunde*,[1] followed by Heyting's formalization of intuitionistic logic[2] and the later Constructivist perspective,[3] logical anti-realism has evolved from a non-classical setting for the formalization of mathematics to a more extended and demanding conceptual framework for formal logical languages. Constructivism, in particular, has strengthened the philosophical orientation of intuitionistic mathematics and has reformulated some of the most important intuitions behind anti-realism. Nowadays, two issues can be located at the heart of the philosophical analysis that constructivism has brought forth in logic: *verificationism* and *epistemic agency*.[4] The importance of these topics for logic and formal epistemology in general is confirmed by the great deal of attention they are recently claiming in different areas of philosophical and mathematical logic, in game and decision theories and in artificial intelligence and social choice theory. Their combination represents probably the very core of Constructivism intended as a philosophical framework, and the aim of the present contribution is to uncover some connections between a philosophical analysis of constructive logic and knowledge. In particular, I shall give some new insights in

---
[1] Brouwer (1907), republished in Brouwer (1975).
[2] See e.g. Heyting (1956).
[3] For a systematic introduction, see Troelstra and van Dalen (1988).
[4] For a recent analysis of the realism and anti-realism debate which focuses on these two issues among others, see Brock and Mares (2007).

the interpretation of epistemic attitudes by knowing subjects in terms of constructive modalities.

The nature of proofs as ideals for the definition of truth, the reinterpretation of truth-values as proof-conditions and the reading of logical connectives in terms of introduction and elimination rules, have all been crucial steps in the development of the intuitionistic/constructivist approach to the philosophical analysis of logic. In particular, the issue of *verifications*, introduced already by Frege in the treatment of identity procedures for canonical definitions, was dealt with by the Positivist in terms of observation protocols at the beginning of the 1920's,[5] and later by Dummett's theory of proof-conditions.[6] In the anti-realistic perspective, the verification principle 'truth = existence of a proof' requires a solid philosophical basis: What does it mean to prove or verify a given propositional content? Under which *conditions* does such a verification become acceptable? Are degrees of certainty allowed within the definition of truth as possession of a proof? The formulation of appropriate answers needs to be given in terms not only of the formal rules that allow the constructively acceptable derivation of a theorem; It rather concerns also the level of assertions of truth for such propositional contents ('proposition A is true'), and therefore it requires an appropriate analysis of assertion conditions for judgements. In the following, this last topic plays a main role in view of the analysis of modal judgements.

On the other hand, the role of the *epistemic agent* for assertions of knowledge can be traced back to Kant's epistemology. Brouwer's theory of two-ity, based on the method of the creating subject, constituted the philosophical background for the intuitionistic reformulation of the method of proof by contradiction.[7] The issue of the knowing subject has recently been restored according to the constructivist point of view in terms of the 'first-person perspective' approach to the judgement-based process of knowledge acquisition. In view of the recent developments of modal and epistemic logics, where the role of the epistemic actor is explicitly formulated in the language,[8] it would be obvious to expect from the anti-realistic approach (and thus from the constructivist perspective in particular) to play a major role in this constantly growing track of research. This means to allow an explicit analysis of the agent's attitude towards the content of the knowledge act ('proposition A is true/proven by agent a'), and to reconsider in this direction the previously mentioned analysis of assertion conditions. What does it mean for an agent to prove a given propositional content, i.e. under which conditions a content is proven by an agent? Which degrees of provability are admitted in the context

---

[5] See e.g. Ayer (1959).
[6] See Dummett (1977), Dummett (1993).
[7] See van Atten (2008).
[8] The explicit formulation of an agent-operator transforms standard formulas in the form $K_a A$, where K is the epistemic/modal operator; Index a stands for the knowing agent; A stands for the propositional content to which knowledge is directed.

of an agent-based judgemental theory of knowledge? How to deal with the *communication* of these knowledge contents, i.e. how to avoid a solipsistic deviation of such a theory of knowledge?[9] The formal and conceptual analyses offered by frameworks such as Constructive Type Theory (CTT) and the Logic of Proofs (LP) are particularly apt to play this role.[10]

In connection with intuitionistic logic and the structure of proofs, the role of modalities has been recently explored in a large number of research areas.[11] Nonetheless, despite their milestone character in the explanation of knowledge attitudes since the debate on the role of hypothetical judgement in intuitionistic logic from Brouwer (1907), and even though the analysis of epistemic acts has become essential for the cognitive act and the meaning of mood,[12] no consistent explanation of epistemic modalities exists which has a non-purely arithmetical interpretation. The main aim of this paper is hence to focus on an epistemic interpretation of modalities and to analyze them formally from a first-person perspective of judgemental knowledge processes. This shall be done especially in view of the nature of *constructive modalities*, their role for epistemic agency and the analysis of assertion conditions for judgements.[13] I will argue that an epistemic interpretation of modalities is given by a crucial attitude towards knowledge contents, by analysing their assertion conditions. This analysis will not be propositional, in the form '(Agent *a* knows that) *A* is necessary/possible;' rather, it will be given with respect to the related judgemental formulation: '(Agent *a* necessarily/possibly knows that)

---

[9]Thanks to Catarina Dutilh Novaes for suggesting an explicit mention of the problem of solipsism in the present framework of modal judgemental constructive knowledge.

[10]Martin-Löf's Type Theory, first introduced in Martin-Löf (1975), has been revised and reformulated in the strongly predicative format of Constructive Type Theory in Martin-Löf (1998). For comparison with Artemov's Logic of Proof, I shall refer in particular to Artemov (1994) and Artemov (2001). The philosophical and foundational perspective offered by Constructive Type Theory has been investigated at length in a number of aspects. B.G. Sundholm has explored a great number of issues that constitute the basis for the present contribution. In the following I will hit upon themes such as the relation between proof-conditions and truth-makers, considered in Sundholm (1994); the formulation of knowledge judgements from the first-person perspective, as in Sundholm (1997); the relation between the constructive and the classical notions of inference, Sundholm (1998); the issue of identity of assertion conditions, Sundholm (1999). In Sundholm (2003), he has offered a constructive semantics for propositions and judgements which spells out the notion of epistemic necessity: I will, in the following, make an extensive comparative analysis of this work, in order to present a more complete account of epistemic modalities.

[11]Among other works, of the greatest importance are the sequential approach from Sambin and Valentini (1982), the general formulation of an intuitionistic modal logic in Bierman and de Paiva (2000), the interpretation under the Curry-Howard correspondence in Bellin et al. (2001) and the natural deduction reading of contextual reasoning in de Paiva (2003).

[12]As it has been recently considered in van der Schaar (2007) for CTT.

[13]Such explanation refers mainly to the epistemic analysis of dependent judgements for CTT introduced in Primiero (2008); it is crucially based on the description of the role of assumptions and presuppositions, following Primiero (2004), and it focuses explicitly on an agent-based definition of information, as shown in Primiero (2007). The proper semantic interpretation of this notion of information is extensively analyzed in Primiero (2009).

'A is true.' In this latter form, modalities require an analysis that considers the judgement at hand both in its apodictic and hypothetical form. This allows to reformulate their constructive interpretation by proof-object and assertion conditions. The here introduced analysis allows moreover for an extension in terms of prioritized structures and multi-agents languages. The implications of these latter issues shall only be sketched towards the end of this contribution.

2 From Constructions to Conditions

The notion of construction – as the formal counterpart to truth-value for connectives in a classical logic setting – has been considered systematically at least since Heyting's work, and its analysis has received a new impulse with Kreisel's interpretation.[14] This led to the translation of the definition of meaning based on alethic conditions to one based on epistemic conditions.

A crucial step in the formulation of epistemic conditions for propositional connectives is represented by the interpretation of hypothetical judgements. The peculiar nature of such form of judgement was obvious already to Brouwer[15] and it can be taken as the basis for his explanation of the relation between intuitionistic logic and (mathematical) knowledge. In the third part of his dissertation, titled "Wiskunde en Logica," Brouwer deals explicitly with the seemingly innate hypothetical nature of logical reasoning, the one where logic seems to proceed ahead of mathematics:

> "There is a special case, where the combination of syllogism has a different nature, that appears to resemble the usual logical figures, and which really seems to presuppose the hypothetical judgement from logic. This occurs when a construction is defined through some relation in a construction, without being directly evident how to provide it. It seems one assumes here that the sought was constructed, and a chain of hypothetical judgements derives from the assumptions."[16]

The kind of construction involved by hypothetical judgements is central to the understanding of the intuitionistic inference relation. The reading of the implication sign from a BHK-style semantics reflects this very same difficulty. Let us mention two standard interpretations:

> Kreisel (1962): The implication $p \to q$ can be asserted, if and only if we possess a construction r, which, joined to any construction proving p (supposing the latter be effected), would automatically effect a construction proving q;

> van Dalen (1979): A proof p of $A \to B$ is a construction which assigns to each proof q of A a proof $p(q)$ [p, provided that q] of B, plus a verification that p indeed satisfies these conditions.

---

[14] See Heyting (1956), Kreisel (1962); see Sundholm (1983) for an overview.
[15] See Brouwer (1907).
[16] Brouwer (1907), pp.124-125. The explanation of the very different interpretations of the nature of hypothetical judgements in the history of intuitionistic and constructive logic is presented in a quite detailed and fascinating way in van Atten (in press); my personal thanks to Mark van Atten for providing me with a preprint.

According to these standard readings, the satisfaction of the epistemic condition on a proof p for the connective $\to$ applied to the couple of propositional contents A and B requires two separate operations to be performed: the formulation of a construction a for A, along with a construction b for B, the latter being deduced from a in terms of explicitly or implicitly given tautologies.[17] The latter condition on tautologies requires that any step from the first construction to the latter can be performed according to definitions and logical laws.

Sundholm (1983) has argued for the basic distinction between the explanation of an implicational relation and the process of constructing such a proof-object: The hypothetical method – one of the allowed construction methods in intuitionistic logic – is the corresponding abstracted process. Under Sundholm's reasoning, the explanation provided by the formula 'for all q: q proves $A \Rightarrow p(q)$ proves B,' cannot itself be regarded as a mathematical object. This is essentially due to the distinction between the dynamic process that is the act of proving, and the resulting mathematical object that is a proof. Hence, in particular for the construction stating that 'A is a proposition,' the assertion condition cannot be propositional itself.[18]

The way out of this impasse is represented by the constructive distinction between act of knowledge and its content: To demonstrate the truth of a proposition A one needs to carry out the construction a which corresponds to a proof-object for A, which will in turn allow to state the judgement 'proposition A is true.' This sets the basic constructive distinction between proposition and judgement. In turn, also the analysis of conditions is extended: *Proof-conditions* formulated for propositional contents are reconsidered as *assertion conditions* for judgements. I aim to show in the following that the former are explicitely identified with the latter only in the case of *categorical judgements*, and in the case of *hypothetical judgements* only under the explicit requirement of closed constructions. This is shown starting from the basic distinction between implication among propositional contents and inference between judgements.

The just mentioned basic distinction between categorical and hypothetical judgement is fully endorsed by Martin-Löf's Type Theory.[19] The formulation of propositional contents A, B for the judgment '$A \to B$ true,' and justified in terms of constructions, provides the following analysis of the implicational relation:

Proof Conditions-interpretation: A proof p of '$A \to B$ true' is given as the pair of proof-objects $<a,b>$, such that one obtains a formal object of a function type $f = <a,b>$, which is the construction for the implicational relation $f:(A \to B)$.

The ground distinction between this standard constructive implicational relation

---

[17] This is called the α-interpretation in van Atten (in press).
[18] See Sundholm (1983), pp.161-165. As I have recalled in my Primiero (2008), this reflects the Russellian distinction between *knowing-that* and *knowing-how*.
[19] See Martin-Löf (1987) and Primiero (2008), especially chapter 2, section 2.

and the inferential hypothetical relation is expressed by the resulting switch from formal constructions to assertion conditions. Whereas by the implicational relation one obtains a categorical object of the function type $f:(A \to B)$, corresponding to a categorical judgement satisfied by the ordered pair of constructions $< a,b >$; In the assertion conditions interpretation of the inferential relation, one requires a dependent object which represents a new functional relation of the form $f:(x:A)B$; The inferential relation is therefore justified by a formal construction for B whose formulation depends on condition $x:A$.

The construction of such formal object $f:(x:A)B$, is given by the implicational relation abstracted with respect to all possible instances of the construction a:

(1) $$\frac{x:A \vdash b:B}{\lambda((x)b):A \to B.}$$

The object $f:(x:A)B$ denotes a dependent function type, that is the type that contains *all* the functions with domain A and range B such that $f(a):B$ for all a of type A. The values satisfying this function define the hypothetical judgement, or logical consequence, 'If A is true, then B is true:'[20]

Assertion Conditions-interpretation: In order to establish 'A true $\Rightarrow B$ true,' one requires that the satisfaction of the conditions that make the proposition A true, can be transformed constructively into the satisfaction of the conditions that make the proposition B true.

The explicit requirement on the constructive transformation of satisfiable conditions is similar to that of the previous implicational relation (definitions plus logical laws).

This distinction reflects the switch from construction (*proof*) to process-of-construction (*proving*), according to the requirement from Sundholm (1983). Hence, assertion conditions for $f:(A \to B)$ rely on the ordered pairs of constructions $a:A$ and $b:B$ defining f, provided that A and B are both of the type of propositions: $A,B:prop$ (the latter represent the due presuppositions allowing for the required constructions); The last condition on type-introductions expresses type-predicability, which corresponds to *knowability* in view of the definition of truth as *knowledge*.[21] On the other hand, the formal expression $f:(x:A)B$ requires two separate such conditions: the first is the type declaration for B ($B:prop$), allowed by the formulation of a construction $b:B$; The second condition (on which such construction b depends) is the assumption $(x:A)$ that declares that a construction of A is given. In other words, provided the *verification* of the assumption on the truth of A, it is possible to formulate the construction b. Also in this case a principle of knowability is at hand for the conditions on constructions.[22] It is at

---

[20]Such an explanation corresponds to what has been formulated in van Atten (in press) as the β-interpretation.

[21]In Primiero (2004) I have suggested the use of the notion of *meaningfulness* as an appropriate counterpart of the standard verificationist definition of *meaning* for type-introductions.

[22]For more on this and the connection to the notion of function see Primiero (2008), pp.47-54.

this second stage that an important step occurs in the constructive perspective: The introduction of the assumption $x:A$ and the generalization thereof is allowed by term introduction, i.e., by the presence of some a which can be substituted for the place-holder x; This means in other words that the assertion conditions interpretation relies ultimately on the proof conditions of the corresponding implicational relation. This interpretation allows therefore the generalization on the application of identical proof objects $a_1, \ldots, a_n$ for the ordered pair $< a_i, b >$ that build all the valid implicational relations $A \to B$. On the other hand, this reading clearly misses the aim of providing an interpretation for hypothetical reasoning under simple assumption on the formulation of such constructions a_i, that is for the pure reason of expressing what one *should* know, in order the conclusion to be inferred.

There is therefore an important distinction in the shift from constructions to assertion conditions. On what Martin-Löf calls the *conceptual order*, this shows knowability to be presupposed by actual knowledge: This principle, which might result trivial for categorical judgements, reveals the requirement of an explicit *verification procedure* in the case of hypothetical judgements. It opens therefore the up-to-now scarcely explored issue of derivations under open assumptions in the construcitve setting. It is intuitive to translate this epistemic relation in terms of provability and knowledge and to show their connection to the notion of logical necessity. In view of the proposed account, it seems appropriate to reconsider this issue from the perspective of the epistemic interpretation of modalities, including both necessity and possibility. In particular, the reading of epistemic modalities in terms of assertion conditions proceeds in two directions:

1. it preserves the usual reading of necessity satisfying the constructive interpretation of the provability predicate (semantically satisfied in a modal language **S4** for factual truth);

2. it provides an embedded reading of possibility that interprets the agent-based perspective of assertion conditions in a weaker frame for open assumptions.

I shall investigate the resulting notion of constructive knowledge under an interpretation that does not consider necessity as arithmetical provability, rather from the perspective of agent-based justified knowledge, where modalities are intended as expressing epistemic attitudes.

3 Constructive modalities as epistemic attitudes

The explanation of the notions of construction (*process of proving*) and proof (*object that shows what is proved*) in the previous section leads to the distinction between proof-conditions for propositions and assertion conditions for judgements: First one goes through the process of satisfying all the conditions needed for proving a proposition A, and then one obtains the object that allows one to assert that

'A is true.' In the case of a categorical judgement, the notion of assertion condition simply reduces to that of proof-object for the propositional content at hand; On the other hand, whereas the standard constructive interpretation of dependent (hypothetical) judgements obtains the same by requiring the substitution of closed construction for assumptions, the extension to conditions becomes enlightening when looking at a formulation of dependency from open assumptions. This last step is essential to provide an interpretation of modalities as epistemic attitudes for the constructive framework.

Martin-Löf (1996) introduces an analysis of the notion of hypothetical judgement based on the comparison with Gentzen style sequents, considering the required presuppositions for the formulation of antecedents (or hypotheses) $(A_1,\ldots,A_n)$. That 'A_1 is true' is a judgement presupposes, according to such explanation, that A_1 is a proposition; and that 'A_2 is true' is a judgement under the assumption that A_1 holds, presupposes that A_2 is a proposition and that A_1 is true, and so on up to A_n:

$$< prop:type >$$
$$(A_1:prop)A_1\ true$$
$$(A_1:true)A_2:prop$$
$$\vdots$$
$$(A_1:true,\ldots,A_{n-1}:true)A_n:true.$$

This shows the intuitionistic formulation of hypothetical judgements based on presuppositions and satisfied assumptions. An interpretation of modalities as epistemic attitudes for the constructive framework – extending the usual interpretation of the necessity operator with a special reading for the possibility operator – aims at illustrating the connected epistemic values of proof-objects and (open) assumptions.

The first step in formulating the connection between epistemic attitudes and constructive modalities is obviously the direct translation of intuitionistic truth as classical provability in terms of necessity, as introduced by Gödel's modal calculus of provability.[23] As recollected in Artemov (2001), Gödel's introduction of the modal reading of intuitionistic provability establishes that a intuitionistically derivable formula F implies a formula $t(F)$ with proof-term t derivable in **S4** such that each subformula of F is boxed:

$$\vdash_{Int} F \Rightarrow\ \vdash_{S4} t(F)\ |\ \forall A \subseteq F, \vdash_{S4} \Box A.$$

The inverse was established by McKinsey and Tarski (1948). Nonetheless, the intended semantics of the provability operator *Provable*(F) with respect to **S4**

---

[23] Gödel (1933).

and for □F were to be found different. Gödel intended the provability predicate to denote the form 'x is a code of a proof of a formula having a code y' for a theory containing Peano Arithmetic (PA), such that a translation is possible where □ means 'it is provable in PA.' But the problem arises with the non-constructive nature of the existential quantifier, which implies that the reflection principle $Provable(F) \rightarrow F$ is not derivable. The appropriate semantics for □F is then given in Artemov (2001) as the derivability of the explicit version of the provability predicate $Provable(n, F) \rightarrow F$ for any natural number n.[24] The corresponding modal logic of the arithmetical provability predicate $Provable(F)$ was given in Solovay (1976). This has an identical counterpart in the formulation '$Proof(A)$ exists' which defines constructively 'proposition A is true:' 'a is a proof-object for A' ($a:A$) justifies A true.

Considering the more recent approach to modal languages to formulate properties of epistemic subjects involved in intelligent processes of knowledge acquisition and exchange, and provided the innate nature of constructivism to deal with the notion of knowledge and judgement from a first person perspective, it seems reasonable to require a reading of justifications (and in turn of modalities) not only in view of arithmetical provability,[25] but also as epistemic attitudes of knowing agents. Such a non-standard interpretation underlines the distinction between the pure provability of a given propositional content A and the description of conditions under which an agent can prove/has proven A. This is the very aim of the present and following sections. In such context, the mentioned switch from formulas to judgements, leading to the reading of assertion conditions along with proof-conditions, plays an important role for the epistemic interpretation of the standard possibility operator.

Sundholm (2003) presents the different interpretations of the necessity operator applied to the judgemental form 'proposition A is true.' The following different readings are provided:

1. Necessarily A is true;

2. A is necessarily true;

3. 'A is true' is necessary.

In the first form, the judgement declares the truth of the proposition □A; In the second form, it is a form of judgement, where necessity is expressed as a form of predicating the truth; In the third form, a judgement 'A is true' is declared to be necessary. Sundholm claims that modal logic accounts only for the first reading;

---

[24] It also provides a complete axiom system for a classical propositional logic with the additional axiom 't is a proof of T' ($t:T$), and the classical BHK semantics for **Int** is formulated.

[25] See e.g. the one given in Fitting (2005).

The contentual approach required by a constructivist's (anti-formalist) perspective suggests the connection to necessity as truth in all possible worlds, and when the semantic truth-conditions are reflected by syntactic proof-terms in the constructive vein, equi-assertability (or identity of assertion conditions) allows for readings 1 and 2 to be identified. The kind of necessity declared by the third form is different, because it applies in a proper sense to a judgement. The judgement $\Box(A\ true)$, and the related possibility version, are the forms of modal judgements I will refer to throughout the rest of this paper.

The meaning of judgemental necessity $\Box(A\ true)$ needs to be interpreted in terms of assertion conditions. Provided that the conditions for having the right to express a judgement are satisfied, the corresponding notion of necessity for a judgement-candidate is that of an *apodictic judgement*: what is known to be so and cannot be known to be otherwise. Hence, the constructive interpretation that identifies provability, truth and knowledge (that a proposition A is true means that a proof for A is known), allows to justify the following extension:[26]

4. 'A is true' is necessary $\Rightarrow$ 'A is true' is known

$$\Box(A\ true) \Rightarrow K(A\ true).$$

Here K can be seen as a *knowledge-operator*, in the style of the mentioned systems of epistemic logic, or as an explicit operator for the knowing agent. Necessity and knowledge relate here in terms of inference under assertion conditions, rather than by equivalence of such conditions: We shall see that necessary knowledge is implied only by knowledge under an empty set of conditions.

For the previously introduced analysis of assertion conditions, the notion of judgemental knowledge needs to be understood as satisfaction of the conditions for the judgement 'A is true.' The basic condition for the truth of A is the construction a that makes it true $(a:A)$; When A presupposes further propositions to be known, these represent the context in which A is known to be true, $\Gamma = (A_1\ true, \ldots, A_n\ true)$; then the notation $(\Gamma)A\ true$ will be used. Thus, when $\Box(A\ true)$ is referred to contextually formulated knowledge, one needs to give explicit satisfaction procedures for each $x_i : A_i \in \Gamma$, in line with the constructivist requirement. We can see this as providing the condition for the reduction to the implicational relation $\bigwedge A_i \to A$, so that it corresponds to knowledge for which no further contextual conditions are needed ($\Gamma = \emptyset$):

---

[26] Sundholm (2003). In the following, when the symbol $\Rightarrow$ occurs among (modal) judgements ('A is true'), such symbol is not to be intended as the propositional connective of implication, rather as a meta-theoretical sign of inferential assertability. It can be read as follows: If the conditions to assert X are satisfied, then the conditions to assert Y must be satisfied as well (where here X and Y are meta-variable for judgements). Correspondingly, the bidirectional arrow $\Leftrightarrow$ expresses identity of assertion conditions for the judgements on the two sides. Thanks to Maria van der Schaar for having pointed out to me the possible confusion.

5. 'A is true' is necessary ⇔ Agent K knows that A, for any knowledge state agent K is in

$$\Box(A\ true) \Leftrightarrow K((\emptyset)A\ true).$$

The latter represents a crucial step: 'alethic'(/modal-theoretical) necessity as truth in all possible worlds corresponds directly to the 'epistemic'(/proof-theoretical) satisfaction of proof-conditions, including satisfaction of assumptions, which means derivation from premises.[27]

The corresponding interpretation of a judgemental possibility operator can now be provided. Let us consider the propositional equivalence $\Box A \leftrightarrow \neg \Diamond \neg A$, then the following is formulated in Sundholm (2003):

6. 'A is true' is possible ⇔ 'A is false' is not known

$$\Diamond(A\ true) \Leftrightarrow \neg \Box(\neg A\ true).$$

This equivalence is based on the fact that the right-hand side formula expresses that it is not known that 'A is false,' which obviously does not allow constructively to say that 'A is true.' But this use of the possibility operator does not give yet a corresponding interpretation on the syntax of judgements, which also shows how under this reading the duality on the two modal operators is partially lost. One can obtain such translation based on the previous remark on assertion conditions as contextual knowledge, by referring to conditions needed for knowledge that can be satisfied, but not necessarily are:

7. Agent K knows that A, for some knowledge state Γ agent K is in

$$\Diamond(A\ true) \Rightarrow K((\Gamma)A\ true).$$

For this to make a difference with respect to the reading of the necessity operator, we need to interpret the context Γ as the set of data or information on which the construction of A depends, without the reduction to a proof-bject for the implication relation being guaranteed. In other words, one needs to keep the reasoning at the level of assertion-conditions rather than bring it at the level of proof-objects. Only with Γ empty this formula will then reduce to the conditions

---

[27]This interpretation appears to me slightly more effective then the one given in Pfenning and Davies (2001). In this latter work, the context of hypotheses corresponds to a description of the knowledge of a given world; A valid judgement A is one of the form 'A is true in a world about which we know nothing.' This interpretation reflects the knowing subject's attitude towards the apodictic judgment. Nonetheless, it is difficult to understand the formulation of the required demonstration in such unknown world. It seems more intuitive to understand emptiness of a context as indifference with respect to the required conditions; Then it follows that the agent *knows everything needed* for the due construction.

for $\Box(A\ true)$, which provides again the translation to the dual operator. Otherwise, it means that truth is preserved under certain knowledge states in which the agent can formulate the appropriate conditions that need to be satisfied (where $\Gamma = (A_1\ true, \ldots, A_n\ true), n \geq 1$), but without expressing the appropriate constructions. This leads directly to a treatment of derivations from open assumptions.[28]

In Pfenning and Davies (2001), the knowledge that A is possibly true corresponds to the existence of a world in which A is true but nothing else is known; this allows to assume that A is true and that (any) C is possible. This let to draw conclusions on the *possibility* of the propositional content C. In section 4, I shall show how appropriate structural rules can be formulated for a judgemental possibility operator that preserves the meaning of conditions for hypothetical judgements.[29] Under such interpretation, one focuses therefore on the different epistemic conditions at the basis of categorical and dependent judgements, justifying the conceptual distinction between *unverified* and *verified assumptions* within the process of proving. An agent who is always able to verify the assumptions in proof-procedures, is an ideal knower; A real knower shall often be in the condi-

---

[28]From this conceptual justification we introduce the role of the possibility operator for judgements. The syntactic justification obviously requires a system in which terms for the type of propositions can be given both as proper constructions and as assumptions. This clearly would weaken the construcitve nature of the system at least for the module of the language that allows truth to be predicated of non properly proved (assumed) contents. This aim, only sketched in a later section, lies outside of the scope of the present contribution and is addressed in complementary research.

[29]The interpretation of modalities in the form of propositions 5 and 7 presents the immediate and intuitive correspondence also with an intuitionistic model for Kripke semantics, requiring a tuple $\langle W, \leq, R, v \rangle$ where R is the usual accessibility relation over the ordered set of worlds $W, \leq$, with worlds being intended as epistemic states, and the function v evaluating necessity and possibility formulas as accessibility respectively in all and in some orderly accessible world. Under the epistemic reading, assertion conditions are reduced to the accessibility relations on epistemic states for the formulas $K_a A$ (agent a knows A) and $B_a A$ (agent a believes A), where epistemic states substitute 'ontological worlds' and the semantic definition on K/B-operators are dictated by the different clauses for defining respectively proven and assumed truths. This gives also a new impulse on the distinction between knowledge and belief, started with Hintikka (1962), where the system $S4$ is chosen to express both knowledge/belief of a propositional content as compatibility with previous knowledge/belief; Compatibility amounts in turn to nothing else than accessibility by an epistemic relation on worlds. Even stronger systems have been proposed: in van der Hoek (1996) the system $S4.3$; in Fagin et al. (1995) and van Ditmarsch et al. (2006) the system $S5$ to include negative introspection. In general, in epistemic logics the meaning of necessity and possibility for epistemic states has been variously interpreted: Sometimes one speaks in terms of the distinction between knowledge and belief; or corresponding notions of hard/soft information are called upon; or one analyzes the persistence of the given contents in possible epistemic alternatives. In turn, knowledge is sometimes interpreted as a strong notion that requires alternatively truth, correctness or verification on propositional contents; whereas belief asks for some sort of individual, weak confirmation attached to contents. The standard interpretation of accessibility on propositional contents is preserved in the judgemental interpretation from Pfenning and Davies (2001), where modal propositional operators are still present, and no full explanation of the corresponding epistemic attitudes for the first-person perspective approach is required. Such explanation is here obtained by an analysis of the verification principle of truth defining knowledge under conditions.

tion of formulating contents only assuming their conditions being satisfied (e.g., because so a reliable source says), but in practice being unable to provide appropriate verifications.[30] These are the different epistemic attitudes analyzed by modal judgements. I shall in the following unfold this interpretation and show how this allows to formulate appropriate definitions of constructive epistemic attitudes.

4 The meaning of satisfied conditions

The aim of the present section is to complete the process of translating the meaning of epistemic attitudes into constructive modalities. To formulate such an interpretation, I shall focus on the notion of conditions on constructions for propositional contents (as suggested by the act/object of knowledge distinction); This will provide an appropriate syntactic reading for judgements of the form $\Box(A\ true)$ and $\Diamond(A\ true)$.

Let us start again from the judgemental reading of the necessity operator suggested in Sundholm (2003):

4. 'A is true' is necessary $\Rightarrow$ 'A is true' is known.

Let us recall that this translation is based on the identity between $A\ true$ and $Proof(A)$. The corresponding epistemic attitude for constructive necessity is given by the apodictic judgement in which the truth of A is asserted under an empty context of conditions $(\Gamma = \emptyset)A\ true$, where one has reduced to implication (and thus also to an empty context of assumptions) the case of hypothetical reasoning. This means that conditions for the truth of the propositional content are already satisfied by the formulation of the judgement, which in turn provides an appropriate proof-object.

In correspondence with the previous reading in 5, I shall now focus on the relation between possibility and conditions via knowledge (rather than necessity and conditions): If necessity corresponds to knowledge under no extra conditions (than proof), possibility shall be related to knowledge under *some* assertion conditions to be satisfied. In both cases, we have a relation among knowledge and conditions, and the epistemic attitude of knowing agents in the first person perspective can be introduced, so that K is no longer a knowledge-operator, rather an agent-operator:

5'. 'A is true' is known $\Leftrightarrow$ Agent K has satisfied all conditions to know that A is true.

---

[30]The issue of reliability of sources is a very important one in this context. Obviously, e.g. in a mathematical context, researchers rely on each other's knowledge very often, for example in assuming the content of a given theorem proven by someone else. The generalization here presented, is based on the import of the first-person perspective principle, where the ideal knower may be a given epistemic agent or the whole of an epistemic community. In each case, it is relevant to stress the different values and roles that contents have in the process of knowledge. A next stage in this research is the analysis of prioritized structures of information sources, by which it is possible to formalize the relation of dependency between the knowing agent and his sources in a particular relevance order.

Under this reading, necessity is reflected as *actual knowledge* in the following syntactic representation (where J is the judgement stating 'A is true' and Γ a set of assumptions $x_n : A_n$):[31]

(2) $\Box - Rule \qquad \dfrac{\Gamma \vdash J}{\Box \Gamma \vdash \Box J}$

By this rule, one accounts for judgements whose conditions have been verified, therefore allowing provability of the conclusion. It says that, given the derivability of the judgement 'A is true' under the list of (minimal) conditions expressed by Γ, the satisfaction of proof conditions for each element in Γ (interpreted as substitutions of proof-variables: $([x_1/a_1]:A_1,\ldots[x_n/a_n]:A_n)$) allows to formulate proof condition a such that it makes A known.[32] This $\Box$-Rule is justified in a different way than the necessitation rule introduced in Pfenning and Davies (2001). In the latter, an introduction rule for (contextual) boxed formulas has the following form:[33]

(3) $\dfrac{\Delta; \cdot \vdash J}{\Delta; \Gamma \vdash \Box J}$

it means that propositional necessity is implied by judgmental validity: If J is justified by any context, then it is necessarily justified. The corresponding elimination rule allows to use a judgement J as an assumption in context once $\Box J$ has been derived:

(4) $\dfrac{\Delta; \Gamma \vdash \Box J \quad \Delta, J, \Gamma \vdash J_1}{\Delta; \Gamma \vdash J_1}$

which means to extend contexts in terms of proved formulas. Obviously, this is valid also in our interpretation: By the previous $\Box$-Rule in equation 2, the derivation of the formula $\Box J$ means that related judgmental assertion conditions (in contexts) have all been boxed (proved), therefore the content of J can be safely used in a context of assumptions.[34] The extension of context by new formulas still preserves derivability of $\Box J$ if its minimal conditions have been satisfied, provided that no contradictory extension can be allowed of the initial context.

---

[31] See Bellin et al. (2001).

[32] For the problem of composition of boxed formulas see de Paiva (2003).

[33] In the following, the original notation from Pfenning and Davies (2001) has been modified to conform to ours. J is equivalent to A *true* expressed on empty context, from which one can infer A *valid*, i.e. judgmental necessity; their $\Box J$ is equivalent to $(\Box A)$ *true* and it refers to the propositional form of necessity – whereas we always look at judgements of the form $\Box(A\ true)$. The notation '·' refers to an empty context.

[34] In the following section I shall introduce hypotheses by a formal rule in the style of Pfenning and Davies (2001) but with the basic difference of formulating it both for the $\Box$ and the $\Diamond$ version (reflecting the verified/unverified distinction).

An important issue at this stage is the iteration of the necessity operator, which is usually interpreted as positive introspection by axiom 4 in epistemic logic, and as exponentiation in the arithmetical interpretation of provability. According to the previous analysis introducing the first-person perspective on knowing acts, one can easily substitute justified truth by satisfaction of categorical constructions by means of the necessity operator and to extend it by adding explicitly an index that expresses the knowing agent:

4'. $\Box(A\ true) \Leftrightarrow \Box_K(A\ true)$.

This formulation says that 'A is true' is necessary if and only if there is an agent K satisfying all the due conditions in order to formulate a proof for A. Provided $\Box(A\ true)$ is justified by a formula $a:A$, the identity with the operator $\Box_K$ allows the reduction to the standard possible iteration of modalities. That is, the following derivation is sound:

1	$a:A$	PREM
2	$A\ true$	verification principle of truth
3	$\Box(A\ true)$	(5)
4	$K(A\ true)$	(4)
5	$\Box_K(A\ true)$	(4')
6	$\Box_K \Box(A\ true)$	(4,5)
7	$\Box_K \Box_K(A\ true)$	(4')

The formula $\Box_K \Box(A\ true)$, as the basic formulation for the iteration of modalities in the perspective of epistemic attitudes, says that 'Agent K knows that she has a proof object for A.' The formula 4' allows to reduce this to the usual positive introspection, being existence of a proof-object admissible only if an agent K is able to formulate it. Even though the reduction is admissible for proven contents (necessity judgements), the meta-theoretical structure is crucial with respect to the iteration of the $\Diamond$-operator: When switching to knowledge of the conditions that make a hypothetical judgement true, one cannot express them – as in the previous case – by iteration of $\Box_K$, which implies direct satisfaction (as a provability operator). From the present perspective, if one wants to preserve the different attitudes expressed as 'knowing that' and 'assuming that' (otherwise also expressible as: 'receiveing the information that'), it is crucial to express the difference between conditions for knowing and conditions for the possibility of a propositional content.[35]

---

[35] This argument is particularly relevant in view of the development of a multi-agent system, by which one wants to formulate knowledge of a propositional content possessed by one agent and transmitted to another (which might not possess the corresponding verification object). This point is obviously linked

As I hope to make clear in what follows, there are two sorts of *desiderata* that one wants to satisfy when the interpretation of necessity is connected to the verification procedure of assertion conditions:

- to make the necessitation of the consequent dependent on the verification of the assumptions; This is formally obtained by the introduction rule for $\Box J$;

- consequently, to provide a $\Diamond$-introduction rule as an appropriate counterpart of the $\Box$-elimination rule: Not only one wants to be able to extend contexts by valid judgments, rather one also wants to make explicit extensions of contexts by *unverified assumptions*, preserving their epistemic value in the conclusion.

5 Interpreting possibility as satisfiable conditions

In correspondence with the list $1 - 3$ of interpretations of necessity given in Sundholm (2003), the following list of meanings for possibility are formulated:

1'. Possibly A is true;

2'. A is possibly true;

3'. 'A is true' is possible.

As in the previous analysis, let us focus on the judgemental form 3', to show how it corresponds to a reading in terms of assertion conditions. The proper counterpart to the notion of *apodictic judgement* – covering also the notion of judgement formulated under satisfied conditions – is obviously the one of *hypothetical judgement* with open assumptions: what is known to be so, but can be known to be otherwise; In particular, it can be otherwise if its conditions are not satisfied. The final aim is therefore to explicitate the appropriate epistemic attitude formulated by the constructive interpretation of hypothetical judgement under the formulation of the possibility modality.

The sentence in 5' from the previous section, implicitly expressing the condition on the formula '$Proof(A)$ exists' as the categorical judgement of the form 'a is a proof for A' $(a:A)$, turns explicitly as follows in the interpretation of truth under assumptions:

5''. 'A is true' is known, provided knowledge of contents $(A_1,\ldots,A_n)$ $\Leftrightarrow$ Agent K knows that A is true, *if* K satisfies conditions $(A_1,\ldots,A_n)$.

---

to the already mentioned problem of solipsism and to the resulting epistemic attitude of 'becoming informed,' see Primiero (2009). More on this in the concluding section. Thanks to Bjørn Jespersen, for urging an exposition of the problem of introspection at this particular stage of the discussion on epistemic modalities.

The case of 5″ refers (explicitly) to a hypothetical judgement of the form 'provided that (all and only) constructions a_1 for A_1 up to a_n for A_n are satisfied, a is a proof for A.' In this last reading, one implicitly refers to a minimality property on the conditions needed to satisfy the verification of A; More to the point, deviating from the strictly constructive meaning of hypothetical judgements, one allows here the formulation of construction a assuming that constructions $a_1, \ldots, a_n$ are given. The constructive interdefinability of possibility and necessity is correspondingly translated as follows:

6′. $\Diamond(A\ true) \Leftrightarrow$ in some minimal world the conditions for $A\ true$ are satisfied

$$\Diamond(A\ true) \Leftrightarrow \exists \Gamma(\neg\Box((\Gamma)\neg A\ true)).$$

The previous means that, considering conditions for hypothetical judgements, there is a list of assumptions $\Gamma = (A_1, \ldots, A_n)$ such that if these are verified, no pair composed with them and a construction a can be formulated such that $\neg A$ holds true. Consider that, according to this translation of possibility under assertion conditions and for the forthcoming analysis, our $\Diamond A$ collapses into the standard possibility operator of modal logic only when $\Gamma = \emptyset$ (that is only for the case of categorical apodictic judgement).

The corresponding positive reading under assertability conditions is as follows:

7′. If (all and only) conditions $\Gamma = (A_1, \ldots, A_n)$ are satisfied, a proof can be formulated such that agent K knows that A

$$\exists \Gamma, s.t.\ (\Gamma)a:A\ \text{and}\ K(\Gamma) \Leftrightarrow K(A\ true).$$

The explanation of assertion conditions for possibility finally amounts to:

8. 'A is true' is knowable $\Leftrightarrow$ Conditions for A are satisfiable

$$\Diamond(A\ true) \Leftrightarrow \exists \Gamma, s.t.\ (\Gamma)a:A\ \text{and}\ \Diamond(\Gamma).$$

The latter in turn means that a list of assumptions can be formulated such that its knowledge makes A true, and thus A is *coherently assertable* in a hypothetical context (again, only in the case $\Gamma = \emptyset$ the previous reduces to the case of the apodictic judgement). Under the first person perspective interpretation and the reading of conditions on constructions for judgements, knowability of 'A is true' means that A can be satisfied, and therefore it can be used to satisfy conditions on further constructions, as explained in the conceptual order among constructions and conditions in the previous section.

The definition given in Pfenning and Davies (2001) of possibility judgments valid under assumptions is again a useful starting point for a better understanding of our notion of possible knowledge. Their definition preserves possibility under validity (which is intuitive, because it allows verified formulas to be formulated in hypothetical contexts), and it defines possibility with necessity as a substitution principle. The calculus in Pfenning and Davies (2001) defines a propositional operator ($\Diamond$) along with a judgemental predicate (*poss*):

- If $\Gamma \vdash A$ *true* then $\Gamma \vdash A$ *poss*;

- If $\Gamma \vdash A$ *poss* and A *true* $\vdash C$ *poss* then $\Gamma \vdash C$ *poss*.

In this definition, the *poss* predicate is judgemental in the same sense the *true* predicate is judgemental, that is, it forms a judgement A *poss*. Its interpretation based on necessity is obtained by allowing assumptions about validity (where, again, validity is interpreted as necessity):

- If $\Delta; \Gamma \vdash A$ *poss* and $\Delta; A$ *true* $\vdash C$ *poss*, then $\Delta; \Gamma \vdash C$ *poss*.

In the present context I focus instead on the judgemental possibility intended as possibility applied to a judgement: $\Diamond(A\ true)$. The main property in common among the two analyses is that assumptions of validity (i.e. formulas A *true* in contexts) are extended in order to derive further possible contents (formulas C *poss* in the analysis from Pfenning and Davies (2001); formulas of the form $\Diamond(A\ true)$ in the present context[36]). This desirable property is complementary to the $\Box$-elimination rule in the previous analysis of the necessitation procedure. The validity of assumptions considered in Pfenning and Davies (2001) corresponds to our explicit formulation of *verified assumptions*, namely the derivation-tree that goes from Γ to $\Box\Gamma$, where $\Gamma = (x_1:A_1,\ldots,x_n:A_n)$. The extension that preserves *unverified assumptions*, inducing possibility on the conclusion, is given as follows (as in the previous rule, J stands for a formula of the form A *true*):

(5) $\Diamond - Rule$ $\quad \dfrac{\Gamma, J_1 \vdash J_2}{\Box\Gamma, \Diamond J_1 \vdash \Diamond J_2.}$

This reading extends the previous interpretation of necessity as proof-conditions to the assertability conditions of hypothetical judgements: It preserves the formulation of knowledge contents epistemically weaker than strictly proved ones. Possibility expresses thus – in the present context – the knowability of contents

---

[36] It is not my aim here to investigate to which extent the judgemental formulation C *poss* is relevantly different from the propositional $\Diamond C$ in Pfenning and Davies (2001).

under *assumption of appropriate assertion conditions*.[37] This second rule accounts therefore for the *transmission* of contents in the knowledge frame without explicit proof/verification, by referring to the epistemic modality expressed by the ◇-operator.[38] This formulation has the corresponding agent-based abbreviation of the formula at 7:

7″. $\Diamond(A\ true) \Leftrightarrow \Diamond_K(A\ true)$

saying that 'A is true' is possible if and only if there is an agent K who can formulate the due conditions needed to assert that A is true. This condition expresses the ability to tell what would be needed in order to know that A (without the verification procedure required by the corresponding necessitation attitude); The object of knowledge for the agent's epistemic state is obviously given by the formulation of construction a under conditions in Γ.[39]

To sum up, consider the third reading of the □ operator in its application to the judgemental form ('A is true' is necessary) from the previous section. It satisfies the notion of apodictic judgement: A proposition known to be true and which cannot be otherwise, being its assertion conditions necessarily satisfied. The relation here described is of an analytical nature. The role of the ◇-rule is complementary. It refers to knowledge assertions formulated on the basis of a set of assumptions: If knowing the truth of a formula depends on the validity of a set of assumptions, then the instantiation of these assumptions constraints to the knowledge of the given formula. The epistemic value of open assumptions expresses the notion of acquired information, where contents that might not be explicitly verified are used to coherently extend a knowledge base. If the derivability of a judgement is valid under extension of its assertion conditions by a further judgement, then the inferred judgement becomes dependent on the verification of the new assumptions. The relation here described is of a synthetic nature.

---

[37] This rule is introduced also for the calculus presented in Bellin et al. (2001). The modal analysis presented in Primiero (2009) extends the application to the syntactic-semantic method of CTT in the same direction, and as such it provides an adequate modal reading of the constructive calculus of dependent judgements. It shows how to interpret axiom **B** as the core of **S4** in which the logic of dependent proofs can be formulated.

[38] The procedure of transmission of contents (already mentioned in relation with the problem of solipsism and introspection) is not analyzed in the present paper, but it describes in an intuitive way the various processes of communication based on reliability, trust, signed messages. An extension of the present framework on the basis of prioritized contexts shall cover this important aspect of multi-agent knowledge processes. See the final section for some further remarks.

[39] It is maybe useful to mention that the identification of the notion of dependent condition with the **B**-axiom from Primiero (2009), referred to in footnote 37, is at this stage easily justified: $A \rightarrow \Box\Diamond A$ means that the truth of A implies that the assertion conditions for A (i.e. formally $\Diamond A$) have been entirely verified (i.e. the necessitation imposed by the □-operator).

6 The formal system

I shall now briefly formulate the formal language for the calculus of modal dependent judgements. Standard types for propositions and formulas in contexts are introduced as axioms. Constructed formulas are standardly given by propositional closure. Judgments formulated within a *context* are considered assumptions or hypotheses are introduced in terms of elements of the set of proof-variables ($Var = \{x_1,\ldots,x_n\}$); judgments for the $prop : type$ are justified in terms of elements of the set of proof-constants ($Con = \{a_1,\ldots,a_n\}$). In order to satisfy the requirement on non-reducibility of the former to the latter and thus to the implicational relation, one needs to restrict the truth relation for categorical judgements and to allow hypotheses to be of the type of propositions without appropriate construction being already provided. The standard judgemental grammar is then extended by using a different truth predicate when introduced by non-contradictory assumptions; finally, modal judgements are introduced:[40]

$prop : type$
$context : type$
$A : prop$
$J ::= A\ true \mid (A_1 \wedge A_2)\ true \mid (A_1 \vee A_2)\ true \mid (A_1 \supset A_2)\ true \mid (A \supset \bot)\ true;$
$a : A := A\ true$
$(\neg(A \supset \bot)) \supset x : A := A\ true^*$
$\Gamma, \Delta : context$
$\Gamma ::= (x_1 : A_1, \ldots, x_n : A_n); \Delta ::= (x_{n+1} : A_{n+1});$
$mod(J) ::= \Box J \mid \Diamond J.$

Whereas Introduction and Elimination Rules for the propositional connectives are standard, let us mention here only the formulation of rules for modal judgments. The standard hypothesis rule will be extended by rules for introducing modal judgments as verified/unverified assumptions; I shall accordingly change the nomenclature as follows:

(6) Premise Rule $\quad \dfrac{}{\Gamma, \Box J, \Delta \vdash \Box J}$

(7) Hypothesis Rule $\quad \dfrac{}{\Gamma, \Diamond J, \Delta \vdash \Diamond J}$

The $\Box$- and $\Diamond$-Rules are finally given as follows:

(8) $\Box$ – Rule Introduction $\quad \dfrac{\Gamma \vdash J}{\Box \Gamma \vdash \Box J}$

---

[40] An extended version of the following formal analysis is contained in Primiero (Technical Report 1/09).

(9) □ − Rule Elimination $\dfrac{\Box\Gamma \vdash \Box J}{\Gamma, \Delta \vdash J}$

(10) ◇ − Rule Introduction $\dfrac{\Gamma, J_1 \vdash J_2}{\Box\Gamma, \Diamond J_1 \vdash \Diamond J_2}$

(11) ◇ − Rule Elimination $\dfrac{\Gamma, \Delta \vdash J_1 \quad \Box\Gamma, \Diamond J_1 \vdash \Diamond J_2}{\Gamma, \Delta \vdash J_2}$

7 Sensible extensions for the epistemic attitudes framework

The analysis of epistemic possibility carried out in the present contribution lies crucially on the understanding of the notion of knowledge and the related necessitation rule. In particular, it refers to the role of assumptions and the explanation of the notion of dependent derivation. The introduction of the latter (which in CTT is completely satisfied by the formulation of dependent types with substitution rules on place-holders) requires that the implicational relation from truth to necessary truth be translated in terms of the corresponding epistemic version: From known content to necessarily known content. This relation is weakened in view of the formulation of proven contents under open assumptions.[41]

A calculus that includes categorical and dependent (open) judgements needs to be appropriately tuned with respect to the mentioned epistemic modalities. The standard constructivist view on hypothetical reasoning makes knowledge of the conclusion dependent on knowledge of the premises, thus preserving knowability.[42] In the present context, the role of assumptions is taken in a more strict sense: By the use of modalities, one is able to import in the language the distinction between *verified* premises and *unverified* assumptions, reflected by the corresponding epistemic attitudes formulated by the knowing agent. Moreover, it is well-known that dependent knowledge in this sense describes the kind of effective knowledge processes advocated e.g. by natural deduction systems and derivations by contexts. The introduced modal rules are equivalent to those in Bellin et al. (2001) for the calculus of constructive modal logic **IK**, extended by the needed hypothesis rules. In such a system, axioms for intuitionistic logic hold, plus an axiom that allows for the distribution of the □-operator on implication; On the other hand, distribution of the ◇-operator on disjunction is discarded. The further step of this research is obviously represented by the formulation of a corresponding Constructive Kripke semantics, that be sound and complete with respect to the syntactic representation here introduced. It seems safe at this stage to suggest that such semantics might be defined by a set of different accessibility relations on a subset of the worlds

---

[41] See Hakli and Negri (2008) for the role of this distinction in the formulation of the Deduction Theorem. In Primiero (2009) I have considered the relevance of this distinction for the problem of logical omniscience.

[42] See Martin-Löf (1996), Sundholm (1997).

representing those where contents would hold, were the appropriate conditions be satisfied.[43]

A recent result in Kramer (2008) has shown the reducibility of provability to knowledge, referring to the identity between provability and "a combination of individual knowledge (knowledge of messages), plain propositional knowledge, common knowledge (propositional knowledge shared in a community of agents) and a new kind of knowledge, namely adductive knowledge (propositional knowledge contingent on the adduction of certain individual knowledge, e.g. through oracle invocation)."[44] This shows that, in the context of information exchange for multi-agent systems, the notion of knowledge requires transfer of (signed) messages and (signed) proofs. This reflects the very same distinction here underlined between the epistemic attitudes towards proved and assumed contents. In turn, this suggests that a modal type-theoretical framework, including appropriate judgments for necessity and possibility, can be extended in view of a contextual dynamics to a multi-modal version. In this way, the role of messages is played by update dynamic operations with assumed formulas in contexts. The corresponding semantic interpretation would be designed by a set of indexed accessibility relations (for the agents) on the subset of possible worlds, on whose basis respectively distributed and common knowledge can be modelled.

8 Acknowledgments

In the course of the last years, during my PhD and later on, Göran Sundholm has been constantly a source of inspiration and challenge on topics that were interesting for us both. I will remain grateful for what I have learned from him, and for the help received in learning to do philosophy in an independent and autonomous way. Catarina Dutilh Novaes, Bjørn Jespersen, Hans Lycke, Dagmar Provijn and Maria van der Schaar have all been very kind in reading previous drafts of this paper, everyone suggesting improvements and requiring clarifications that surely made this text a better one. Everything left unclear or incorrect, is the author's own fault.

BIBLIOGRAPHY

Ayer, A.J. (1959). *Logical Positivism*. The Free Press, New York.

Artemov, S. (1994). "Logic of proofs", *Annals of Pure and Applied Logic*, 67(2), pp.29–59.

Artemov, S. (2001). "Explicit provability and constructive semantics", *Bulletin of Symbolic Logic*, 7(1), pp.1–36.

---

[43] See Primiero (Technical Report 2/09)
[44] Kramer (2008), p.1.

van Atten, M. (2008). *The development of intuitionistic logic*, in Edward N. Zalta, editor, *The Stanford Encyclopedia of Philosophy*. Fall 2008.

van Atten, M. (in press). *On the hypothetical judgement in the history of intuitionistic logic*, in *Logic, Methodology and Philosophy of Science XIII*, College Publications, in press.

Bellin, G., de Paiva, V. and Ritter, E. (2001). "Extended Curry-Howard correspondence for a basic constructive modal logic, " Preprint; presented at M4M-2, ILLC, UvAmsterdam.

Bierman, G.M. and de Paiva, V. (2000). "On an intuitionistic modal logic", *Studia Logica*, (65), pp.383–416.

Brock, S. and Mares, E. (2007). *Realism and Anti-Realism*. Acumen, Stocksfield.

Brouwer, L.E.J. (1907). *Over de grondslagen der Wiskunde*. Maas & van Suchtelen, Amsterdam. Republished by Mathematisch Centrum, Amsterdam, D. van Dalen (ed.) 1981; 276 pp.; and by Epsilon, Utrecht, D. van Dalen (ed.) 2001, 2005.

Brouwer, L.E.J. (1975). *L.E.J. Brouwer. Collected Works, vol.I*. North-Holland, Amsterdam.

van Dalen, D. (1979). "Interpreting intuitionistic logic", in Baayer, v.Dulst, and OOsterhoof, editors, *Proc. Bicentenical Congree Wiskundig Genootschap*, volume 1 of *Mathematical Centre Tracts*, pp.133–148. Mathematisch Centrum, Amsterdam.

van Ditmarsch, H., van der Hoek, W. and Kooi, B. (2006). *Dynamic Epistemic Logic*. Synthese Library Series, Springer, Dordrecht.

Dummett, M. (1977). *Elements of Intuitionism*, Oxford University Press, Oxford.

Dummett, M. (1993). *The Seas of Language*, Oxford University Press, Oxford.

Fagin, R., Halpern, J.Y., Moses, Y. and Vardi, M.Y. (1995). *Reasoning about Knowledge*, MIT Press, Cambridge.

Fitting, M. (2005). "A logic of explicit knoweldge", in *The Logica Yearbook 2004*, pp.11–22, Filosofia Publisher, Institute of Philosophy, Czech Academy of Sciences, Prague.

Gödel, K. (1933). "Eine interpretation des intuitionistischen aussagenkalküls", *Ergebnisse des Mathematischen Colloquium*, 4, pp.39–40.

Hakli, R. and Negri, S. (2008). "Does the deduction theorem fail for modal logic?", manuscript.

Heyting, A. (1956). *Intuitionism: An Introduction*. North-Holland, Amsterdam.

Hintikka, J. (1962). *Knowledge and Belief: An Introduction to the Logic of the two Notions*. Cornell University Press, Cornell.

van der Hoek, W. (1996). "Systems for knowledge and belief", *Journal of Logic and Computation*, 3, pp.173–195.

Kramer, S. (2008). "Reducing provability to knowledge in multi-agent systems", in *Proceedings of the Workshop on Intuitionistic Modal Logic and Applications (IMLA08)*, Technical Report MSR-TR-2008-90. Microsoft Research, Elsevier Science, Amsterdam.

Kreisel, G. (1962). *Foundations of intuitionistic logic*, in E. Nagel, P. Suppes, and A. Tarski, editors, *Proceedings of the Logic, Methodology and Philosophy of Science Congress, 1960*, pp.198–210, Stanford University Press, Palo Alto CA.

Martin-Löf, P. (1975). "An intuitionistic theory of types: predicative part", In H.E. Rose and J C. Shepherdson, editors, *Proceedings of the Logic Colloquium '73*, volume 80 of *Studies in Logic and the Foundations of Mathematics,*, pp.73–118, North-Holland, Amsterdam.

Martin-Löf, P. (1987). "Truth of a proposition, evidence of a judgement, validity of a proof", *Synthese*, 73(3), pp.407–420.

Martin-Löf, P. (1996). "On the meanings of the logical constants and the justification of the logical laws", *Nordic Journal of Philosophical Logic*, 1(1), pp.11–60.

Martin-Löf, P. (1998). "An intuitionistic theory of types", in G. Sambin and J. Smith, editors, *Twenty-five Years of Constructive Type Theory*, pp.127–172. Clarendon Press, Oxford.

McKinsey, J.C.C. and Tarski, A. (1948). "Some theorems about the sentential calculi of Lewis and Heyting", *The Journal of Symbolic Logic*, 13, pp.1–15.

de Paiva, V. (2003). *Natural Deduction and Context as (Constructive) Modality*, volume 2680 of *Lecture Notes in Computer Science*, page 1025, Springer, Dordrecht.

Pfenning, F. and Davies, R. (2001). "A judgmental reconstruction of modal logic", *Mathematical Structures in Computer Science*, 11, pp.511–540.

Primiero, G. (2004). *Presuppositions, assumptions, premises*, Master's thesis, Universiteit Leiden.

Primiero, G. (2007). "An epistemic constructive definition of information", *Logique & Analyse*, 50(200), pp.391–416.

Primiero, G. (2008). *Information & Knowledge*, volume 10 of *Logic, Epistemology and the Unity of Sciences*. Springer, Dordrecht.

Primiero, G. (2009). "An epistemic logic for becoming informed", *Synthese*, 167(2), pp.363–389.

Primiero, G. (TR1/09). "A constructive judgmental modal calculus for provability up to refutation", Technical report, Centre for Logic and Philosophy of Science, Ghent University.

Primiero, G. (TR2/09). "A constructive modal semantics for contextual verification", Technical report, Centre for Logic and Philosophy of Science, Ghent University.

Sambin, G. and Valentini, S. (1982). "The modal logic of provability. The sequential approach", *Journal of Philosophical Logic*, 11, pp.311–342.

van der Schaar, M. (2007). "The assertion-candidate and the meaning of mood", *Synthese*, 159(1), pp.61–82.

Solovay, R.M. (1976). "Provability interpretations of modal logic", *Israel Journal of Mathematics*, 25, pp.287–304.

Sundholm, B.G. (1983). "Constructions, proofs and the meaning of logical constants", *Journal of Philosophical Logic*, 12, pp.151–172.

Sundholm, B.G. (1994). "Existence, proof and truth-making", *Topoi*, 13, pp.117–126.

Sundholm, B.G. (1997). "Implicit epistemic aspects of constructive logic", *Journal of Logic, Language and Information*, 6, pp.191–212.

Sundholm, B.G. (1999). "Identity: Propositional, criterial, absolute", in *The Logica 1997 Yearbook*, pp.26–35, Filosofia Publisher, Czech Academy of Science, Prague.

Sundholm, B.G. (1998). "Inference, consequence, implication: a constructivist's approach", *Philosophia Mathematica*, 6, pp.178–194.

Sundholm, B.G. (2003). "Mind your P's and Q's. On the proper interpretation of modal logic", in *The Logica 2002 Yearbook*, pp.101–111, Filosofia Publisher, Czech Academy of Science, Prague.

Troelstra, A.S. and van Dalen, D. (1988). *Constructivism in Mathematics: An Introduction*, 2 vols. North-Holland, Amsterdam.

Kripke Completeness Revisited

SARA NEGRI

ABSTRACT. The evolution of completeness proofs for modal logic with respect to the possible world semantics is studied starting from an analysis of Kripke's original proofs from 1959 and 1963. The critical reviews by Bayart and Kaplan and the emergence of Henkin-style completeness proofs are detailed. It is shown how the use of a labelled sequent system permits a direct and uniform completeness proof for a wide variety of modal logics that is close to Kripkes' original arguments but without the drawbacks of Kripke's or Henkin-style completeness proofs.

1 Introduction

The question about the ultimate attribution for what is commonly called Kripke semantics has been exhaustively discussed in the literature, recently in two surveys (Copeland 2002 and Goldblatt 2005) where the rôle of the precursors of Kripke semantics is documented in detail.

All the anticipations of Kripke's semantics have been given ample credit, to the extent that very often the neutral terminology of 'relational semantics' is preferred. The following quote nicely summarizes one representative standpoint in the debate:

"As mathematics progresses, notions that were obscure and perplexing become clear and straightforward, sometimes even achieving the status of 'obvious.' Then hindsight can make us all wise after the event. But we are separated from the past by our knowledge of the present, which may draw us into 'seeing' more than was really there at the time."[1]

We are not going to treat this issue here, nor discuss the parallel development of the related algebraic semantics for modal logic,[2] but instead concentrate on one particular and crucial aspect in the history of possible worlds semantics, namely the evolution of completeness proofs for modal logic with respect to Kripke semantics.

Kripke published in 1959 a proof of completeness for first-order **S5** and in 1963[3] an extension of the method to cover the propositional modal systems **T**,

---
[1] Goldblatt (2005), section 4.2.
[2] See Jonsson and Tarski (1951).
[3] The results of Kripke's 1963 paper had already been announced in an abstract published in 1959.

S4, **S5**, and **B**. His method employed a generalization of Beth tableaux and completeness was established in a direct and explicit way by showing how a failed search for a countermodel gives a proof.

Kripke's proof was criticized in a review by Kaplan as lacking in rigor and as making excessive use of 'intuitive' arguments on the geometry of tableau proofs. Kaplan suggested a different, more 'mathematical' and more elegant approach based on an adaptation of Henkin's completeness proof for classical logic. Indeed, a Henkin-style completeness proof for **S5** had already been published in 1959 by Bayart and other proofs were published at the time of the review or shortly after.[4]

Henkin-style completeness proofs were since then preferred in the literature on modal logic, even for labelled systems. The explicit character of Kripke's original proof, that constructs countermodels for unprovable formulas, is lost with the Henkin approach.

The purported mathematical elegance of a proof of completeness in the Henkin style resembles a well-designed trick. The proof gives no way to obtain derivability from validity, nor does it show how to construct a countermodel for underivable propositions. Other specific problems arise for modal logic. For instance, in systems with an irreflexive accessibility relation, as those needed in temporal logic, the canonical accessibility relation need not be irreflexive. Some extra devices, such as the one called bulldozing have to be used to obtain an irreflexive frame from the canonical one.[5]

The criticism of insufficient formalization in Kripke's original argument can be overcome by the use of a system that embodies Kripke semantics in an explicit way, through the use of a labelled syntax. In Kripke, the ramified structure of systems of sets of alternative tableaux contains the semantics in the form of geometric conditions on tree-structures in proofs. We show how this structure can be replaced by a simple labelled sequent system. Completeness is established with a Schütte-style construction of an exhaustive proof search in the labelled system: Either a proof or a countermodel is found. The countermodel is extracted directly from the labels used in a non-conclusive branch in the search tree. The problems mentioned above with the treatment of negative properties of the accessibility relation, such as irreflexivity, simply do not arise.

The contents of the paper are as follows: Section 1 presents a background on modal logic and its Kripke semantics; It can be skipped by readers already familiar with modal logic. In Section 2 we present a re-reading of Kripke's original completeness proofs, as published in the papers from 1959 and 1963, respectively, and the reviews to these papers by Bayart and Kaplan. In Section 3 we give a sketch of a Henkin-style completeness proof for modal logic. In Section 4 we present our method for obtaining labelled sequent calculi with good structural properties for

---

[4]Makinson (1966), Cresswell (1967).
[5]Cf. Bull and Segerberg (1984).

all modal logics characterized by a relational semantics. The completeness proof is presented is Section 5.

2 Background on modal logic and its Kripke semantics

Traditional Kripke completeness is concerned with systems of normal modal logics, that is, systems obtained as extensions of basic modal logic. In this Section we shall recall the basic definitions and the standard notions of what is nowadays regarded as 'Kripke semantics.'

2.1 The language and axioms of modal logic

In modal logic, we start from the language of propositional logic and add to it the two modal operators $\Box$ and $\Diamond$, to form from any given formula A the formulas $\Box A$ and $\Diamond A$. These are read as "necessarily A" and "possibly A," respectively.

A system of modal logic can be an extension of intuitionistic or classical propositional logic. In the latter, the notions of necessity and possibility are usually connected by the equivalence $\Box A \supset\subset \sim \Diamond \sim A$.

It is seen that necessity and possibility behave analogously to the quantifiers: In one interpretation, the necessity of A means that A holds in all circumstances, and the possibility of A means that A holds in some circumstances. The definability of possibility in terms of necessity is analogous to the definability of existence in terms of universality.

The system of **basic modal logic**, denoted by **K** in the literature, adds to the axioms of classical propositional logic the following:

Table 1. The system of basic modal logic

1. Axiom: $\Box(A \supset B) \supset (\Box A \supset \Box B)$,

2. Rule of necessitation: From A to infer $\Box A$.

One axiom and one rule is added to the axioms and rules of propositional logic. The rule of necessitation requires that the premiss be derivable in the axiomatic system, i.e., its contents are that if A is a theorem, also $\Box A$ is a theorem. If instead of axiomatic logic we start from a system of natural deduction, the following rules are added:

Table 2. Natural deduction for basic modal logic

$$\frac{\Box(A \supset B) \quad \Box A}{\Box B} \qquad \frac{A}{\Box A}$$

The second rule, called "necessitation" or "box introduction," requires a restriction: if from any formula A one could conclude $\Box A$, by first assuming A and then applying necessitation and implication introduction, one could conclude $A \supset \Box A$.

Anything implies its own necessity, which clearly is wrong. In the axiomatic formulation, the premiss of necessitation was a theorem. In a natural deduction system, one requires that A be derivable with no open assumptions. If one thinks of the analogy between necessity and universal quantification, it appears that the restriction is analogous to the variable condition in the rule for introducing the universal quantifier. Inappropriate formulations of the rule of necessitation have caused considerable confusion in the literature and led many authors to the conclusion that the deduction theorem fails in modal logic.[6]

Other operators besides necessity and possibility can be justified in terms of the quantifiers and their duality. For example, whatever must be done is *obligatory*, whatever can be done is *permitted*. These two notions belong to **deontic** logic. Even more simply, we can read $\Box A$ as "always A" and $\Diamond A$ as "some time A," respectively, which gives rise to **tense logic**.

The early study of modal logic, to the late 1950s, consisted mainly of suggested axiomatic systems based on an intuitive understanding of the basic notions. Certain axiomatizations became standard and are collected here in the form of a table. All of them start with the axioms of classical propositional logic and the axioms of basic modal logic of table 1.

Table 3. Extensions of basic modal logic

	Axiom
T	$\Box A \supset A$
4	$\Box A \supset \Box\Box A$
E	$\Diamond A \supset \Box\Diamond A$
B	$A \supset \Box\Diamond A$
3	$\Box(\Box A \supset B) \vee \Box(\Box B \supset A)$
D	$\Box A \supset \Diamond A$
2	$\Diamond\Box A \supset \Box\Diamond A$
W	$\Box(\Box A \supset A) \supset \Box A$

Well-known extensions of basic modal logic are obtained through the addition of one or more of the above axioms to system **K**, for instance **K4** is obtained by adding 4, **S4** by adding T and 4, **S5** by adding T, 4, and E (or T, 4, and B), deontic **S4** and **S5** are obtained with the addition of axiom D to **S4** and **S5**, respectively. The addition of W gives what is known as the Gödel-Löb system. Axiom 2, also known as axiom M, gives the extensions of **K4** and **S4** known as **K4.1** and **S4.1**, respectively. Axiom 3 is used for instance in the extension **S4.3** of system **S4**.

The study of modal logic was completely changed in the late 1950s through the invention of a **relational semantics** of modal logic to which we now turn.

---

[6]See Hakli and Negri (2008) for a thorough discussion of this issue.

2.2 Kripke semantics

What is known as Kripke semantics, also known under the neutral term relational semantics, was presented by Saul Kripke in 1959 for the modal logic **S5**. It was modified later to accommodate also other modal logics and intuitionistic logic.[7] The idea had several significant anticipations in the work of Arnould Bayart, Rudolf Carnap, Jaakko Hintikka, Stig Kanger, Richard Montague, Arthur Prior, and others. Questions about the originality and ultimate attribution for the invention of Kripke semantics have raised a considerable debate. We shall not take any position on these issues here, but refer to Goldblatt (2005) for an in-depth discussion.

The basic idea of the semantics is that a proposition is necessary if and only if it is true in all "possible worlds." The idea is made precise as follows: a **Kripke frame** is a set W, the elements of which are called **possible worlds**, together with an **accessibility relation** R, that is, a binary relation between elements of W. A Kripke frame becomes a **Kripke model** when a **valuation** is given. A valuation val takes a world w and an atomic formula P and gives as value 0 or 1, to determine which atomic formulas are true at what particular worlds. The notation is

$w \Vdash P$ whenever $val(w, P) = 1$.

It is read as: *formula P is true at world w*, alternatively as *w forces P*. If $val(w,P) = 0$, we write $w \nVdash P$. A valuation is just like a line in a truth table, except that it is indexed by a world. If there is just one world, we have essentially the truth-table semantics of classical propositional logic. Valuations are supposed to be actually given, not just to exist in some abstract sense, so we have $w \Vdash P$ or $w \nVdash P$ for each atom P.

Valuations are extended in a unique way to arbitrary formulas by means of inductive clauses. For the propositional connectives, the inductive extension is straightforward:

Table 4. Valuations for the connectives

$w \Vdash A \& B$	whenever $w \Vdash A$ and $w \Vdash B$,
$w \Vdash A \vee B$	whenever $w \Vdash A$ or $w \Vdash B$
$w \Vdash A \supset B$	whenever from $w \Vdash A$ follows $w \Vdash B$
$w \Vdash \bot$	for no w.

It was assumed above that it is decidable if an atomic formula is forced at a given world. The same property holds then for arbitrary formulas, by the inductive clauses of table 4. Further, if $w \nVdash A$, then $w \Vdash \sim A$. To prove this, assume $w \Vdash A$. We have a contradiction (in fact, 0=1), so that $w \Vdash \bot$. Therefore, by the inductive clause for implication, $w \Vdash \sim A$.

---

[7] Kripke (1963), Kripke (1965).

DEFINITION 2.1 *Given a Kripke frame W, formula A is* **valid in** *W if, for every valuation, $w \Vdash A$ for every world w in W.*

The central idea in Kripke's semantics for modal logic is that a formula of the form $\Box A$ is true at w if A is true at all worlds accessible from w through the relation R:

$w \Vdash \Box A$ if and only if for all o, $o \Vdash A$ follows from wRo.

The second key insight of Kripke semantics is that the axioms of different systems of modal logic correspond to special properties of the accessibility relation. Let us take what is probably the simplest example, namely a **reflexive** frame: we assume the accessibility relation to be reflexive. The condition corresponds to axiom T of table 3:

$w \Vdash \Box A \supset A$ for every world w.

To see this, assume $w \Vdash \Box A$. Then $o \Vdash A$ for every o accessible from w, in particular, by reflexivity, for w itself, so $w \Vdash A$. Therefore $w \Vdash \Box A \supset A$. On the other hand, it is easily seen that a frame that validates $\Box A \supset A$ has to be reflexive, so that reflexivity of the accessibility relation is equivalent to having a modal system with axiom T.

Similarly, it is seen that $\Box A \supset \Box \Box A$ is valid in every transitive frame and that every frame validating it has to be transitive. We say that there is a **correspondence** between a modal axiom and a property of the accessibility relation.

Observe that the defining axiom of the system of basic modal logic **K**, $\Box(A \supset B) \supset (\Box A \supset \Box B)$, is valid in every frame.

Table 6 of Section 4 gives a list of common modal axioms together with their corresponding frame conditions.

3 Kripke's original completeness proofs

In this Section we shall analyze the content of Kripke's original completeness proofs for systems of modal logic, as published in 1959 and 1963, and the criticisms that were raised by Bayart and Kaplan in their reviews, both published in 1966.

3.1 A completeness theorem in modal logic

In his paper of 1959, the young Saul Kripke presented a completeness theorem for 1st-order **S5** with equality.

The starting point is a Hilbert-style axiomatization obtained from Rosser's 1953 first-order predicate calculus with equality, with the addition of the following axiom schemes and rules of inference:

A1: $\Box A \supset A$

A2: $\sim\Box A \supset \Box \sim \Box A$

A3: $\Box(A \supset B) \supset (\Box A \supset \Box B)$

R1: If $\vdash A$ and $\vdash A \supset B$ then $\vdash B$

R2: If $\vdash A$ then $\vdash \Box A$

Given a non-empty domain D and a formula A, a **complete assignment** for A in D is a function which to every free individual variable assigns an element of D, to every propositional variable either T or F, and to n-ary predicates $P(x_1,\ldots,x_n)$ n-tuples of D.

A **model** of A **in** D is a pair (G,K) where G is a complete assignment in a set K of assignments, such that every member of K agrees with G on the assignment of free variables of A. The evaluation of an element H of K on an arbitrary subformula of A is obtained inductively in the usual way from the assignment of individual and propositional variables and of predicates. For example, $P(x_1,\ldots,x_n)$ is true in the model for the assignment H if the values $\alpha_1,\ldots,\alpha_n$ assigned to the variables belong to the subset of n-tuples assigned to the n-ary predicate P; $\forall x B$ is true if $B(x)$ is true for every assignment of x in D; $\Box B$ is true if B is true under every assignment in K.

A formula A is **valid** in (G,K) if it is assigned T by G, valid in D if it is valid in every model in D, **satisfiable** if it is valid for some model based on D, and **universally valid** if it is valid on every non-empty domain.

The intuitive idea here is that a proposition is necessary if and only if it is true in all possible worlds; All possible worlds are just all possible evaluations, the real world being represented by G and the other members of K representing possible worlds:

> "The basis of the informal analysis which motivated these definitions is that a proposition is necessary if and only if it is true in all 'possible worlds.' (It is not necessary for our present purposes to analyze the concept of a 'possible world' any further.) [...] In modal logic, however, we wish to know not only about the real world but about other conceivable worlds; P may be true in the real world but false in some imaginable one, and similarly for $P(x_1,\ldots,x_n)$. Thus we are led not to a single assignment but to a set K of assignments, all but one of which represent worlds which are conceivable but not actual; the assignment representing the actual world is singled out as G, and the pair (G,K) is said to form a model of A."[8]

Kripke shows, using an adaptation to modal logic of Beth's method of semantical tableaux, that a formula is derivable in **S5** if and only if it is universally valid. The tableau method is presented as a test of semantical entailment from $A_1 \& \ldots \& A_n$ to B through a systematic search for a countermodel in which $A_1,\ldots,A_n$ are valid but B is not. The construction produces a system of alternative sets of tableaux, each set containing a main tableau and subsidiary tableaux.

---

[8] Kripke (1959), p.2.

The rules for tableaux are the familiar ones, with the splitting into alternative tableaux in the case of conjunction in the right (and disjunction in the left if a full language is used). The rules for necessity are:

Yl. If $\Box A$ appears in the left column of a tableau, then we put A in the left column of every tableau of the set.

Yr. If $\Box A$ appears in the right column of a tableau, then we introduce a new auxiliary tableau which is started out by putting A in its right column.

A tableau is **closed** if and only if either a formula occurs in both of its columns, or $a = a$, for some variable a, occurs in its right column. A set of tableaux is closed if and only if at least one of its members is closed. A system is closed if and only if all its alternative sets are closed.

THEOREM 1 B is semantically entailed by $A_1 \& \ldots \& A_n$ if and only if the construction beginning with $A_1 \& \ldots \& A_n$ in a left column and B in a right column is closed.

The proof is divided in two parts, the first part (Lemma 1, validity) shows that if B is not semantically entailed by $A_1 \& \ldots \& A_n$, then the tableau construction cannot be closed.

If B is not semantically entailed by $A_1 \& \ldots \& A_n$, there is a model (G, K) on D such that $A_1, \ldots, A_n$ are true and B false in it. The inductive clauses for valuations match the tableaux rules in such a way that they preserve countermodels, so in the end, since the construction is closed, every alternative set contains a tableau which either has a formula in both columns or has $a = a$ in the right column. So there would be some formula which is at the same time true and false in the model, or $a = a$ would be false, a contradiction.

The information on the existence of a countermodel is directly transferred to a constraint on the tableau.

The second part (Lemma 2, completeness) shows that if the tableau construction is not closed, then a countermodel is found.

LEMMA 2 If the construction starting with $A_1, \ldots, A_n$ on the left and B on the right is not closed, then B is not semantically entailed by $A_1, \ldots, A_n$.

The proof of Lemma 2 is obtained by choosing one of the alternative sets which is not closed and by defining a suitable countermodel on the basis of that. As domain D the set of free variables in the latter set is taken. The assignment is defined as follows: Every free variable except those eliminated by the rule of substitution for identity (*Il*) is assigned to itself. Free variables eliminated by rule *Il* are assigned the variable that replaces them. Propositional variables occurring on the left of the tableaux of the chosen set are assigned T, those occurring on the right are assigned F. Predicates P^n of n variables are assigned the set of n-tuples $(x_1,\ldots,x_n)$ of variables such that $P^n(x_1,\ldots,x_n)$ appears on the left in the tableaux of the set. It is then shown by induction on formulas that every formula occurring on the left (resp. right) is assigned T (resp. right).

After proving a Löwenheim-Skolem result, by which a formula is satisfiable in a finite or denumerable domain if it is satisfiable in a nonempty domain, Kripke proceeds with the proof of completeness with respect to the original Hilbert-type system. For importing the completeness result proved for the tableau system, a form of a deduction theorem is proved. If we start a construction with $A_1,\ldots,A_n$ in the left column and B on the right of a tableau (initial stage), after m applications of the rules there are finitely many tableaux, and this is called the $(m+1)$th stage of the construction. Each stage is put in correspondence with an equivalent *characteristic formula*: The characteristic formula of a given tableau with $A_1,\ldots,A_m$ in the left and $B_1,\ldots,B_n$ in the right is $A_1\&\ldots\&A_n\&\sim B_1\&\sim B_m$; The characteristic formula of any of the alternative sets at a given stage is $\exists x_1 \ldots \exists x_p (A \& \Diamond B_1 \& \ldots \& \Diamond B_q)$ where A is the characteristic formula of the main tableau of the set and $B_1,\ldots,B_q$ are the characteristic formulas of the auxiliary tableaux of the set and $x_1,\ldots,x_p$ are the free variables of $A \& \Diamond B_1 \& \ldots \& \Diamond B_q$. Finally, the characteristic formula of a stage is $D_1 \vee \ldots \vee D_r$, where $D_1,\ldots,D_r$ are the characteristic formulas of the alternative sets of the stage. Then it is shown (Lemma 4) that if A is the characteristic formula of the initial stage and B is the characteristic formula of any stage, then $A \supset B$ is provable in the given Hilbert system $\mathbf{S5}^{*=}$ ($\mathbf{S5}$ with equality and quantifiers).

The proof of completeness for $\mathbf{S5}^{*=}$ (Theorem 5) can be summarized as follows: If A is universally valid, then the tableau construction beginning with A in the right column is closed. If B is the characteristic formula of the earliest stage at which the closure holds, by Lemma 4 the implication between the closure formula of the initial stage and B is derivable in $\mathbf{S5}^{*=}$, that is, $\vdash \exists a_1 \ldots a_p \sim A \supset B$. It is detailed how the fact that B is the characteristic formula of a stage of closure gives $\vdash \sim B$, from which $\vdash \sim \exists a_1 \ldots a_p \sim A$, and therefore $\vdash A$ follows.

Theorem 6 establishes validity: If $\vdash A$ in $\mathbf{S5}^{*=}$, A is universally valid. The proof consists in observing that the axioms of $\mathbf{S5}^{*=}$ are universally valid and that modus ponens preserves universal validity. As for the rule of necessitation, if A is universally valid, the tableau construction starting with A on the right closes, hence also the tableau construction starting with $\Box A$ closes, so universal validity of $\Box A$ follows by Theorem 1.

Kripke proceeds with defining truth tables for $\mathbf{S5}$ and establishing that a formula is a tautology if and only if it is universally valid. In the final part, second order quantification is treated.

3.2 Semantical analysis of modal logic I. Normal modal propositional calculi

In Kripke (1963) the results of Kripke (1959) were extended to various systems of modal logic: Gödel-Feys-von Wright's **M** (**T**), the *Brouwersche* system **B**, and Lewis' **S4** and **S5**.

The core of normal modal systems is taken to consist of axioms A1 and A3 and rules R1 and R2 (see previous Section), which give the system **M**, alternatively called system **T**. System **S4** is obtained by the addition of

A4: $\Box A \supset \Box\Box A$,

the Brouwersche system by the addition of

$A \supset \Box \Diamond A$,

and **S5** by the addition of

A2: $\sim\Box A \supset \Box \sim\Box A$.

The novelty here is the explicit appearance of the accessibility relation:

> "A *normal model structure* (n.m.s) is an ordered triple $(G, K, \mathbf{R})$ where K is a non-empty set, $G \in K$, and **R** is a reflexive relation defined on K. If **R** is transitive, we call the n.m.s. an **S4** *model structure*; if **R** is symmetric, we call it a BROUWER*sche model structure*; if **R** is an equivalence relation, we call it an **S5** *model structure*."[9]

An **M** (**S4**, **S5**, Brouwersche) model for a formula A is given by a binary function Φ that has as arguments the propositional variables P of A and the elements H of K, and as range the truth values T, F. The function is extended in a unique way to all subformulas of A by the following inductive clauses:

$\Phi(B\&C, H) = T$ if and only if $\Phi(B, H) = T$ and $\Phi(C, H) = T$;
$\Phi(\sim B, H) = T$ if and only if $\Phi(B, H) = F$;
$\Phi(\Box B, H) = T$ if and only if $\Phi(B, H') = T$ for all H' such that HRH'.

Truth (falsity) of A in a model is defined as truth (resp. falsity) in the real world G, $\Phi(A, G) = T$ (resp. $\Phi(A, G) = F$); validity as truth in all models, and satisfiability as truth in at least one of them.

The relation of the new definitions to those in Kripke (1959) is given in section 2.1 (Informal explanation). Two main differences arise: Whereas in the previous work worlds were identified with complete assignments, here the two notions are distinct, to the effect that there can be worlds in which the same truth values are assigned to the atomic formulas. The second novelty is the relation **R**, for which the following informal explanation is given:

---

[9]Kripke (1963), p.8.

"Intuitively we interpret the relation **R** as follows: given any two worlds $H_1, H_2 \in K$, we read '$H_1 \mathbf{R} H_2$' as 'H_2 is possible relative to H_1,' 'possible in H_1,' or 'related to H_1;' that is to say, every proposition true in H_2 is to be possible in H_1."[10]

On the basis of this reading, modal axioms are related to properties of the accessibility relation: Transitivity is shown to correspond to $\Diamond\Diamond A \supset \Diamond A$, and symmetry to $A \supset \Box\Diamond A$

In section 2.2 a *connected* model structure is defined as one in which all the possible worlds are related through the transitive closure $\mathbf{R}^*$ (the *ancestral*, in Kripke's terminology) of $\mathbf{R}$. Here Kripke shows that every satisfiable formula has a connected model, or equivalently, that every non-valid formula has a connected countermodel. Given a model with valuation Φ on $(G, K, \mathbf{R})$, a connected countermodel is defined by the restriction of Φ and $\mathbf{R}$ to the set of worlds accessible form the 'real world' G through the transitive closure of $\mathbf{R}$. By the restriction to connected models an equivalence relation gives the same models as a total relation, so the treatment of Kripke (1959) for **S5**, without the accessibility relation, can be seen as a special case of the new one.

A further reduction is prepared for with the definition of a *tree* as a triple (G, K, S) where S is a binary relation on K, G has no predecessor with respect to S, and every other element of K has a unique predecessor.

In section 3, semantic tableaux are presented as a generalization of the tableaux of Kripke (1959). A tableau construction gives at each stage a system of *alternative* sets of tableaux, each containing a *main* tableau and *auxiliary* tableaux.

The completeness proof is given by a systematic search of a countermodel; If no countermodel is found, the formula is valid. The procedure for a formula of the form $A_1 \& \ldots \& A_m \supset B_1 \vee \ldots \vee B_n$ starts by imposing $A_1, \ldots, A_m$ to be true in the model and $B_1, \ldots, B_n$ false, that is in putting $A_1, \ldots, A_m$ to the left and $B_1, \ldots, B_n$ to the right of the main tableau. The rules for the tableau construction transform the requirement into equivalent conditions on more elementary formulas. The rules can either produce a continuation of the same tableau (rules for negation, conjunction, and necessity on the left) or, in the case of right conjunction, the splitting of the tableau into alternative sets of tableaux $t_1, t_2, \ldots$. The splitting corresponds to the fact that in order to falsify a conjunction it is enough to falsify one of the conjuncts. For necessity on the right, the tableau construction proceeds by creating from the given tableau t that contains a formula $\Box A$ on the right, a new auxiliary tableau t' such that tRt'. For specific modal systems, additional properties are assumed on R. These properties are not made a formal part of the syntax in the tableau construction.

Kripke did not devise a formal notation to fully describe his tableau construction and wrote in fact:

---

[10] Kripke (1963), p.70. Observe that this informal reading imposes the definition of the canonical accessibility relation if possible worlds are Henkin sets (see Section 3 below).

"I hope that this explanation makes the process clear intuitively; the formal statement is rather messy [...]."[11]

A tableau is closed if the same formula appears on both sides of the tableau, a set of tableaux is closed if some tableau in it is closed, and a system is closed if each of its alternative sets is closed. A construction is closed if at some stage a closed system of alternative sets appears.

Some additional restrictions are posed in order to facilitate the tableau construction, for instance, a rule need not be applied if it produces a formula that is already in the tableau.[12] Another optimization in the tableau construction is the irrelevance of the order of application of the rules, so that no special strategy is needed.[13]

The procedure is clarified by an example that shows how the tableau procedure works for the formula $\Box(A \& B) \supset \Box(\Box A \& \Box B)$: A closed S4 tableau construction is obtained, that is, the search for a countermodel fails and therefore the formula is valid.

In section 3.2, Kripke proves the completeness of the tableau procedure with respect to the semantics.

First (Lemma 1) he shows validity: If the construction for A is closed, then A is valid. This is proved by contradiction: If A is not valid, there exists a valuation Φ on a model structure $(G, K, \mathbf{R})$ with $\Phi(A, G) = F$. It is then shown by induction on the stages of the tableau construction, started with A on the right, that each stage can be put in correspondence with worlds in the model and the relation that links auxiliary tableaux with the relation in the model, and that formulas on the right are false and those on the left true under the assignment Φ. The construction is closed, so a contradiction follows because there would be a formula both true and false under Φ.

Lemma 2 proves completeness, again by contradiction: If the construction for A is not closed, then A is not valid. The proof can be summarized as follows: The tableau construction has a tree structure that can either be finite or infinite. If the construction is finite it has, because it is not closed, at least one alternative set S_0 that is not closed. Then a countermodel $(G, K, \mathbf{R})$ is defined by taking for K the set S_0, for $\mathbf{R}$ the relation R between elements of S_0, and for G the main tableau of S_0. The valuation Φ is defined by putting $\Phi(P, H) = T$ if P appears on the left side of H, and $\Phi(P, H) = F$ otherwise. It is then shown by induction on formulas that for arbitrary formulas A we have $\Phi(A, H) = T$ if A appears on the left of H, and $\Phi(A, H) = F$ otherwise. Therefore A is false under Φ.

In case the tableau construction is infinite, the proof is a bit more complex

---

[11] Kripke (1963), p.73.

[12] Observe the analogy to the search for minimal derivations in Gentzen systems with height-preserving contraction, where a rule need not be applied if it leads to a duplication of formulas in a sequent.

[13] But see Bayart's objection in Section 3.3 below for the predicate case.

and König's lemma is used to extract an infinite path that is used to define the countermodel.

By the completeness proof, the previously established reduction to connected structures is strengthened to a reduction to trees. The reduction is used for obtaining a reformulation of the tableau rules in which relation R is replaced by a unary successor relation S. The tableau rules are thus modified so that the properties of the relation R become part of the rules. For example, the left rule for $\Box$ that subsumes transitivity is as follows:

> "Y1: If $\Box A$ appears on the left of a tableau t_1, we put A on the left of t_1 and put $\Box A$ on the left of any tableau t_2 such that $t_1 R t_2$."[14]

A similar modification is given for the Brouwerische system, which incorporates symmetry of R.

Kripke defines two tableaux t_1 and t_2 to be *contiguous* if $t_1 S t_2$ or $t_2 S t_1$ and observes that, because of the new formulation, application of a rule to a tableau affects only the tableaux contiguous with it.[15]

In section 4, completeness of the Hilbert systems is proved. The proof of validity (consistency) reduces to the immediate task of verifying that the axioms are valid and that the rules preserve validity.

For completeness, first the definition of a characteristic formula of a tableau is given as in Kripke (1959), and the lemma already proved in Kripke (1959) for **S5** (Lemma 4) is extended to the systems here considered. If the tableau procedure started with A is closed, we find for each of the alternative sets S_j a stage of closure, that is, with the same formula both on the left and on the right of the tableau, so the characteristic formula D_j contains the conjuction of a formula and its negation. By the above mentioned lemma, we have $\vdash \sim A \supset D_1 \vee \ldots \vee D_m$ and since for all $j, \vdash \sim D_j$, we have $\vdash \sim (D_1 \vee \ldots \vee D_m)$, and therefore, because all the systems are extensions of classical logic, we get $\vdash A$.

The rest of the paper contains proofs of decidability for all the systems considered, obtained by means of a bound in the tableau proof search procedure, a section on matrices to establish independence results, and a proof, by the method later called 'glueing of Kripke models,' of the modal disjunction property. The last one had already been proved by McKinsey and Tarski (1948) and Lemmon (1960) using algebraic semantics.

3.3 Reviews

Both of Kripke's papers were carefully reviewed. In 1966, a review of Kripke (1959) by Arnould Bayart appeared. After a detailed summary of the

---

[14] Kripke (1963), p.81.
[15] This system of rules thus enjoys the remarkable property nowadays called *locality*.

basic definitions and results of the paper, the reviewer observed a lack of determinism in the tableau rules for the quantifiers and suggested the introduction of a control mechanism to avoid dead-ends:

> "An objection against both the proof and the statement of theorem 1 is that, at each step of the construction of a system of tableaux, several possibilities generally occur so that different end results can be reached. If one starts with bx and $\sim bx$ in the left column and with $(x)bx$ in the right column, one obtains a closed tableau by working on $\sim bx$, and one obtains an infinite not closed tableau by working on $(x)bx$. The rules for constructing tableaux should be supplemented by a rule imposing some definite choice at each step and guaranteeing that each formula appearing in a not closed branch of the construction will at some moment become the object of an application of a construction rule and that the rules for universal quantification in a left column will be applied for all individual variables appearing in the branch."[16]

The same year, Kripke (1963) was reviewed by David Kaplan who, even if he praised Kripke's result, found that the development lacked in rigor:

> "Although the author extracts a great deal of information from his tableau constructions, a completely rigorous development along these lines would be extremely tedious. As a consequence a number of small gaps must be filled by the reader's geometrical intuition [...]. The dangers inherent in relying on intuition are illustrated by the author's need to correct a fallacious proof in a completeness theorem in modal logic [...] the author criticizes another writer's faulty version of the rule; but his own formulation also requires amendment [...]. The proofs of the decision procedures seemed to the reviewer excessively intuitive even within the allowable space, and the proof for S4 contains an error (which can be corrected [...])."[17]

Kaplan went on suggesting an alternative approach to the completeness proof for modal logic:

> "The reviewer believes that future research will bring considerably simpler more rigorous proofs which avoid the tableau technique. In fact the interesting half of the main theorem can be established by using the technique of Henkin [...]."[18]

After a sketch of the idea of the Henkin-style proof of completeness for modal logic, Kaplan observed that the proof was suggested to him by Dana Scott and that the argument was already foreshadowed in Kanger (1957). The review also witnesses the already existent debate on the ultimate attribution of the possible world semantics. The contributions of Carnap, Kanger, Hintikka, and Montague are mentioned as important anticipations. The review ends with words of praise for the paper as one "among the most important contributions to the study of modal logic," followed by a list of corrections to about 20 misprints.

4 Henkin-style completeness proof

Kaplan's review of 1966 gave the guiding ideas of the adaptation of the proof of Henkin (1949) to modal logic as suggested to him by Lemmon and Scott.

---

[16] Bayart (1966).
[17] Kaplan (1966).
[18] Kaplan (1966).

Henkin-style completeness proofs for various systems of modal logic with respect to the relational semantics were published at the same time,[19] or shortly after.[20] Indeed, Cresswell refers to the papers by Arnould Bayart (1958, 1959), who was apparently the first to have given a proof of completeness in this style for modal logic. Bayart considered second-order **S5**. In the first of the two papers, he proved validity (in French, *correction*), in the second quasi-completeness (in French, *quasi-adequation*), the quasi being referred to the limitations for the second order case. The papers by Bayart are difficult to read; They use Polish notation and an archaic style of exposition, which may explain why they were little known. Instead of tableaux, they use a system of sequent calculus with invertible rules. The use of the possible worlds semantics is independent of Kripke's, and declared by the author to have been inspired by Leibniz. Bayart himself did not mention his alternative approach to completeness for **S5** in his review of Kripke (1959).

Henkin-style completeness proofs seem to have been unanimously considered superior to the proofs originally devised by Kripke. Kaplan's criticism of Kripke (1963) was confirmed in Makinson's review of Kripke (1965).

> "It should be remarked that alternative and simpler proofs have since been constructed for these completeness theorems, in which maximal consistent set constructions do the work of Kripke tableaux."[21]

In addition, a reviewer of Makinson's paper wrote:

> "The proof, which makes no use of the axiom of choice, and which is nicely worked out in detail, is Henkin-style and thus avoids the Beth-Hintikka technique of semantic tableaux employed by Kripke [...]."[22]

Before discussing the relative advantages of the two approaches, we sketch the structure of Henkin-style completeness proofs for modal logic.

We recall that a frame $\mathcal{F}$ is a non-empty set S endowed with a binary relation R. A model $\mathcal{M}$ is given by a frame together with an assignment V of atomic formulas to subsets of S. Often the forcing relation notation $\mathcal{M} \Vdash_v P$ is used for $v \in V(P)$. The assignment V is extended to arbitrary formulas by the standard inductive clauses, for example:

$$\mathcal{M} \Vdash_v A \supset B \text{ if from } \mathcal{M} \Vdash_v A, \mathcal{M} \Vdash_v B \text{ follows.}$$
$$\mathcal{M} \Vdash_v \Box A \text{ if } \mathcal{M} \Vdash_w A \text{ for all } w \text{ such that } vRw.$$

Validity in a model is defined by truth in every world:

---

[19] Makinson (1966).
[20] In 1967 a paper by Cresswell with a completeness proof for T and S with the Barcan formula appeared. See Cresswell (1967).
[21] Makinson (1970).
[22] Åqvist (1970), p.136.

$$\mathcal{M} \Vdash A \text{ if } \mathcal{M} \Vdash_v A \text{ for all } v \in S.$$

Validity in a frame is defined by validity in every model based upon the frame:

$$\mathcal{F} \Vdash A \text{ if } \mathcal{M} \Vdash A \text{ for all } V.$$

If C is a class of frames, A is valid in the class of frames, $C \Vdash A$, if it is valid in every frame of the class, that is, $\mathcal{F} \Vdash A$ for all $\mathcal{F}$ in C. For example, $\Box(A \supset B) \supset (\Box A \supset \Box B)$ is valid in all frames; $\Box A \supset \Box\Box A$ is valid in all transitive frames.

Let L be a normal modal logic. L is usually defined by the set of propositional tautologies plus the axiom $\Box(A \supset B) \supset (\Box A \supset \Box B)$ and closure under the rules of modus ponens and necessitation.

The deducibility relation is defined implicitly by: $\vdash_L A$ if $A \in L$.

L is *sound* with respect to C if from $\vdash_L A$, $C \Vdash A$ follows.

L is *complete* with respect to C if from $C \Vdash A$, $\vdash_L A$ follows.

Soundness is proved by a straightforward induction on the derivation of A in L and we need not go into the details here. If L has additional axioms, then it is proved that they are valid in the class of frames considered.

Completeness is proved by the canonical model construction. From L a special model is built in which validity and derivability coincide. We start with the proof for classical logic.

A set of formulas Δ is a *maximal set* if it is consistent and has no consistent extension. We recall that a set of formulas is consistent if for no finite subset Δ_0 of Δ, $\Delta_0 \vdash_L \bot$. An equivalent characterization for a maximal set Δ requires that Δ is consistent and for every A, either A or $\sim A$ is in Δ.

By Lindenbaum's Lemma, every consistent set of formulas Γ can be extended to a maximal consistent set: One starts from an enumeration of the formulas in the language, $A_0, \ldots, A_n, \ldots$, and defines inductively a chain of sets of formulas as follows:

$$\Delta_0 \equiv \Gamma.$$
$$\Delta_{n+1} \equiv \Delta_n \cup \{A_n\} \text{ if } \Delta_n \vdash_L A_n.$$
$$\Delta_{n+1} \equiv \Delta_n \cup \{\sim A_n\} \text{ otherwise.}$$

It is not difficult to verify that $\Delta \equiv \bigcup_{n \geq 0} \Delta_n$ is a maximal consistent set that contains Γ.

By the construction of a maximal set containing a set of formulas Γ, the following hold:

1. Maximal sets are deductively closed.

2. If $\Gamma \nvdash A$, then there exists a maximal set that contains Γ but not A.

The valuation in the *canonical model* is defined by putting $\Delta \Vdash P$ if $P \in \Delta$. It is then shown by induction on formulas that also for arbitrary formulas A we have $\Delta \Vdash A$ if $A \in \Delta$. By taking the contrapositive of 2. above, we have:

Completeness. If $\Gamma \models A$, then $\Gamma \vdash A$.

The argument is augmented as follows to cover modal logic: The canonical model is a Kripke model in which the nodes are maximal consistent sets of formulas, the accessibility relation is such that two nodes Γ, Δ are related if all the necessary truths in the former are in the latter, and a formula is forced at a node if it belongs to that node. The notation is:

$$\mathcal{M}^{\mathcal{L}} \equiv (S^{\mathcal{L}}, R^{\mathcal{L}}, V^{\mathcal{L}}).$$

Here

$$S^{\mathcal{L}} \equiv \{\Gamma : \Gamma \text{ is } \mathcal{L}\text{-maximal consistent}\}$$
$$\Gamma R^{\mathcal{L}} \Delta \text{ if for all } A, \Box A \text{ in } \Gamma \text{ implies } A \text{ in } \Delta$$
$$V^{\mathcal{L}}(P) \equiv \{\Gamma : P \in \Gamma\}$$

We have:

Truth Lemma. $\mathcal{M}^{\mathcal{L}} \Vdash_\Gamma A$ if and only if $A \in \Gamma$.

The proof is by induction on A, the only non-trivial case being the one of a modalized formula. The case follows from the definition of validity in the model, from the definition of $R^{\mathcal{L}}$, and from the fact that maximal consistent sets are deductively closed.

To prove that validity and derivability coincide in the canonical model, that is,

$$\mathcal{M}^{\mathcal{L}} \Vdash A \text{ if and only if } \vdash_{\mathcal{L}} A$$

it is enough to prove the more general:

Lemma. $\Gamma \models_{\mathcal{M}^{\mathcal{L}}} A$ if and only if $\Gamma \vdash_{\mathcal{L}} A$.

The left-hand side of the latter amounts to the fact that every maximal set that contains Γ also contains A, so the equivalence immediately follows from the properties 1. and 2. above.

Observe that the proof gives no way to obtain derivability from validity, nor does it show how to construct a countermodel for underivable propositions. The explicit character of Kripke's original proof, that constructs countermodels for unprovable formulas, is lost with the Henkin approach.

In the next two Sections we review the definition and properties of our labelled calculi, and show how they can be used to make rigorous and generalize Kripke's original completeness proof.

5 A sequent system with internalized Kripke semantics

The development of structural proof theory has led to the remarkable class of sequent calculi, called **G3**-calculi, in which all of the structural rules–weakening, contraction, and cut–are admissible. We shall present a method for obtaining similarly behaving labelled sequent calculi for modal logics. In these, all the structural rules are admissible; They support, whenever possible, proof search, and have a simple and uniform syntax that allows easy proofs of metatheoretic results, such as those reviewed in the previous Sections.

We shall present in this Section a sequent system for the basic modal logic **K** with rules for the modalities $\Box$ and $\Diamond$. These rules are obtained through a meaning explanation in terms of the possible worlds semantics and an inversion principle. The modal logic **K** is characterized by arbitrary frames and restrictions on the class of frames that characterize a given modal logic amount to the addition of certain frame properties to our sequent calculus. These properties are added in the form of mathematical rules, following the method of extension of sequent calculus presented in chapter 6 of Negri and von Plato (2001). All the extensions are thus obtained in a modular way. As a consequence, the structural properties of the resulting calculi can be established in one theorem for all systems. A basic knowledge of sequent calculus, for example chapter 3 in Negri and von Plato (2001), is sufficient for what follows.

5.1 Basic modal logic

Basic modal logic is formulated as a labelled sequent calculus through an internalization of the possible worlds semantics within the syntax. First we enrich the language so that sequents are expressions of the form $\Gamma \to \Delta$ where the multisets Γ and Δ consist of **relational atoms** wRo and **labelled formulas** $w : A$, the latter corresponding to the forcing $w \Vdash A$ in Kripke models. Here w, o range over a set W of labels/possible worlds and A is any formula in the language of propositional logic extended by the modal operators of necessity and possibility, $\Box$ and $\Diamond$.

The rules for each connective/modality are obtained from its meaning explanation in terms of the relational semantics. The inductive definition of forcing for a modal formula is:

$w \Vdash \Box A$ *whenever for all* o, *from* wRo *follows* $o \Vdash A$.

The definition gives:

If $o : A$ can be derived for an arbitrary o accessible from w, then $w : \Box A$ can be derived.

Formally, we have the rule

$$\frac{wRo, \Gamma \to \Delta, o : A}{\Gamma \to \Delta, w : \Box A} \, R\Box$$

In the rule, the arbitrariness of o becomes the variable condition that o must not occur in Γ, Δ.

Reading the semantical explanation in the other direction, we have that $w \Vdash \Box A$ and wRo give $o \Vdash A$. A corresponding rule for the antecedent side is:

$$\frac{o : A, w : \Box A, wRo, \Gamma \to \Delta}{w : \Box A, wRo, \Gamma \to \Delta} \, L\Box$$

The rules for $\Diamond$ are obtained similarly from the semantic explanation

$w : \Diamond A$ whenever for some o, wRo and $o : A$.

The rules of sequent calculus for the propositional connectives are obtained by a labelling of the active formulas with the same label in the premisses and conclusion of each rule of the calculus **G3cp**.[23] The following sequent calculus **G3K** for basic modal logic is thus obtained:

---

[23] Cf. Negri and von Plato (2001), p.49.

Table 5. The sequent calculus G3K

Initial sequents:

$$w : P, \Gamma \to \Delta, w : P \qquad\qquad wRo, \Gamma \to \Delta, wRo$$

Propositional rules:

$$\frac{w : A, w : B, \Gamma \to \Delta}{w : A\&B, \Gamma \to \Delta} L\& \qquad \frac{\Gamma \to \Delta, w : A \quad \Gamma \to \Delta, w : B}{\Gamma \to \Delta, w : A\&B} R\&$$

$$\frac{w : A, \Gamma \to \Delta \quad w : B, \Gamma \to \Delta}{w : A \vee B, \Gamma \to \Delta} L\vee \qquad \frac{\Gamma \to \Delta, w : A, w : B}{\Gamma \to \Delta, w : A \vee B} R\vee$$

$$\frac{\Gamma \to \Delta, w : A \quad w : B, \Gamma \to \Delta}{w : A \supset B, \Gamma \to \Delta} L\supset \qquad \frac{w : A, \Gamma \to \Delta, w : B}{\Gamma \to \Delta, w : A \supset B} R\supset$$

$$\frac{}{w : \bot, \Gamma \to \Delta} L\bot$$

Modal rules:

$$\frac{o : A, w : \Box A, wRo, \Gamma \to \Delta}{w : \Box A, wRo, \Gamma \to \Delta} L\Box \qquad \frac{wRo, \Gamma \to \Delta, o : A}{\Gamma \to \Delta, w : \Box A} R\Box$$

$$\frac{wRo, o : A, \Gamma \to \Delta}{w : \Diamond A, \Gamma \to \Delta} L\Diamond \qquad \frac{wRo, \Gamma \to \Delta, w : \Diamond A, o : A}{wRo, \Gamma \to \Delta, w : \Diamond A} R\Diamond$$

In the first initial sequent, P is an arbitrary atomic formula. In $R\Box$ and in $L\Diamond$, o is a fresh label. Observe that atoms of the form wRo in the right-hand side of sequents are never active in the logical rules nor in the rules that extend the logical calculus. Moreover, the modal axioms that correspond to the properties of the accessibility relation are derived from their rule presentations alone. As a consequence, initial sequents of the form $wRo, \Gamma \to \Delta, wRo$ are needed only for deriving properties of the accessibility relation, namely, the axioms that correspond to the rules for R given below. Thus such initial sequents can as well be left out from the calculus without impairing the completeness of the system.

5.2 Extensions

Our aim is to extend the above basic calculus so that the structural properties of the extensions are automatically guaranteed. This will follow from the form of the axioms that characterize the extensions. The following table continues table 3 with the frame properties of modal axioms:

Table 6. Modal axioms with corresponding frame properties

	Axiom	Frame property
T	$\Box A \supset A$	$\forall w \; wRw$ reflexivity
4	$\Box A \supset \Box\Box A$	$\forall wor(wRo \& oRr \supset wRr)$ transitivity
E	$\Diamond A \supset \Box \Diamond A$	$\forall wor(wRo \& wRr \supset oRr)$ euclideanness
B	$A \supset \Box \Diamond A$	$\forall wo(wRo \supset oRw)$ symmetry
3	$\Box(\Box A \supset B) \vee \Box(\Box B \supset A)$	$\forall wor(wRo \& wRr \supset oRr \vee rRo)$ connectedness
D	$\Box A \supset \Diamond A$	$\forall w \exists o \; wRo$ seriality
2	$\Diamond \Box A \supset \Box \Diamond A$	$\forall wor(wRo \& wRr \supset \exists l(oRl \& rRl))$ directedness
W	$\Box(\Box A \supset A) \supset \Box A$	no infinite R-chains + transitivity

The frame properties in the first group (T, 4, E, B, 3) are universal axioms, those in the second group are what are known as geometric implications,[24], whereas the last one is not expressible as a first-order property.

The systems **T, K4, KB, S4, B, S5,** ... are obtained by adding one or more axioms to the system **K**. Sequent calculi are obtained by adding to the system **G3K** the rules that correspond to the properties of the accessibility relation that characterize their frames. For instance, a sequent calculus for **S4** is obtained by adding to **G3K** the rules that correspond to the axioms of reflexivity and transitivity of the accessibility relation:

$$\frac{wRw, \Gamma \to \Delta}{\Gamma \to \Delta} \; Ref \qquad \frac{wRr, wRo, oRr, \Gamma \to \Delta}{wRo, oRr, \Gamma \to \Delta} \; Trans$$

A system for **S5** is obtained by adding also the rule that corresponds to symmetry:

$$\frac{oRw, wRo, \Gamma \to \Delta}{wRo, \Gamma \to \Delta} \; Sym$$

The rule for euclideanness is

$$\frac{oRr, wRo, wRr, \Gamma \to \Delta}{wRo, wRr, \Gamma \to \Delta} \; Eucl$$

If o is substituted for r in *Eucl*, a duplication wRo, wRo is produced in the premiss and in the conclusion. The same happens in *Trans* if $w \equiv o \equiv r$. Contracted instances of these rules must be added to the system:

$$\frac{wRw, wRw, \Gamma \to \Delta}{wRw, \Gamma \to \Delta} \; Trans^* \qquad \frac{oRo, wRo, \Gamma \to \Delta}{wRo, \Gamma \to \Delta} \; Eucl^*$$

Both of the contracted rules are instances of rule *Ref*, therefore, in order to have the rule of contraction admissible, they have to be added into systems that do not

---
[24] Cf. Negri (2003).

contain rule *Ref*. Similar additions must be made for all extensions by rules that have instances with two occurrences of the same relational atom in the conclusion. The condition we require to be satisfied by this addition is called *closure condition*[25] The closure condition is unproblematic because it requires only a bounded, very small number (usually one or two) of rules to be added. This is general and uniform even if, as seen above, there are contracted rules that may turn out to be superfluous in some systems.

Extensions are obtained in a modular way for all possible combinations of properties:

G3T = G3K + *Ref*

G3K4 = G3K + *Trans*

G3KB = G3K + *Sym*

G3S4 = G3K + *Ref* + *Trans*

G3TB = G3K + *Ref* + *Sym*

G3S5 = G3K + *Ref* + *Trans* + *Sym*

A system for **deontic logic** is obtained by the addition of the geometric rule *Ser*:

$$\frac{wRo, \Gamma \to \Delta}{\Gamma \to \Delta} \; Ser$$

Here the variable condition is $o \notin \Gamma, \Delta$.

Directedness is another property that follows the pattern of a geometric implication, and it is converted into the rule

$$\frac{oRl, rRl, wRo, wRr, \Gamma \to \Delta}{wRo, wRr, \Gamma \to \Delta} \; Dir$$

The variable condition is $l \notin wRo, wRr, \Gamma, \Delta$.

The property that corresponds to axiom W, needed for provability logic, can be incorporated in the system through a modification of the rules for $\Box$.[26]

---

[25] Cf. section 6.1 of Negri and von Plato (2001).
[26] Cf. section 5 of Negri (2005).

5.3 Structural properties

Let **G3K*** be any extension of **G3K** by rules for the accessibility relation that follow the *regular rule scheme* for extensions of sequent calculus[27] or the more general *geometric rule scheme*.[28] The following properties can be established uniformly for all systems that belong to the class **G3K***.[29]

LEMMA 5.1 Sequents of the form

$$w : A, \Gamma \to \Delta, w : A$$

with A an arbitrary modal formula are derivable in **G3K***.

To prove the correspondence between our systems and their Hilbert-style presentations, it is necessary to show that the characteristic axioms are derivable and the systems closed under the rules of necessitation and *modus ponens*. The latter will be a consequence of admissibility of cut.

LEMMA 5.2 For arbitrary A and B, the sequent

$$\to w : \Box(A \supset B) \supset (\Box A \supset \Box B)$$

is derivable in **G3K***.

The rule of necessitation,

$$\frac{\to w : A}{\to w : \Box A}$$

is a context-dependent rule, as it requires both the antecedent and succedent contexts to be empty. As an explicit rule, it would impair the flexibility of the systems in the permutations that are needed for proving cut elimination. However, we do not need to add any such rule because we can show that it is admissible. To prove this, we exploit the first-order features of the system to show a lemma about substitution.

Substitution of labels is defined in the obvious way for relational atoms and labelled formulas and is extended to multisets componentwise. We have

LEMMA 5.3 If $\Gamma \to \Delta$ is derivable in **G3K***, then $\Gamma(o/w) \to \Delta(o/w)$ is also derivable, with the same derivation height.

---

[27] As in chapter 6 of Negri and von Plato (2001).
[28] As in Negri (2003).
[29] We refer to Negri (2005) for the proofs.

THEOREM 5.4 The rules of weakening

$$\frac{\Gamma \to \Delta}{w:A,\Gamma \to \Delta} LW \qquad \frac{\Gamma \to \Delta}{\Gamma \to \Delta, w:A} RW$$

$$\frac{\Gamma \to \Delta}{wRo,\Gamma \to \Delta} LW \qquad \frac{\Gamma \to \Delta}{\Gamma \to \Delta, wRo} RW$$

are height-preserving admissible in **G3K***.

COROLLARY 5.5 The necessitation rule is admissible in **G3K***.

We also obtain a very useful property of a sequent calculus, namely:

LEMMA 5.6 All the rules of **G3K*** are height-preserving invertible.

The most important structural property of our calculi, besides cut-admissibility, is height-preserving admissibility of contraction. First observe that there are, *a priori*, four contraction rules, namely left and right contraction for expressions of the forms $w:A$ and wRo. Explicitly stated, the rules of left and right contraction are:

$$\frac{w:A,w:A,\Gamma \to \Delta}{w:A,\Gamma \to \Delta} LC \qquad \frac{wRo,wRo,\Gamma \to \Delta}{wRo,\Gamma \to \Delta} LC_R$$

$$\frac{\Gamma \to \Delta, w:A, w:A}{\Gamma \to \Delta, w:A} RC \qquad \frac{\Gamma \to \Delta, wRo, wRo}{\Gamma \to \Delta, wRo} RC_R$$

Observe that rule RC_R is not needed in case we use the calculus without the initial sequent $wRo, \Gamma \to \Delta, wRo$.

THEOREM 5.7 The rules of contraction are height-preserving admissible in **G3K***.

Also cut can take two forms, namely

$$\frac{\Gamma \to \Delta, w:A \quad w:A, \Gamma' \to \Delta'}{\Gamma, \Gamma' \to \Delta, \Delta'} Cut$$

and

$$\frac{\Gamma \to \Delta, wRo \quad wRo, \Gamma' \to \Delta'}{\Gamma, \Gamma' \to \Delta, \Delta'} Cut_R$$

However, Cut_R is not needed if the variant of **G3K** without the initial sequent $wRo, \Gamma \to \Delta, wRo$ is used. We have:

THEOREM 5.8 The cut rule is admissible in **G3K***.

6 Kripke completeness revisited

Kripke's original proof of completeness for modal logic used a direct construction of a Beth tree from a failed proof search. In later proofs, Kripke countermodels had nodes built from Henkin sets of formulas and extra devices that impose additional properties on the accessibility relation that are not automatically captured by the Henkin construction.[30] Kripke used tableaux in which the semantical element was hidden in their tree structure and therefore had to use some not fully formalized arguments in his completeness proofs. We show that, for the labelled calculus introduced in the previous Section, we can give a completeness proof close to Kripke's original argument but without any appeal to geometric intuition. For every sequent, the proof search either ends in a proof or fails, and the failed proof tree gives a Kripke countermodel.

6.1 Soundness

We reformulate first the semantical notions of Section 2 so that they apply to our labelled calculi:

DEFINITION 6.1 Let K be a frame with an accessibility relation $\mathcal{R}$ that satisfies the properties $*$. Let W be the set of variables (labels) used in derivations in **G3K***. An *interpretation* of the labels W in frame K is a function $[\![\cdot]\!] : W \to K$. A *valuation* of atomic formulas in frame K is a map $\mathcal{V} : AtFrm \to \mathcal{P}(K)$ that assigns to each atom P the set of nodes of K in which P holds; the standard notation for $k \in \mathcal{V}(P)$ is $k \Vdash P$.

Valuations are extended to arbitrary formulas by the following inductive clauses:

$k \Vdash \bot$ for no k,

$k \Vdash A \& B$ if $k \Vdash A$ and $k \Vdash B$,

$k \Vdash A \vee B$ if $k \Vdash A$ or $k \Vdash B$,

$k \Vdash A \supset B$ if $k \Vdash A$ implies $k \Vdash B$,

$k \Vdash \Box A$ if for all k', from $k\mathcal{R}k'$ follows $k' \Vdash A$,

$k \Vdash \Diamond A$ if there exists k' such that $k\mathcal{R}k'$ and $k' \Vdash A$.

DEFINITION 6.2 A sequent $\Gamma \to \Delta$ is *valid for an interpretation and a valuation* in K if for all labelled formulas $w : A$ and relational atoms oRr in Γ, whenever $[\![w]\!] \Vdash A$ and $[\![o]\!]\mathcal{R}[\![r]\!]$ in K, then for some $l : B$ in Δ, $[\![l]\!] \Vdash B$. A sequent is *valid* if it is valid for every interpretation and every valuation in a frame.

---

[30] Such devices include for example 'bulldozing' methods for imposing irreflexivity.

THEOREM 6.3 If the sequent $\Gamma \to \Delta$ is derivable in **G3K***, then it is valid in every frame with the properties $*$.

Proof. By induction on the derivation of $\Gamma \to \Delta$ in **G3K***. If it is an initial sequent, then there is a labelled atom $w : P$ both in Γ and in Δ so the claim is obvious, and similarly if the sequent is conclusion of $L\bot$ since for no valuation can $\bot$ be forced at any node.

If $\Gamma \to \Delta$ is a conclusion of a propositional rule, assume the rule is $L\&$ with the premiss $w : A, w : B, \Gamma' \to \Delta$. Assume that for an arbitrary assignment and interpretation, all the formulas in Γ are valid. Since $[\![w]\!] \Vdash A\&B$ is equivalent to $[\![w]\!] \Vdash A$ and $[\![w]\!] \Vdash B$, the inductive hypothesis, i.e., validity of $w : A, w : B, \Gamma' \to \Delta$ for every interpretation, gives the desired conclusion.

If $\Gamma \to \Delta$ is a conclusion of a modal rule, say $R\square$, with the premiss $wRo, \Gamma' \to \Delta', o : A$, assume by the induction hypothesis that the premiss is valid. Let $[\![\cdot]\!]$ be an arbitrary interpretation that validates all the formulas in Γ'. We claim that one of the formulas in Δ' or $w : \square A$ is valid under this intepretation. Let k be an arbitrary element of K such that $[\![w]\!]Rk$; Let $[\![\cdot]\!]'$ be the interpretation identical to $[\![\cdot]\!]$ except possibly on o, where we set $[\![o]\!]' \equiv k$. Clearly $[\![\cdot]\!]'$ validates all the formulas in the antecedent of the premiss, so it validates a formula in Δ' or $o : A$ (the alternative being independent of the choice of $[\![o]\!]'$). In the former case we have that also $[\![\cdot]\!]$ validates a formula in Δ', in the latter that $[\![\cdot]\!]$ validates $w : \square A$.

If the sequent is a conclusion of a mathematical rule without eigenvariables, let the rule be for instance *Trans*:

$$\frac{wRr, wRo, oRr, \Gamma \to \Delta}{wRo, oRr, \Gamma \to \Delta}$$

Let $[\![w]\!]\mathcal{R}[\![o]\!]$ and $[\![o]\!]\mathcal{R}[\![r]\!]$. Since $\mathcal{R}$ satisfies transitivity by assumption, we have $[\![w]\!]\mathcal{R}[\![r]\!]$, so validity of the premiss gives validity of the conclusion.

If the sequent is a conclusion of a mathematical rule with eigenvariables, let the rule be for instance *Directedness*:

$$\frac{oRl, rRl, wRo, wRr, \Gamma \to \Delta}{wRo, wRr, \Gamma \to \Delta}$$

Here l is an eigenvariable. Since by hypothesis the frame is directed, if $[\![w]\!]\mathcal{R}[\![o]\!]$ and $[\![w]\!]\mathcal{R}[\![r]\!]$, there exists d such that $[\![o]\!]\mathcal{R}d$ and $[\![r]\!]\mathcal{R}d$. The premiss is valid for all interpretations, in particular for one that coincides with $[\![\cdot]\!]$ on all labels, except possibly on l where it is assigned value d (this choice is possible because l is an eigenvariable). It follows that one of the formulas in Δ holds under this interpretation. **QED.**

6.2 Completeness

The proof of completeness follows the pattern of the proof of completeness for predicate logic.[31]

The idea we pursue with the labelled system is the same as in Kripke's proof, but instead of looking for a failed search of a countermodel, we look directly for a proof: To see whether a formula is derivable, we check if it is universally valid, that is, valid at an arbitrary world for an arbitrary valuation, $w \Vdash A$. This is translated to a sequent $\rightarrow w : A$ in our calculus. The rules of the calculus applied backwards give equivalent conditions until the atomic components of A are reached. It can happen that we find a proof, or that we find that a proof does not exist either because we reach a stage where no rule is applicable, or because we go on with the search forever. In the two latter cases the attempted proof itself gives a countermodel.

THEOREM 6.4 Let $\Gamma \rightarrow \Delta$ be a sequent in the language of **G3K***. Then either the sequent is derivable in **G3K*** or it has a Kripke countermodel with properties $*$.

Proof. We define for an arbitrary sequent $\Gamma \rightarrow \Delta$ in the language of **G3K*** a reduction tree by applying the rules of **G3K*** root first in all possible ways. If the construction terminates we obtain a proof, else the tree becomes infinite. By König's lemma an infinite tree has an infinite branch that is used to define a countermodel to the endsequent.

1. *Construction of the reduction tree.* The reduction tree is defined inductively in stages as follows: Stage 0 has $\Gamma \rightarrow \Delta$ at the root of the tree. Stage $n > 0$ has two cases:

    Case I: If every topmost sequent is an initial sequent or a conclusion of $L\bot$ or of a zero-premiss mathematical rule, the construction of the tree ends.

    Case II: If not every topmost sequent is an initial sequent or a conclusion of $L\bot$ or of a zero-premiss mathematical rule, we continue the construction of the tree by writing above those topsequents that are not initial, nor conclusions of $L\bot$ or of a zero-premiss mathematical rule, other sequents that are obtained by applying root-first the rules of **G3K*** whenever possible, in a given order.

    There are $10 + r$ different stages, 10 for the rules of the basic modal systems, r for the mathematical rules. At stage $n = 10 + r + 1$ we repeat stage 1, at stage $n = 10 + r + 1$ we repeat stage 2, and so on for every n.

    We start, for $n = 1$, with $L\&$: For each topmost sequent of the form

$$w_1 : B_1 \& C_1, \ldots, w_m : B_m \& C_m, \Gamma' \rightarrow \Delta$$

---
[31] As in Negri and von Plato (2001), section 4.4.

where $B_1\&C_1,\ldots,B_m\&C_m$ are all the formulas in Γ with a conjunction as the outermost logical connective, we write

$$w_1 : B_1, w_1 : C_1,\ldots,w_m : B_m, w_m : C_m, \Gamma' \to \Delta$$

on top of it. This step corresponds to applying root first m times rule $L\&$.

For $n = 2$, we consider all the sequents of the form

$$\Gamma \to w_1 : B_1\&C_1,\ldots,w_m : B_m\&C_m, \Delta'$$

where $w_1 : B_1\&C_1,\ldots,w_m : B_m\&C_m$ are all the labelled formulas in the succedent with a conjunction as the outermost logical connective. We write on top of them the 2^m sequents

$$\Gamma \to w_1 : D_1,\ldots,w_m : D_m, \Delta'$$

where D_i is either B_i or C_i and all possible choices are taken. This is equivalent to applying $R\&$ root first successively with principal labelled formulas $w_1 : B_1\&C_1, \ldots, w_m : B_m\&C_m$.

For $n=3$ and $n=4$ we consider $L\vee$ and $R\vee$ and define the reductions symmetrically to the cases $n = 2$ and $n = 1$, respectively.

For $n = 5$, for each topmost sequent that has the labelled formulas $w_1 : B_1 \supset C_1, \ldots, w_m : B_m \supset C_m$ with implication as the outermost logical connective in the antecedent, Γ' the other formulas, and succedent Δ, we write on top of it the 2^m sequents

$$w_{i_1} : C_{i_1},\ldots,w_{i_k} : C_{i_k}, \Gamma' \to w_{j_{k+1}} : B_{j_{k+1}},\ldots,w_{j_m} : B_{j_m}, \Delta$$

Here $i_1,\ldots,i_k \in \{1,\ldots,m\}$ and $j_{k+1},\ldots,j_m \in \{1,\ldots,m\} - \{i_1,\ldots,i_k\}$. This step, perhaps less transparent because of the double indexing, corresponds to the root-first application of rule $L\supset$ with principal formulas $w_1 : B_1 \supset C_1$, $\ldots, w_m : B_m \supset C_m$.

For $n = 6$, we consider all the labelled sequents that have implications in the succedent, say $w_1 : B_1 \supset C_1,\ldots,w_m : B_m \supset C_m$, and Δ' the other formulas, and write on top of them

$$w_1 : B_1,\ldots,w_m : B_m, \Gamma \to w_1 : C_1,\ldots,w_m : C_m, \Delta'$$

that is, apply root first m times rule $R \supset$.

For $n = 7$, we consider all topsequents with modal formulas $w_1 : \Box B_1, \ldots, w_m : \Box B_m$ and relational atoms $w_1 R o_1, \ldots, w_m R o_m$ in the antecedent, and write on top of these sequents the sequents

$$o_1 : B_1, \ldots, o_m : B_m, w_1 : \Box B_1, \ldots, w_m : \Box B_m, w_1 R o_1, \ldots, w_m R o_m, \Gamma' \to \Delta$$

that is, apply m times rule $L\Box$.

For $n = 8$, let $w_1 : \Box B_1, \ldots, w_m : \Box B_m$ be all the formulas with $\Box$ as the outermost connective in the succedent of topsequents of the tree, and let Δ' be the other formulas. Let $r_1, \ldots, r_m$ be fresh variables, not yet used in the reduction tree, and write on top of each sequent the sequent

$$w_1 R r_1, \ldots, w_m R r_m, \Gamma \to \Delta, r_1 : B_1, \ldots, r_m : B_m$$

that is, apply m times rule $R\Box$.

For $n = 9$, let $w_1 : \Diamond B_1, \ldots, w_m : \Diamond B_m$ be all the formulas with $\Diamond$ as the outermost connective in the antecedent of topsequents of the tree, and let Γ' be the other formulas. Let $l_1, \ldots, l_m$ be fresh variables, and write on top of each sequent the sequent

$$w_1 R l_1, \ldots, w_m R l_m, l_1 : B_1, \ldots, l_m : B_m, \Gamma' \to \Delta$$

that is, apply m times rule $L\Diamond$.

For $n = 10$, consider all topsequents with modal formulas $w_1 : \Diamond B_1, \ldots, w_m : \Diamond B_m$ in the succedent and relational atoms $w_1 R o_1, \ldots, w_m R o_m$ in the antecedent, and write on top of these sequents the sequents

$$w_1 R o_1, \ldots, w_m R o_m, \Gamma \to \Delta', w_1 : \Diamond B_1, \ldots, w_m : \Diamond B_m, o_1 : B_1, \ldots, o_m : B_m$$

that is, apply m times rule $R\Diamond$.

Finally, for $n = 10 + j$, we consider the generic case of a mathematical rule, that is, a rule for the relation R. For systems with the *subterm property*,[32] the mathematical rules need to be instantiated only on terms in the conclusion or on eigenvariables. Thus, if the system contains rule *Ref*, instances of that rule consist in adding to the antecedent all the relational atoms wRw for w in $\Gamma \to \Delta$; With a rule with eigenvariables, such as seriality, the step for that rule adds all the atoms of the form wRo for w in

---

[32] Cf. section 6 of Negri (2005).

$\Gamma \to \Delta$ and o a fresh variable. Observe that because of height-preserving substitution and height-preserving admissibility of contraction, once a rule with eigenvariables has been considered, it need not be instantiated again on the same principal formulas. If it is a rule such as *Trans*, consider all the sequents with a pair of atoms of the form wRo, oRr in the antecedent and write on top of them the sequents with the atoms wRr added.

For any n, for each sequent that is neither initial, nor conclusion of $L\bot$, nor of a zero-premiss mathematical rule, nor treatable by any one of the above reductions, we write the sequent itself above it.

If the reduction tree is finite, all its leaves are initial or conclusions of $L\bot$, or of zero-premiss mathematical rules, and the tree, read from the leaves to the root, yields a derivation.

2. *Construction of the countermodel:* If the reduction tree is infinite, it has an infinite branch. Let $\Gamma_0 \to \Delta_0 \equiv \Gamma \to \Delta, \Gamma_1 \to \Delta_1 \ldots, \Gamma_i \to \Delta_i, \ldots$ be one such branch. Consider the sets of labelled formulas and relational atoms

$$\Gamma \equiv \bigcup_{i \geq 0} \Gamma_i \quad \Delta \equiv \bigcup_{i \geq 0} \Delta_i$$

We define a Kripke model that forces all the formulas in Γ and no formula in Δ and is therefore a countermodel to the sequent $\Gamma \to \Delta$.

Consider the frame K the nodes of which are all the labels that appear in the relational atoms in Γ, with their mutual relationships expressed by the wRo's in Γ. Clearly, the construction of the reduction tree imposes the frame properties of the countermodel, for instance, in the system **G3S4**, the constructed frame is reflexive and transitive. The model is defined as follows: For all atomic formulas $w : P$ in Γ, we stipulate that $w \Vdash P$ in the frame, and for all atomic formulas $o : Q$ in Δ we stipulate that $o \nVdash Q$. Since no sequent in the infinite branch is initial, this choice can be coherently made, for if there were the same labelled atom in Γ and in Δ, then, since the sequents in the reduction tree are defined in a cumulative way, for some i there would be a labelled atom $w : P$ both in the antecedent and in the succedent of $\Gamma_i \to \Delta_i$.

We then show inductively on the weight of formulas that A is forced in the model at node w if $w : A$ is in Γ and A is not forced at node w if $w : A$ is in Δ. Therefore we have a countermodel to the endsequent $\Gamma \to \Delta$.

If A is $\bot$, it cannot be in Γ because no sequent in the branch contains $w : \bot$ in the antecedent, so it is not forced at any node of the model.

If A is atomic, the claim holds by the definition of the model.

If $w : A \equiv w : B\&C$ is in Γ, there exists i such that $w : A$ first appears in Γ_i, and therefore, for some $l \geq 0$, $w : B$ and $w : C$ are in Γ_{i+l}. By the induction hypothesis, $w \Vdash B$ and $w \Vdash C$, and therefore $w \Vdash B\&C$.

If $w : A \equiv w : B\&C$ is in Δ, consider the step i in which the reduction for A applies. This gives a branching, and one of the two branches belongs to the infinite branch, so either $w : B$ or $w : C$ is in Δ, and therefore by the inductive hypothesis, $w \not\Vdash B$ or $w \not\Vdash C$, and therefore $w \not\Vdash B\&C$.

If $w : A \equiv w : B \vee C$ is in Γ, we reason similarly to the case of $w : A \equiv w : B\&C$ in Δ.

If $w : A \equiv w : B \vee C$ is in Δ, we argue as with $w : A \equiv w : B \vee C$ in Γ.

If $w : A \equiv w : B \supset C$ is in Γ, then either $w : B$ is in Δ or $w : C$ is in Γ. By the inductive hypothesis, in the former case $w \not\Vdash B$, and in the latter $w \Vdash C$, so in both cases $w \Vdash B \supset C$.

If $w : A \equiv w : B \supset C$ is in Δ, then for some i, $w : B \in \Gamma_i$ and $w : C \in \Delta_i$, so by the inductive hypothesis $w \Vdash B$ and $w \not\Vdash C$, so $w \not\Vdash B \supset C$.

If $w : A \equiv w : \Box B$ is in Γ, we consider all the relational atoms wRo that occur in Γ. If there is no such atom, then the condition that for all o accessible from w in the frame, $o \Vdash B$ is vacuously satisfied, and therefore $w \Vdash \Box B$ in the model. Else, for any occurrence of wRo in Γ we find, by the construction of the reduction tree, an occurrence of $o : B$ in Γ. By the inductive hypothesis, $o \Vdash B$, and therefore $w \Vdash \Box B$ in the model.

If $w : A \equiv w : \Box B$ is in Δ, consider the step at which the reduction for $w : A$ applies. We then find $o : B$ in Δ for some o with wRo in Γ. By the induction hypothesis, $o \not\Vdash B$, and therefore $w \not\Vdash A$.

The cases of $w : A \equiv w : \Diamond B$ in Γ and of $w : A \equiv w : \Diamond B$ in Δ are symmetric to those of $w : A \equiv w : \Box B$ in Δ and of $w : A \equiv w : \Box B$ in Γ, respectively. **QED.**

COROLLARY 6.5 *If a sequent $\Gamma \to \Delta$ is valid in every Kripke model with the frame properties $*$, then it is derivable in the system* **G3K***.

In case the system has an irreflexive accessibility relation, we have a zero-premiss mathematical rule of the form

$$wRw, \Gamma \to \Delta.$$

A sequent of this form cannot appear in the infinite branch, and therefore the countermodel will be irreflexive by construction. The problem with properties such as irreflexivity in the Henkin-style completeness proof thus disappears with our approach.

7 Conclusion and further work

We have reviewed here two main styles of completeness proofs for modal logic, Kripke's original proofs and Henkin-style proofs, and discussed their relative merits. Although Kripke's original proofs were more informative, Henkin-style proofs have been preferred in the literature on modal logic because of the difficulties in formalizing Kripke's original proof.

There are two main trends in the recent literature on the proof theory of modal logic: one that enriches the language of sequents by the use of labels (cf. Negri 2007 for references to the vast literature), another that avoids the use of labels. Recent variants of the latter approach include the systems of *nested sequents* (Kashima 1994), *tree-sequents* (Cerrato 1996), *deep sequents* (Brünnler 2006 and Stouppa 2007), and *tree-hypersequents* (Poggiolesi 2008). These works can be regarded as formalizations of Kripke's original approach even if they do not explicitly refer to Kripke's own contributions. Also, the treatment of modal systems with geometric frame conditions has so far remained out of their scope.

Section 1.5 of Boretti (2008) contains a useful methodological discussion of labelled and unlabelled systems. In her words, "whereas the semantic notions are explicitly internalised into the labelled calculi in the form of the syntactical counterparts of forcing $(x:A)$ and accessibility relation (xRy), tree-hypersequents and deep sequent systems hide their relational semantics under a more complex syntax."

We have presented here a labelled sequent system that simplifies Kripke's tableau method thanks to the fact that the accessibility relation is an explicit part of the syntax and not an implicit property of proof-trees. A wide class of modal systems is covered and a uniform, simple, and direct proof of completeness obtained that does not present the shortcomings of the original Kripke proofs, nor the limitations of Henkin-style proofs. Completeness proofs for first-order modal logic and for provability logic, along the lines of the method presented here, appear in Negri and von Plato (2008). A similar treatment for non-normal modal logics should not present any extra difficulty, and is left to future work.

8 Acknowledgements

The question that gave origin to the research done in this work, about the possibility of a direct completeness proof for the labelled systems introduced in Negri (2005), was first posed to me by Erik Palmgren during a conference at Benediktbeuern (Baviera) in 2005. A completeness proof for systems of temporal logic along the lines presented in this work was first presented in a seminar in Uppsala in 2007. The contents of this paper have been presented in seminars during 2008 at the Universities of Helsinki and Pisa. Parts of it have also been presented the same year in talks at the "Workshop on Proof Theory" in Bern and at the workshop on "Advances in Constructive Topology and Logical Foundations" in Padua.

Comments to my presentations and to the manuscript, in particular for the latter by Raul Hakli and Giuseppe Primiero, are gratefully acknowledged. In 2005, Ilpo Halonen gave a course on the "Birth and development of possible worlds semantics" at the University of Helsinki. The unpublished course material that he has maintained available through the web has been very useful to me, especially in respect to the contributions of Finnish philosophers to the rise of Kripke semantics.

BIBLIOGRAPHY

Bayart, A. (1958). "La correction de la Logique de Modale du 1er and 2eme ordre", *Logique et analyse*, vol. 1, pp.28–45.

Bayart, A. (1959). "Quasi-adequation de la logique modale de 2eme ordre", *Logique et analyse*, vol. 2, pp.99–121.

Bayart, A. (1966). Review of Kripke (1959), *The Journal of Symbolic Logic*, vol. 31, pp.276–277.

Boretti, B. (2008). *Proof analysis in temporal logic*, Ph.D. thesis, University of Milan.

Brünnler, K. (2006). "Deep sequent systems for modal logic", in G. Governatori, I. Hodkinson, and Y. Venema (eds), *Proceedings of the Sixth Conference on Advances in Modal Logic*, AiML 2006, College Publications, vol. 6, pp.107–119.

Bull, R. and Segerberg, K. (1984). *Basic modal logic*, in D. Gabbay and F. Guenther (eds) *Handbook of Philosophical Logic*, vol. 2, pp.1–88, Kluwer, Dordrecht. Second edition 2001.

Cerrato, C. (1996). "Modal tree-sequents", *Mathematical Logic Quarterly*, vol. 42, pp.197–210.

Copeland, B.J. (2002). "The Genesis of Possible Worlds Semantics", *Journal of Philosophical Logic*, vol. 31, pp.99–137.

Cresswell, M. J. (1967). "A Henkin completeness for T", *Notre Dame Journal of Formal Logic*, vol. 8, pp.186–190.

Goldblatt, R. (2005). *Mathematical modal logic: A view of its evolution*, in *Handbook of the History of Logic*, vol. 6, D. Gabbay and J. Woods (eds), Elsevier, Amsterdam.

Hakli, R. and Negri, S. (2008). "Does the deduction theorem fail for modal logic?", manuscript.

Halonen, I. (2005). *Mahdollisten maailmojen semantiikan synty ja kehitys*, slides in Finnish for a course given at the University of Helsinki, available at http://www.helsinki.fi/hum/fil/filosofia/

Henkin (1949). "The completeness of the first-order functional calculus", *The Journal of Symbolic Logic*, vol.14, pp.159-166.

Jonsson, B. and Tarski, A. (1951). "Boolean algebras with operators I", *American Journal of Mathematics*, vol. 23, pp.891–939.

Kanger, S. (1957). *Provability in Logic*, Almqvist & Wiskell, Stockholm.

Kaplan, D. (1966). Review of Kripke (1959), *The Journal of Symbolic Logic*, vol. 31, pp.120–122.

Kashima, R. (1994). "Cut-free sequent calculi for some tense logics". *Studia Logica*, vol. 53(1), pp.119–136.

Kripke, S. (1959). "A completeness theorem in modal logic", *The Journal of Symbolic Logic*, vol. 24, pp.1–14.

Kripke, S. (1959a). "Semantical analysis of modal logic" (abstract), *The Journal of Symbolic Logic*, vol. 24, pp.323–324.

Kripke, S. (1963). "Semantical analysis of modal logic I. Normal modal propositional calculi", *Zeitschrift f. math. Logik und Grund. d. Math.*, vol. 9, pp.67–96.

Kripke, S. (1965). *Semantical analysis of modal logic II. Non-normal modal propositional calculi*, in J. W. Addison, L. Henkin and A. Tarski (eds.) *The Theory of Models*, pp.206–220. North-Holland, Amsterdam.

Lemmon, E.J. (1960). "An extension algebra and the modal system T", *Notre Dame Journal of Formal Logic*, vol. 1, pp.3–12.

McKinsey, J.C.C. and Tarski, A. (1948). "Some theorems about the sentential calculi of Lewis and Heyting". *The Journal of Symbolic Logic*, vol.13, pp.1–15.

Makinson, D. (1966). "On some completeness theorems in modal logic", *Zeitschrift f. math. Logik und Grund. d. Math.*, vol. 12, pp.379–384.

Makinson, D. (1970). Review of Kripke (1965), *The Journal of Symbolic Logic*, vol. 35, p.135.

Negri, S. (2003). "Contraction-free sequent calculi for geometric theories, with an application to Barr's theorem", *Archive for Mathematical Logic*, vol. 42, pp.389–401.

Negri S. (2005). "Proof analysis in modal logic", *Journal of Philosophical Logic*, vol. 34, pp.507–544.

Negri S. (2007). *Proof analysis in non-classical logics*, in C. Dimitracopoulos, L. Newelski, D. Normann, and J. Steel (eds), *ASL Lecture Notes in Logic*, vol. 28, pp.107–128.

Negri, S. and von Plato, J. (2001). *Structural Proof Theory*, Cambridge University Press, Cambridge.

Negri, S. and von Plato, J. (2008). *Proof Analysis: A Contribution to Hilbert's Last Problem*, book manuscript.

Poggiolesi, F. (2008). "A cut-free simple sequent calculus for modal logic S5", *The Review of Symbolic Logic*, vol. 1, pp.3–15.

Stouppa, P. (2007). "A deep inference system for the modal logic S5", *Studia Logica*, vol. 85, pp.199–214.

Åqvist, L. (1970). Review of Makinson (1966), *The Journal of Symbolic Logic*, vol. 35, pp.135–136.

Judgement, Belief and Acceptance

MARIA VAN DER SCHAAR

ABSTRACT. The notion of judgement plays a central role in constructive type theory. It nearly has disappeared, though, from modern logic and philosophy. The empiricist, naturalist term 'belief' became the substitute for the rational term 'judgement.' L.J. Cohen has argued that there is a two-fold ambiguity involved in the term 'belief,' of which the act of acceptance is the notion that comes closest to the notion of act of judgement. It is argued here that the two notions cannot be identified. Acceptance is rather what is left of judgement after the metamathematical turn: it has become an act of arbitrary choice. A four-fold ambiguity of the term 'belief' has to be acknowledged, instead. 'Belief' may refer to the act of judgement, conviction, opinion, or faith. This four-fold ambiguity has its origin in the philosophy of John Locke. For Locke, these four different types of acts and states fall under one category: They can all be explained in terms of rational judgement. It is argued here that these acts and states do not have a genus in common, and that there is thus an important reason to disambiguate the term 'belief,' and to consider the act of judgement to be *sui generis*.

> *Tout a été dit. Sans doute.*
> *Si les mots n'avaient pas changé de sens;*
> *et les sens, de mots.*
> Jean Paulhan

1 Introduction

Within Constructive Type Theory the notion of judgement plays a central role, and Göran Sundholm has written on the topic since his inaugural address in 1988. Since that time I have had the opportunity to talk with him about the notion, and its two counterparts in modern analytic philosophy: assertion and belief. I leave the topic of assertion for another occasion, and I will focus here on judgement and belief. In section 2 I will use Constructive Type Theory to show that logic and epistemology are in need of the notion of judgement. In section 3, an analysis is given of the different meanings of the modern term 'belief.' In these sections the question is raised: What has made it possible that the term 'judgement' has disappeared from logic, and that in modern analytic philosophy the term has been replaced by the term 'belief'? This change in modern thought goes back to John Locke, and I need to devote a separate section to the philosophical reasons Locke has had for introducing the notion of belief with all its ambiguities (section 4). In section 5, I show how the constructivist account of judgement presented in section 1, the systematic analysis of the term 'belief' presented in section 2, and the historical analysis of that term presented in section 3, can be used to evaluate Jonathan Cohen's proposal to disambiguate the term 'belief.'

2 Judgement and Knowledge in Constructive Type Theory

The premises and conclusions in our reasoning are judgements and assertions. The notion of judgement is thus central to logic insofar as it takes our reasoning to be a central object of study. The notion of judgement has disappeared, though, from modern logic. It is often claimed that Frege's anti-psychologism has been responsible for the disappearance of the notion of judgement from logic. According to Frege, the validity of an inference is founded upon truth-relations between objective propositions. By making a distinction between the subjective act of judging and an objective proposition that functions as judgemental content, logic and psychology could be separated. The separation of the psychological (*das Psychologische*) from the logical (*das Logische*) alone cannot explain, though, the banishment of the notion of judgement from logic. In his *Begriffsschrift* (1879), Frege considers logic to be an epistemological project. The axioms of the *Begriffsschrift* are preceded by both the horizontal stroke, as a sign that what follows is a judgeable content, and the judgemental stroke, a sign showing that the content has actually been judged. The axioms are pieces of knowledge in which insight into such concepts as negation and implication are made manifest. The logical theorems that are flawlessly derived from these axioms by an insight-preserving inference rule are pieces of knowledge, too. The judgement stroke that precedes the axioms and theorems of the *Begriffsschrift* is thus also a sign showing that the judgemental content is known. Frege's logic is thus foundationalist in character[1], and is in need of the epistemic notion of judgement.

In the nineteenth century, another development started with Boole's conception of logic as an uninterpreted calculus.[2] The idea of an uninterpreted calculus is also present in Hilbert's conception of mathematics, to which both Frege and Brouwer opposed their conception of mathematics, the one on realist, the other on intuitionist terms. Frege and Brouwer shared the idea that mathematics and logic are to be understood as having content. For Hilbert, only metamathematics has content; The objects of its study are formal axiomatic systems without content. And the axioms of a logical system are a matter of choice. As soon as the axioms are set,

---
[1] Cf. Sundholm (2003), p.110.
[2] Cf. van Heijenoort (1967) and Sundholm (1988).

they become 'true,' as Hilbert writes to Frege in a reaction to his foundationalist conception of axioms:

> "Wenn sich die willkürlich gesetzten Axiome nicht einander widersprechen mit sämtlichen Folgen, so sind sie wahr."[3]

Although not all sets of axioms are equally good - the axioms should not contradict each other - , the axioms are essentially the result of an act of arbitrary choice. The notion of judgement has thus disappeared from logic. What is left is an act of arbitrary choice. I will show that this notion of act of choice plays an important role in modern philosophy under the name of 'acceptance' (see section 5). On the Hilbertian conception of logic, it is no longer possible to conceive of logic as providing a norm for our reasoning, and the philosophical notions of knowledge and truth fall outside the scope of logic.

Constructive Type Theory (CTT) has reconsidered the question what logic is, and its answer is: Logic is the theory of demonstrative science, aiming at the conceptual foundations of the demonstrative sciences. One of the central questions in CTT is therefore: What are the inference rules, and to what do these rules apply? The premises and conclusions in our reasoning are judgements, and the inference rules apply to them, according to CTT; Judgement is thus a central concept in logic, again.

An inference rule, such as &-Elimination, may be applied to a judgement of the form *A&B is true*, thus obtaining the judgement *A is true* (or *B is true*) as a conclusion, where the judgement that *A&B is true* presupposes that *A* is a proposition and that *B* is a proposition. If a judging agent is entitled to judge that *A&B is true*, and applies the rule for &-Elimination to the left side, he is entitled to judge that *A is true*.

One may speak of the judgement *A is true* before it is actually judged, for example, when one wonders under what condition one is entitled to judge that *A is true*. A judgement before it is made is called a *judgement candidate*.[4] Within CTT, the judgement candidate is explained by what one has to *know* in order to be entitled to make the relevant judgement. It is essential to CTT that one is entitled to judge that *A is true*, only if one has constructed a proof-object a for the proposition A, that is, only if one is entitled to judge that a is a proof-object for the proposition A ($a:A$). For example, one is entitled to judge that *A&B is true*, only if one has obtained a proof-object for the proposition *A&B*, which may be a pair consisting of a proof-object for *A* and a proof-object for *B*. Because one is entitled to judge only if the epistemic condition is fulfilled, the notion of judgement in CTT may be called normative, where the norm is an epistemic one. To capture this normative, epistemic aspect of the judgment I will speak of the act of judgement as *rational*:

---

[3] Hilbert (1899), p.66.
[4] For more on the notion of assertion candidate or judgement candidate, see van der Schaar (2007).

The fact that one has obtained a proof-object for the relevant judgement gives one an (epistemic) reason to make that judgement.

The judgement candidate is to be distinguished both from the act of judging and the judgement made that results from the act of judging. The judgement made is necessarily constituted in the act of judgement: no judgement made without a prior act of judgement, and there is no act of judgement that does not result in a judgement made. The act of judgement has a temporary existence, whereas the judgement made is valid after the act of judgement has been made, being an abstract entity, created at a certain place and time, but existing outside space and time. The judgement candidate, though, is independent of both these notions, for it may never be judged. The term 'judgement' thus has a three-fold ambiguity: it may mean the *act of judgement*, the *judgement made*, or the *judgement candidate*. We can now be more specific about the question what the premises and conclusions of our reasonings are: They are judgements made rather than acts of judgement, because the act in which the premise is judged has passed away by the time we have reached the conclusion; and they are judgements made rather than judgements candidate, because the premises and conclusions in our reasoning have been judged.[5]

The three meanings of the terms 'judgement' are to be distinguished from the meaning of the term 'proposition.' Whereas the judgement is explained in epistemic terms, the proposition is explained in non-epistemic terms. A proposition can be considered as a set of its proof-objects, and it is defined by its canonical proof-objects; The proposition $A \& B$ is thus a set, a non-epistemic entity, and it is defined by the pair of proof-objects for A and for B (in this order), which pair is the canonical proof-object for the proposition $A \& B$.

The judgement made, resulting from the act of judging, being an abstract entity, cannot be identified with the capacity to judge that the judging agent may have obtained through an act of judgement. Such a capacity is not necessarily constituted in the act of judging, for it depends, for example, on the fact that the judger is still alive, and has memory. A capacity to judge can be explained as the psychological result of an act of judgement, or in terms of its actualisations, the corresponding acts of judgement. A capacity to judge therefore stands under the same epistemic norm as the act of judgement.

In CTT, (a piece of) knowledge is explained as the justified judgement.[6] If the judgements *A is true* and *B is true* are justified, the judgement *A&B is true* becomes justified by applying &-Introduction on the former two judgements. The act that justifies the judgement that *A&B is true* is thus an act of inference from known premises. Ultimately, the first premises are justified by a non-inferential

---

[5]The question whether the inference rules apply to judgement products or to judgement candidates I have dealt with in van der Schaar (2007), pp.74-75.

[6]Martin-Löf (1998), p.110.

act. In mathematics, such a non-inferential act is an act of intuitive insight. Both the inferential and the non-inferential act may be called a *cognitive act*; in both cases the result is a justified or evident judgement, that is, knowledge. The act of intuitive insight results in an axiom, and the act of inference results in a theorem. It is essential to knowledge that it is the result of a cognitive process; Without a cognitive act, there is no justified judgement, and therefore no knowledge; The judgement is made evident or justified by a cognitive act, thus resulting in a piece of knowledge, which is called the *knowledge product*.

A judgement's being evident or justified consists of two aspects. A judgement's being evident primarily means that it is grounded through a cognitive act; This is the objective side of its being evident. Standardly, the judgement's being grounded is accompanied by a certain degree of *conviction*, which is its subjective side. Without a clear memory of the first cognitive act, the degree of conviction will be weakened. There are deviant cases in which we have grounded our judgement through a cognitive act, but have not obtained a degree of conviction that standardly accompanies our knowledge. The act of inference that makes one's judgement grounded may be a complicated one, so that we become convinced of the correctness of the conclusion only if we have seen the approval of colleagues. Although a judgement's being evident has an objective side, this does not imply that 'evident' means *infallibly evident*. The cognitive act through which our judgement becomes evident is a fallible act in the sense that it may turn out that we have made a mistake in what counted as an act of inference.

If the judgement is made evident, it can be called *correct* insofar as it *can* be justified or made evident. As soon as the judgement is made evident, the judgement can be said to have been correct before it was made evident; It is thus the judgement candidate that is the bearer of correctness. Judgemental correctness is an epistemic notion insofar as it is explained in terms of justification, or being evident.

Although a judgement is always evident to a judging agent, and although it is the judging agent's cognitive act that makes the judgement evident, and thus correct, what counts as an entitlement to make a certain judgement is not to be determined by the judging agent. Whether a certain cognitive act makes the judgement evident is determined by the explanation of the judgement simpliciter. The judgement A *is true* is explained in terms of what one has to know in order to be entitled to make that judgement, and one is entitled to judge that A is true, if one has obtained a canonical proof-object for A, or if one has obtained a non-canonical proof-object for A, which is a method for obtaining a canonical proof-object for A, and there is no other way to be entitled to make that judgement. The condition under which one is entitled to make a judgement is thus not to be settled by the judging agent.

The relation between cognitive act and knowledge product can be understood as a special case of the relation between act of judgement and judgement made. It is to be doubted, though, that the cognitive act is to be defined as the act of judgement that is justified, for there is not a concept that can be used to distinguish the cognitive act from the non-cognitive judgemental act. One cannot use the concept of justification as the specific difference for the cognitive act, for the notion of justification is understood in terms of the cognitive act: What makes a judgement justified is precisely the cognitive act. The notion of correctness can equally not be used to distinguish the cognitive act from other judgemental acts, because this notion is explained in terms of justification. The relation between act of judgement and cognitive act is thus not a *genus-species* relation. Acts of judgement are standardly cognitive acts; A non-cognitive judgemental act is to be understood as a privation.

Knowledge as it is defined by modern epistemologists is knowledge as a mental state that a person has whether asleep or awake, and such knowledge is standardly defined as justified or warranted true belief. The account of knowledge given in CTT suits a justified true belief account only to a certain extent. In the first place, the term 'belief' is to be understood in the sense of *judgement* as explained above, that is, judgement is already an epistemic, rational notion. In CTT, the judgement is correct, that is, knowable, if it is known, but this does not imply that correctness is part of the definition of knowledge. There is also an important difference between the notion of epistemic correctness and that of truth in standard accounts of knowledge, because the latter truth-notion is generally taken to be non-epistemic.

3   Judgement, Conviction, Opinion and Faith

In modern thought, the term 'belief' is highly ambiguous, and because not everyone is always conscious of these ambiguities, the discussions related to the following questions often lead to confusion: Is belief essentially a 'propositional attitude,' or can belief be non-linguistic? Is belief voluntary or involuntary? Is belief a normative notion, or can it be explained completely in naturalist terms? Are there degrees of belief, or is belief an all or nothing affair? Is faith a special case of belief, namely religious belief? Is belief part of our cognitive nature, or of our sensitive nature? What is the relation between dispositional belief and actual or occurrent belief? In all these questions, we have to make up our mind what is meant by 'belief.' In this section, I first give the different meanings of 'belief' in a systematic, non-diachronic way, in such a way that they cohere with related concepts in CTT, introduced in the former section. At the end of the section, I deal with the historical development that has lead to the different meanings and roles of the term 'belief,' and it will turn out that a separate section is needed on Locke's theory of judgement and belief to explain the philosophical background of the ambiguities of that term.

> 1. The term 'belief' may refer to (a) *the act of judgement* or to (b) *the capacity to make a certain judgement.* Judgement is the mental counterpart to the

speech act of assertion, and it is thus an all or nothing affair: We may affirm, deny, or withhold our judgement. The act of judgement as it is introduced in the former section is an epistemic notion: In order to be entitled to judge (or assert) one must have an epistemic reason or ground for one's judgement (or assertion). It is thus understood to be a rational, normative notion. Not every philosopher is willing to explain the notions of judgement and assertion in terms of epistemic grounds, but most of them agree that there is a norm for assertion, and its mental equivalent, the act judgement. The *capacity to judge* can then be explained as the psychological result of a rational act of judgement, presupposing a form of memory and interest on the part of the judger. In the first section, it has been proposed to explain knowledge in terms of 'judgement' rather than in terms of 'belief,', because of the ambiguity of the latter term. The term 'belief' in the explanation of knowledge is generally understood as translation of the term *doxa* in Plato's proposal of the definition of knowledge at the end of the *Theaetetus*. It is convincingly argued, though, that 'judgement' is a better translation of the term in the dialogue, because it is primarily the act of judgement that Plato has in mind.[7]

2. 'Belief' may also refer to a certain degree of *conviction*, a mental state rather then an act of the mind. Belief in this sense is a specific notion on the same level as doubt, disbelief and assurance. It is sometimes understood as a disposition to have a certain feeling, a feeling of assurance, and it is therefore no surprise when Hume understands belief to be part of our sensitive, rather than our cognitive nature. Belief in this sense is the conviction mentioned in the first section as the subjective side of a judgement's being evident. A certain conviction, though, may be the result of non-intellectual passion or custom as well. We profess certain beliefs in order to belong to a group, and conviction may then easily follow.[8] Our passions and our will may have at least an indirect influence on our convictions. 'Belief' may also have the generic sense of conviction, which comes in degrees. One may have such a low degree of belief (conviction) that one does not act upon it. When Bayesians capture the degree of belief of an ideally rational person in terms of the mathematical principles of probability theory, it is this meaning of the term they have in mind.

3. 'Belief' may further have the meaning of *opinion*. We call something *a matter of opinion*, and not a matter of fact, because we consider it to be a personal matter, and not something for which we may be able to give an objective ground. The term 'objective ground' might be misleading: all that

---
[7]Cf. Burnyeat (1990), p.69.
[8]Cf. Losonsky (2000).

is meant is the notion of epistemic ground as it is explained in section 1, that is, a ground that is determined by the explanation of the judgement in question, and thus counts as such for everyone who considers the matter. It is in this sense that knowledge and opinion are conceived as opposites. One is entitled to vote, that is, to express one's opinion concerning a political question, independent of the question whether one has grounds for one's opinion; One is simply entitled to vote when one has reached a certain age, and is citizen of the relevant country or town. Opinion may be explained as *privative judgement*: The *Oxford English Dictionary* (OED) has it that opinion is a "judgement resting on grounds insufficient for complete demonstration." And Kant explains opinion (*Meinung*) as a holding true that is both objectively and subjectively *insufficient* (in contrast to knowledge,*Wissen*, that is both objectively and subjectively sufficient, *Kritik der reinen Vernunft*, B 850). Our opinions may be accompanied by such a weak degree of conviction that we do not act upon them; Still, they may have a value in obtaining the truth:

> "Truth [...] has to be made by the rough process of a struggle between combatants fighting under hostile banners [...] [O]nly through diversity of opinion is there [...] a chance of fair play to all sides of the truth,"

as J.S. Mill put it in his argument for the freedom of expression of opinion.[9] Although the OED does acknowledge the verb 'to opine,' the term 'opinion' is generally understood as standing for a mental state or disposition. One may use the verb 'to assent' for the act in which one's opinion is actualised, either in overt terms or in silence.

4. 'Belief' may also be used in the sense of *faith*. 'Faith' primarily stands for trust, especially trust in a religious object or truth, but it may also be used in the sense of what is sometimes called 'animal faith,' a basic, unquestioned trust that is presupposed in all questioning and judgement. Faith is thus an attitude (*Einstellung*) we may have not only towards God, but, perhaps, also towards "eine unwankende Grundlage (s)einer Sprachspiele," as Wittgenstein put it in *Über Gewissheit* (§403).

The important question to be answered is whether these four different notions have a genus in common. If that were true, one would be entitled to use one term, such as 'belief,' for the different notions. I doubt, though, that this is the case. Suppose that they are all species of the genus *holding true*.[10] Do we really have a clear concept of holding true? Does it have degrees as conviction has, or is it an

---

[9]Mill (1859), p.254.

[10]Bolzano explains these notions in terms of *the act of judgement*. Bolzano, *Wissenschaftslehre*, III, §306. Compare van der Schaar (2007a).

all or nothing affair as is the case with judgement? Is it an *act* of the mind or a mental, dispositional *state*? Is it a normative notion? What do we hold true when we have faith in something? I am afraid that 'holding true' has to be confusingly vague in its meaning, if it were to function as genus in terms of which the concepts introduced above can be defined.

The English language has two terms, 'belief' and 'faith,' where the German language has only one term, 'Glaube' (a translator into Dutch equally has to use for both 'belief' and 'faith' the term 'geloof'). According to the OED, "*Belief* was the earlier word for what is now commonly called *faith*," that is, 'belief' originally meant "the mental action, condition or habit of trusting to or confiding in a person or thing," which is given as sense 1 of 'belief' in the OED. The term 'faith' "began in the 14th c. to be used to translate [the L. *fides*], and in course of time almost superseded 'belief,' esp. in theological language, leaving 'belief' in great measure to the merely intellectual process or state in sense 2." Sense 2 of 'belief' in the OED is: "mental acceptance of a proposition [...] as true on the ground of authority or evidence; assent of the mind to a statement [...], the mental condition involved in this assent." If we interpret these comments in accordance with the four-fold scheme mentioned above, the term 'belief' originally meant what is now called 'faith;' later, it became a substitute for both the term 'judgement' and 'opinion.' The fact that the English language has a preference for the term 'belief' instead of the term 'judgement' thus goes partly back to the fourteenth century.

In logic, the term 'judgement' was the standard term until the nineteenth century. At that time, there came a division between the empiricists, who use the term 'belief' under the influence of J.S. Mill's *A System of Logic* (1843), and the idealists, such as F.H. Bradley, who uses the term 'judgment' in *The Principles of Logic* (1883). Mill's use of the term 'belief' is in accordance with a frequent use of the term in Hume's writings.

Hume's use of the term 'belief' instead of the term 'judgement' enforced the non-intellectual character of the notion of belief, as it became part of an empiricist, naturalist program. Belief is something we partly share with animals, in which the central question is: 'How do we come to have the beliefs we have?', rather than: 'How can we justify our judgements?'. There is, according to Hume, no answer to the latter question, if it concerns the belief that the future will resemble the past, or any empirical belief that is based on this belief.

Hume's use of the term 'belief' is prepared by the use of the terms 'judgement' and 'belief' in John Locke's *Essay Concerning Human Understanding*. Locke made an exclusive distinction between the categories of knowledge and judgement, and he put all the relevant notions introduced above under the category of judgement. The terms 'judgement,' 'belief,' 'opinion,', 'assent,' 'conviction,' and 'faith' are sometimes used with a specific meaning, at other times they are used with a general meaning, thus becoming synonymous with the other terms in their general meaning. Locke consciously used these ambiguities to suit his philosophy, as I hope to make clear in the next section.

4 John Locke on Judgement and Belief

As we have seen in the first section, the notion of judgement is an epistemic one, and it derives its norm from the epistemic ground in terms of which the judgement is explained. In modern analytic philosophy, though, the notion of belief is not explained in epistemic terms; Knowledge is explained in terms of belief, and belief is understood as prior in the order of explanation to knowledge.

In Locke's *Essay*, judgement is not explained in terms of knowledge, but neither is knowledge explained in terms of judgement or belief. The two notions are explained independently of each other, and the reason Locke has for this separate explanation is that he treats knowledge and judgement as exclusive categories. Knowledge we have of infallible certainties, as those in mathematics; Knowledge concerns relations between ideas only. Our judgements relate to probabilities, and are always fallible: Newtonian science does not consist of pieces of knowledge, but of judgements, according to Locke. Our faculty of judgement is needed to account for error:

> "*Errour* is not a Fault of our Knowledge, but a Mistake of our Judgment giving Assent to that, which is not true."[11]

Locke's reason for separating the categories of knowledge and judgement is thus that knowledge is infallibly certain, whereas judgement is fallible.

For Locke, there is also an important connection between knowledge and judgement insofar as both are rational. Demonstrative knowledge is rational because it is the product of a rational act of inference; in knowledge, reason is infallible. Judgement is rational in the sense that the two ideas connected in judgement are always mediated by a third idea, the epistemic ground or reason for the judgement[12]. For Locke, all our beliefs are the product of a rational act of judgement, an act of non-demonstrative inference. Reason as we use it in our judgements is fallible, though, in the sense that it is possible that we make a perfect use of our reason, and still be mistaken. Insofar as the act of judgement is rational it is appropriate to speak of 'judgement' as the term is introduced in the former section. Insofar as Locke understands judgement and knowledge to be exclusive categories, it is rather the term 'opinion' that suits his concept of judgement. It is for this reason that Locke

---

[11] Locke (1690), iv.xx.1, p.706.

[12] "*Rational Knowledge*, is the perception of the certain Agreement, or Disagreement of any two *Ideas*, by the intervention of one or more other *Ideas*. *Judgment*, is the thinking or taking two *Ideas* to agree, or disagree, by the intervention of one or more *Ideas*, whose certain Agreement, or Disagreement with them it does not perceive, but hath observed to be frequent and usual," Locke (1690), iv.xvii.17, p.685. I have given an analysis of Locke's concepts of judgement and knowledge in, respectively, van der Schaar (2008) and van der Schaar (2009).

uses *Of Knowledge and Opinion* as title for the fourth book of the *Essay* that is devoted to these topics.[13]

Although the two categories knowledge and judgement are to be understood as exclusive, Locke does explain judgement in analogy with rational knowledge, as one can judge from the quotation given in footnote 12. Rational knowledge is 'the perception' of the (dis)agreement of two ideas by the mediation of a third idea; Judgement is the mere taking the two ideas to (dis)agree on the basis of a third idea. 'Perception' in Locke's explanation of knowledge is an act of perception, which may be an act of intuitive insight, an act of sensory perception, or an act of demonstrative inference. The act of perception is a cognitive act, and is called actual knowledge by Locke, which he distinguishes from habitual knowledge, a potentiality to make certain cognitive acts, resulting from a prior cognitive act. The analogy between knowledge and judgement explains that important theses that apply to knowledge also apply to judgement. For example, the distinction between actual and habitual knowledge, that is, between the cognitive act and knowledge as state, applies to judgement, too: There is a distinction between the rational act of judgement and a capacity to judge resulting from such an act. Further, knowledge is partly voluntary partly necessary, according to Locke, and the same holds for assent or judgement. Knowledge and judgement are voluntary in the sense that we may start an investigation into the truth of a proposition as the result of wanting to know whether it is true; Without our wanting to know there would have been no knowledge or judgement on our part. At the final stage of our research, though, the act of knowledge or judgement is involuntary, because the object insofar as it is discovered by us determines our knowledge and judgement. It is in this sense that knowledge "has a great Conformity with our Sight,"[14], and that knowledge and judgement are primarily involuntary acts[15]. The topic of the involuntariness of belief plays an important role in modern discussions, and I will come back to it in the final section.

According to Locke, the act of judgement is an all or nothing affair. It is thus like our modern notion of assertion and its mental counterpart. Regarding the act of judgement, there are only three possibilities: assent, dissent or rejection, and suspense of judgement.[16] Belief, though, may also involve degrees. 'Assent' is used ambiguously either for the (linguistic) act of judgement in the all or nothing sense, or for a degree of assent in accordance with a degree of belief or conviction. Because judgement is concerned with probabilities, Locke acknowledges different

---

[13] Locke (1690), p.16.

[14] Locke (1690), iv.xiii.1, p.650.

[15] "[A]ll that is *voluntary* in our Knowledge, is the *employing*, or with-holding any of *our Faculties* from this or that sort of Objects, and a more, or less accurate survey of them: But they being employed, *our Will hath no Power to determine the Knowledge of the Mind* one way or other; that is done only by the Objects themselves, as far as they are clearly discovered." Locke (1690), iv. xiii.2, pp.650-651.

[16] Locke (1690), iv.xx.15, p.716.

degrees of conviction:

> "*the Mind if it will proceed rationally, ought to examine all the grounds of Probability*, and see how they make more or less, *for or against* any probable Proposition, before it assents to or dissents from it, and upon a due ballancing the whole, reject, or receive it, with a more or less firm assent, proportionably to the preponderancy of the greater grounds of Probability on one side or the other."[17]

A man proceeds rationally, if he examines all the evidence available for and against a proposition, and if he proportions his degree of conviction with which he affirms or denies that proposition, to the evidence. Those probabilities can be counted and weighed in a rational way, although it is the rationality of economy, not the rationality of a mathematical proof that applies to these probabilities[18]. Some judgements may thus be accompanied by a strong degree of conviction, others by a weak degree of conviction. Belief as conviction reaches from full assurance and confidence, quite down to conjecture, doubt, and distrust.[19] And 'belief' may also be used for a special degree of conviction:

> "[I]n the Mind [there are] such different Entertainments, as we call *Belief, Conjecture, Guess, Doubt, Wavering, Distrust, Disbelief,* etc."[20]

There is in Locke no confusion of the notions judgement and conviction. In the quote given above, there is an act of receiving or rejecting (the first occurrence of the term 'assent'), which is the act of judgement as an all or nothing affair, and this judgemental act is accompanied by a more or less firm degree of assent (the second occurrence of 'assent') that expresses the degree of conviction that accompanies our judgemental act[21].

Although Locke's notion of judgement is contrasted with knowledge, which means that the terms 'judgement' and 'opinion' may be used synonymously, Locke also uses the term 'opinion' in a more restricted sense for a belief that is not obtained in a rational way. Where the mind does not apprehend the probable connexion of the ideas, "there Men's Opinions are not the product of Judgment, or the Consequence of Reason; but the effects of Chance and Hazard, of a Mind floating at all Adventures, without choice, and without direction."[22] It is important for Locke that this concept of opinion is not a separate category besides knowledge

---

[17] Locke (1690), iv.xv.5, p.656.

[18] Locke uses an economic metaphor for the act of judgement: "Judging is, as it were, balancing an account, and determining on which side the odds lies." Locke (1690), ii.xxi.67, p.278.

[19] Locke (1690), iv.xv.2, p.655.

[20] Locke (1690), iv.xvi. 9, p.663.

[21] Cardinal Newman (1870), p.106-116, has criticised Locke's thesis that there are degrees of assent: "We might as well talk of degrees of truth as of degrees of assent." (p.115). Newman himself understands 'assent' in the sense of act of judgement, sometimes in combination with the sense of act of faith, Newman (1870), p.98: conviction and its degrees thus fall outside of what he calls 'assent.'

[22] Locke (1690), iv.xvii.4, p.669.

and judgement: Opinion is the result of a deprived act of judgement, that is, of an act of judgement that is deprived of something that is essential to it, namely rationality.

The final move that Locke makes is that he puts faith under the category of judgement. Faith, according to Locke, "is the Assent to any Proposition [...] upon the Credit of the Proposer, as coming from GOD, in some extraordinary way of Communication. This way of discovering truths to Men we call *Revelation*."[23] As assent to a proposition, faith falls under the category of judgement, and this means that everything that is said about the general category of judgement, also holds for faith. Faith is thus fallible and stands under a norm of rationality. Locke admits that a revealed truth may go beyond the discovery of our reason, and may contradict a high probability (though it may never contradict our knowledge). Still, faith is answerable to reason, because we always have to give an interpretation of the revelation, and we have to judge whether it is coming from God: "But whether it be a divine Revelation, or no, *Reason* must judge."[24] In this sense, faith cannot go beyond reason. By putting faith under the category of judgement, and thereby making faith answerable to reason, Locke is able to criticise both the Roman Catholics, and the 'Enthusiasts' within the Protestant movement. The Enthusiasts base their religious beliefs on their inner feelings and convictions. They have no reason to think that these feelings and convictions come from God, but their strong feeling that it is true that they do so.[25] The Roman Catholics often base their faith upon Authority alone, instead of using their own reason.

It is thus that the term 'belief' has come to mean *judgement* (as act or capacity), *opinion*, *conviction* (as genus or as specific degree of conviction) and *faith*. It is not to be concluded, though, that Locke was confused about these notions. His aim was to bring these different mental acts and attitudes under one category of judgement, which, though opposed to certain knowledge, is answerable to reason. Reason in judgement will never give us, though, the reason why the judged proposition is true, and it is fallible in the sense that there is no conceptual connection between reason and truth. One has a certain responsibility concerning one's judgements, and each judger is individually responsible for his judgements, when it comes to the final Judgement. The man who makes use of his God-given faculty of reason, "in doing his Duty as a rational Creature, that though he should miss Truth, he will not miss the Reward of it."[26] Using one's reason is the only way to ward off the influence of an authority that puts forward his private prejudices as innate and universal, but it is no guarantee for truth. That the God-given faculty of reason may lead us into error is a fact of human life that should not prevent us

---

[23] Locke (1690), iv.xviii.2, p.689.
[24] Locke (1690), iv.xviii.10, p.695.
[25] Locke (1690), iv.xix.8,9, p.700.
[26] Locke (1690), iv.xvii.24, p.688.

from making judgements.

5 Belief and Acceptance in Modern Analytic Philosophy

Jonathan Cohen has noticed that the term 'belief' hides an ambiguity, and he proposes in his (1992) to distinguish between belief and acceptance. Although Cohen's distinction captures some important points, neither his notion belief nor his notion acceptance, I will argue, captures the rational notion of judgement needed for logic. Cohen's distinction has been influential, and I will also discuss the proposals for improvement of the distinction in Ullmann-Margalit and Margalit (1992) and Bratman (1992).

Cohen defines *belief* as a disposition normally to feel it true that p, when one is attending to issues raised by the proposition that p.[27] Belief is a passive state that we may share with the pre-linguistic infant and animals: it is non-linguistic and involuntary; It varies in strength, depending on the intensity of the feeling, that is, there are degrees of belief. Beliefs are explained as resulting from the operation of causal factors.[28] In contrast, to *accept* the proposition that p, Cohen says, is to treat it as given that p, that is, to include that proposition among one's premises for deciding what to do or think in a particular context. In contrast to belief, acceptance is an active policy, the execution of a choice. It is linguistically structured, voluntary, and may be relative to a context; It is not a matter of degree. Acceptance is something for which we may be held responsible, because an acceptance ought to be based upon an evidential or prudential reason: "for acceptance that p to be justifiable you normally need to have reasons in favour of it."[29] Acceptance is therefore inherently motivated towards the elimination of inconsistency.[30] The notion of acceptance is closely related to the group of 'indicative-mood speech acts,'[31] that is, speech acts that are made by the utterance of an affirmatively intoned indicative sentence: it may be understood as the interiorization of such a speech act.[32] According to Cohen, such a speech act that p does not imply the belief that p, but acceptance that p. Because assertion is the typical speech act that we make by uttering a declarative sentence, assertion, according to Cohen, implies acceptance, not belief.

Cohen argues that the two concepts are independent of each other in the sense that there may be belief without acceptance, and acceptance without belief. And he uses the distinction to elucidate the controversy between a Cartesian theory of judgement that has it that judgement is an act of the will and Hume's theory

---

[27] Cohen (1992), p.4.
[28] Cohen (1992), p.23.
[29] Cohen (1992), p.13, cf. also p.23.
[30] Cohen (1992), p.36.
[31] Cohen (1992), p.72.
[32] Cohen (1992), p.79.

that judgement is involuntary. According to Cohen, Descartes was aiming at an elucidation of acceptance, whereas Hume gave an account of the notion belief. I come back to the involuntariness issue at the end of the section.

An interesting application of the notion of acceptance that Cohen proposes is that a jury's verdict declares what its members accept. Relevant is not what the members of the jury believe, but what they accept on the basis of the evidence. It may happen that the members of the jury believe the defendant to be guilty, whereas they have to accept that he is not guitly, given the available evidence. Besides, juries tend to reflect the prevailing values of the community in their decisions, which means that they may declare the defendant not to be guilty, although the evidence shows him to be guilty: "Jurors would then be accepting on ethical or pragmatic grounds, rather than cognitive ones."[33] The reasons for acceptance may thus be cognitive or pragmatic.

According to Cohen, although knowledge may imply either belief or acceptance, or both, a scientist's knowledge that p ideally implies his acceptance that p, but not his belief that p: "There is a danger that possession of a belief that p might make him less ready to change his mind about accepting that p if new evidence crops up."[34] Especially of interest for the notion of judgement in Constructive Type Theory is Cohen's remark about the attitude with respect to mathematical knowledge. According to Cohen, because the axioms in mathematics are "a matter of choice," "the mental attitude implicit in mathematical knowledge, when classically construed, is the attitude of acceptance."[35] First a remark about Cohen's terminology 'mathematics classically construed:' standardly, this is mathematics understood in realist terms. On a realist account, though, axioms are not a matter of choice; They contain real insights (cf. Frege's conception of axioms in the *Begriffsschrift* presented in section 2). Cohen's description fits, though, Hilbert's metamathematical program, in which an axiom is considered to be the result of an act of choice, as we have seen in section 2. Acceptance is what is left of judgement after the metamathematical turn. In general, the notion of acceptance suits a positivist and instrumentalist conception of science, which conceives of science not as giving insight into nature; Cohen's notion of acceptance has more in common with Popper's notion of conjecture than with the notion of rational judgement, as it is explained in the former three sections.

Cohen makes an interesting caveat concerning Bouwer's intuitionism. According to Cohen's interpretation of Brouwer, the intuition that is central to intuitionism is a kind of belief.[36] Cohen's interpretation that Brouwer has a non-linguistic intuition in mind is not far from the truth, but if belief is defined in Cohen's terms

---

[33]Cohen (1992), p.120.
[34]Cohen (1992), p.88.
[35]Cohen (1992), p.100.
[36]Cohen (1992), p.100.

as a disposition to feel it true that p, Brouwer's intuitionism would collapse into a form of psychologism. Brouwer's intuitionism is not psychologistic, though, as Brouwer scholars have shown.[37] Within Constructive Type Theory, what is essential to the demonstrative sciences is the rational act of judgement, and as I will show below, it is neither Cohen's notion of belief nor his notion of acceptance that captures this notion.

Cohen's notion of belief is what is called 'conviction' in the former sections: it has degrees, it is a mental state, and it is part of our cognitive and our sensitive nature as well. Regarding Cohen's notion of acceptance, there are some agreements with the notion of act of judgement introduced in the former sections: The act of judgement is linguistically structured, an all or nothing affair, it is a conscious activity of the mind, and the judgement made may function as premises and conclusions in our reasoning. One may thus be inclined to identify the notion of acceptance-for-evidential-reasons with the notion of judgement as explained in the former sections.

There are important reasons that count against such an identification. To accept-for-evidential-reasons is a special case of the generic act of acceptance on Cohen's account, but the act of judgement as it is explained in the former sections is not a species that can be defined by a genus together with a specific difference; The act of judgement is *sui generis*. A similar problem we find in Keith Lehrer's introduction to his *Theory of Knowledge*. Lehrer explains knowledge in terms of acceptance, which he, like Cohen, distinguishes from belief. He asserts that what is needed for knowledge is a special kind of acceptance, namely "accepting something for the purpose of attaining truth and avoiding error with respect to the very thing one accepts."[38] We know, though, that scientists accept certain theories for all kinds of purposes: to belong to a certain group, or to get their research funded. Equally, a juror does not pronounce his sentence for merely prudential reasons: his aim is truth as well, as we may hope. The problem is that the notion of acceptance has to take into account all these different types of reasons and aims, because it is defined so broadly. Because judgement as it is explained in the former sections is the narrow epistemic notion, it does not get muddled by all those other aims. Of course, one may judge for all kinds of purposes, but as the notion is explained above, one is *entitled* to judge only upon epistemic reasons. The problem becomes even clearer when we ask whether we can speak of the correctness of an acceptance. With respect to judgment, the notion of correctness is defined in epistemic terms. Equally, one may call an assertion correct if the asserter is entitled to make it, that is, if he knows what he asserts. Judgement and assertion thus stand under an epistemic norm. It is doubtful, though, what a correct acceptance amounts to, for the acceptance may be correct in one context, and incorrect in another context.

---

[37] Cf. van Atten (2004), ch. 6.
[38] Lehrer (1990), p.11.

Correctness of acceptance thus means nothing but appropriateness given the situation, and it can therefore hardly be called a normative notion (notwithstanding the fact that Cohen wants acceptance to be a normative notion, as we have seen in the second paragraph of this section).

The main problem of Cohen's notion is that it is both a theoretical and a practical notion, whereas one cannot have one notion covering both fields. This problem is noted in Bratman (1992) and Ullmann-Margalit and Margalit (1992). Bratman defends a more restricted notion of acceptance: The demands for acceptance are purely practical. The Ullmann and Margalit paper introduces the concept of holding as true, which is a presumptive notion:

> "The point of a presumption (that p), like the point of holding a sentence as true, is practical, not theoretical [...] One is to enter p as a premise into one's pertinent piece of practical deliberation and to proceed as if it were true."[39]

The difference with Cohen's notion of acceptance is that this notion of holding as true is specifically practical. Further, we do not need to have any reason for our holding as true the proposition that p; It may be a matter of burden of proof:[40] given the situation, you may have license to act as if p were the case, without further ground. Philosophy is in need of a notion of practical acceptance, but it is practical, not theoretical philosophy, that is to profit from it. The main difference is that in theoretical questions we can afford to withhold our judgement, whereas in real life we have to make a decision concerning such questions as: What illness has the patient? Is the accused guilty or not?

I will finally evaluate Cohen's point that belief is involuntary, and that the notion of acceptance is present in Descartes' theory of judgement. It is true that Descartes understands the act of judgement to belong to the active part of the soul, but Descartes' thesis that the act of judgement is an act of the will is not to be understood as though judgement is an act of arbitrary choice. For Descartes, my judgement is most free, when I clearly understand that reasons of truth point in the direction of it.[41] The Cartesian act of judgement is thus not taking a certain proposition among one's premises in the way one may choose one's axioms after the metamathematical turn. According to Descartes, the act of judgement essentially involves the possibility of error. It is to be doubted, though, whether an act of acceptance is thus liable to error. As we have seen above we can at most speak about the appropriateness of an acceptance given a certain context. The reason Cohen gives for belief to be involuntary is that belief is a disposition to have certain feelings, not a disposition to speak or act in certain ways: "No one can be

---

[39] Ullmann-Margalit and Margalit (1992), p.171.

[40] Ullmann-Margalit and Margalit (1992), p.172.

[41] Descartes, *Fourth Meditation*, AT, VII, p.58. In his *Principles of Philosophy*, Part One, article 3, Descartes does introduce the notion of acceptance in practical life, but this notion is clearly distinguished from the epistemic notion of act of judgement, see AT, VIIIA, p.5.

said to decide to be disposed to feel one way or another. You cannot decide to feel joyful or suspicious."[42] The point that judgement and belief are involuntary concerns a completely different question, though. As Locke has pointed out, the reason why judgment is involuntary is that it is concerned with an object that exists independently of our will, in contrast to the object of our imagination.[43]

6 Conclusion

Within Constructive Type Theory, the notion of judgement plays a central role: it is a normative, rational notion, where a judgement being rational or justified implies that it is correct in the epistemic sense, but does not imply that it is infallibly true. Partly due to the metamathematical turn in logic and mathematics, the notion of judgement has nearly disappeared from logic. The modern equivalents of judgement - belief and acceptance - lack the normativity that is needed in logic: Belief as (a certain degree of) conviction is a disposition to feel, at least as Cohen explains it, and feelings or their dispositions do not seem to fall under a norm of rationality. And Cohen's notion of acceptance cannot be conceived as a normative notion, either. The rational act of judgement cannot be identified with the act of acceptance as it is understood by Cohen, because Cohen's act of acceptance is more like an act in which a Hilbertian chooses his axioms than that it is a normative, rational notion.

Cohen, though, has a point when he claims that the term 'belief' should be disambiguated. Using one term 'belief' with the meaning of act of judgement and the capacity to judge; conviction; opinion; and faith suggests that there is a genus of which these different acts and states are species, but as we have seen above these different acts and states cannot be defined by the same genus for all, and a specific difference. If one uses one term 'belief' with all its different meanings, one is implicitly committed to the philosophy that introduced one category for all these different acts and states: in each case Locke had a philosophical reason to put judgement, conviction, opinion and faith under the same category. It is not at all evident, though, that modern philosophers who use the term 'belief' with all its different meanings want to commit themselves to the philosophy of John Locke.

BIBLIOGRAPHY

van Atten, M. (2004). *On Brouwer*. Thomson, Wadsorth, Toronto.

Bratman, M.E. (1992). "Practical reasoning and acceptance in a context", *Mind*, 101, pp.1–15.

---

[42]Cohen (1992), p.21.

[43]Locke's thesis that judgement is not subject to the will goes back to Aristotle's *De Anima* (427b16-21): "judging (*doxazein*) is not up to us, for it must be either true or false." And in relation to perception, Aristotle says: "to perceive is not up to us; for there must be the object of perception," (417b25-26).

Burnyeat, M. (1990). *The Theaetetus of Plato*. Hackett Publishing Company, Indianapolis, Cambridge.

Cohen, L. J. (1992). *An Essay on Belief and Acceptance*. Clarendon Press, Oxford.

Engel, P. (2000) (ed.). *Believing and Accepting*. Kluwer Academic Publisher, Dordrecht.

van Heijenoort, J. (1967). "Logic as calculus and logic as language", *Synthese*, 17, pp.324–330.

Hilbert, D. (1899). Hilbert an Frege 29.12.1899, in *G. Frege, Wissenschaftlicher Briefwechsel*, pp.65–68. Felix Meiner, Hamburg.

Hume, D. (1739). *A Treatise of Human Nature*. Clarendon Press, Oxford, 1978.

Lehrer, K. (1990). *Theory of Knowledge*. Routledge, London.

Locke, J. (1690). *An Essay concerning Human Understanding*. Clarendon Press, Oxford. Fourth edition, 1975. References of the form 'book, chapter, section' all relate to the *Essay*.

Losonsky, M. (2000). *On Wanting to Believe*, in Engel (2000), pp.101–131.

Martin-Löf, P. (1998). "Truth and knowability: on the principles C and K of Michael Dummett", in H.G. Dales and G. Oliveri, editors, *Truth in Mathematics*, pp.105–114, Clarendon Press, Oxford.

Mill, J.S. (1859). *On liberty*, in J. M. Robson, editor, *Collected Works by John Stuart Mill, vol. 18*, pp.213–310, University of Toronto Press, Toronto, 1977.

Newman, J.H. (1870). *An Essay in Aid of a Grammar of Assent*. Clarendon Press, Oxford, 1985.

van der Schaar, M. (2007). "The assertion-candidate and the meaning of mood", *Synthese*, 159, pp.61–82.

van der Schaar, M. (2007a). "Bolzano on Judgement and Error", in O.Tomala and R.Honzik (editors), *The Logica Yearbook 2006*, pp.211-221, Filosofia Publisher, Czech Academy of Science, Prague.

van der Schaar, M. (2008). "Locke and Arnauld on judgment and proposition", *History and Philosophy of Logic*, 29, pp.327-341.

van der Schaar, M. (2009). "Locke on knowledge and the cognitive act" *Grazer Philosophische Studien*, 78, in press.

Sundholm, B.G. (1988). *Oordeel en gevolgtrekking*, Inaugural address, Leiden University.

Sundholm, B.G. (2003). "Tarski and Lesniewski on languages with meaning versus languages without use", in J. Hintikka et al., editors, *Philosophy and Logic in Search of the Polish Tradition*, pp.109–128, Kluwer Academic Publishers, Dordrecht.

Ullmann-Margalit, E. and Margalit, A. (1992). "Holding true and holding as true", *Synthese*, 92, pp.167–187.

PART III

FROM LOGIC TO PHILOSOPHY

Sundholm's Paradox of Knowability: A Novel Paradox?

HELGE RÜCKERT

ABSTRACT. In this paper I take a closer look at a recently published paradox by Göran Sundholm involving the notion of knowability. I point out that this paradox is not a novel, genuine paradox, but rather an important variant of the Knower Paradox. I briefly discuss further variations of the Knower Paradox, and in a final section I try to show that it is not unproblematic to assume that knowability is factive. There are several different notions and conceptions of knowability. The most straightforward ones are non-factive. Anti-realists as Sundholm and Tennant obviously use a different notion of knowability which they assume to be factive. But this notion has not yet been made sufficiently clear to rely on it as common ground in debates between realists and anti-realists.

1 Another Knowability Paradox

The modern debates between realists and anti-realists are often centred on the notion of knowability. Typical anti-realists defend an equivalence claim between truth and knowability, whereas realists deny such a thesis. In recent years one focus of this debate has been the so-called Knowability Paradox or Church-Fitch Paradox,[1] and the literature on it has exploded.[2] This paradox consists in a formal derivation that seems to show that the anti-realistic knowability principle 'Every truth is knowable' implies the seemingly much stronger claim that every truth is actually known, a thesis unacceptable even for all but the hardest anti-realists.

Now, in a short note Göran Sundholm (2008) has presented another paradox

---

[1] Originally the argument was published in a paper by Frederic Fitch (1963), who attributed it to an anonymous referee. So, for a long time the paradox was named after Fitch, until Joe Salerno recently found out that the anonymous referee was Alonzo Church and proposed the name Church-Fitch Paradox. Cf. Salerno (2009b).

[2] For an overview see the online entry in the Stanford Encyclopedia of Philosophy by Berit Brogaard and Joe Salerno, and for a recent collection of papers see Salerno (2009a). Even a monograph has been written on this subject: Kvanvig (2006). I have contributed to this debate also myself, Rückert (2004).

that essentially involves the notion of knowability.[3] Here, Sundholm's argument:[4]

"We consider:

($\star$) This sentence is unknowable.
1   Assume that ($\star$) can be known.   Assumption

Then

2   ($\star$) is true.

whence

3   This sentence cannot be known.   From ($\star$) and 2; T-Schema
4   Contradiction.   From 1 and 3

Therefore

5   ($\star$) cannot be known.   From 1 and 2, without assumptions

That is

6   This sentence is unknowable.   ($\star$) is demonstrated on no assumptions

Therefore

7   ($\star$) is known.   From 7, what is demonstrated is known
8   Contradiction.   From 5 and 7 on no assumptions."

Before moving on, I'd like to point out three decisive features of this argument:

---

[3]Sundholm does not seem to be worried by the original knowability paradox at all, because he thinks that it does not apply to his own conception of knowability, but he is rather silent about why he thinks that it does not:

"Personally, even though I admit of constructivist tendencies, I have never felt in the slightest threatened by the Church/Fitch reasoning. To my mind, its rendering of the knowability of truth simply does no justice to what is involved in constructivist knowability." (Sundholm (2008), pp.375-376).

[4]Sundholm (2008), p.376.

a) *Self-reference*. The starting point of the paradox, the sentence 'This sentence is unknowable,' is self-referential, it speaks about itself, claiming that this very sentence is not knowable. As remarked by Sundholm in a footnote,[5] instead of using an indexical formulation with 'this,' one could also start the paradox with '$S =_{def} S$ cannot be known'. The diagonal lemma guarantees that there is such a sentence.

b) *Factivity of knowability*. The step from (1) to (2) in the reasoning uses factivity of knowability, i.e. that every knowable sentence is also true. So, Sundholm assumes without any further comment that knowability, like knowledge, is a factive notion. I'll come back to this in section 5.

c) *Necessitation*. Here, by necessitation I mean the transition from a sentence having been demonstrated or proved to the sentence being known, and as every actually known sentence is *a fortiori* also knowable, to the sentence being knowable (cf. (7) and (8) in Sundholm's derivation).

2 Naming the Paradox

Sundholm begins and ends his short note with comments about the naming of paradoxes. At the beginning of his text he says:

> "The naming of paradoxes is usually a straight-forward matter: they are named after their inventors."[6]

And he concludes his text with the following footnote:

> "Only the delicate question now remains: by what name should one call this novel (?) paradox ...?"[7]

Of course, taken together these two passages suggest to name the paradox that I have presented in section 1 after Sundholm himself, and as it is a paradox that involves the notion of knowability, I propose to call it Sundholm's Paradox of Knowability from now on.

3 A Novel Paradox?

But, the naming of the paradox certainly wasn't the most 'delicate' question anyway.[8] With the title of his note, "A Novel Paradox?" and, indirectly, with his last sentence (note the question mark after "novel"!) Sundholm formulates a more important question: Is Sundholm's Paradox of Knowability really a new, genuine paradox, or is it rather a mere variant of another paradox.

---
[5] Sundholm (2008), pp.376-377.
[6] Sundholm (2008), p.375.
[7] Sundholm (2008), p.375.
[8] Sundholm's use of 'delicate' in the quote above is at least partly ironic, of course.

Maybe Sundholm himself considered his own paradox to be a variant of the Liar. Or better, a variant of the Strengthened Liar.[9] This is suggested by his second footnote:

> "From a constructivist point of view this is very close to the Liar since, in some suitable version or other, truth is nothing but knowability."[10]

Indeed, for philosophers with constructivist views who equate truth with knowability or replace truth by knowability, Sundholm's Paradox is a variant of the Liar paradox. On the other hand, the derivation of Sundholm's Paradox is slightly more complex than the one of the Liar.

Although, in his paper, Sundholm mentions more than a dozen of other paradoxes, it is surprising that the most pertinent other paradox is missing: the so-called Knower Paradox which can be attributed to Montague,[11] and which itself is a generalisation of the Liar Paradox. Here, the most simple version of the Knower Paradox:[12]

Let T be a theory extending Robinson Arithmetic, and K a unary predicate of $L(T)$ that is factive ($K\phi \to \phi$) and satisfies necessitation (if $T \vdash \phi$ then $T \vdash K\phi$). Then T is inconsistent.

Proof.

1	$T \vdash \phi \leftrightarrow \neg K\phi$	[There is such a ϕ because of the diagonal lemma]
2	$T \vdash K\phi \to \phi$	[Factivity of K]
3	$T \vdash K\phi \to \neg K\phi$	[By (1) and (2)]
4	$T \vdash \neg K\phi$	[By (3)]
5	$T \vdash \phi$	[By (1) and (4)]
6	$T \vdash K\phi$	[By (5) and Necessitation for K]
7	Contradiction	[Between (4) and (7)]

This paradox is called the Knower Paradox because K might well be the knowledge predicate, but it need not be. Any predicate that satisfies Factivity and Necessitation will do. Égré (2005), p.18, calls factive predicates 'knowledge predicates.' This might be a bit misleading as not only epistemic predicates like 'knowledge' satisfy Factivity. Among others, the truth predicate itself is trivially factive.

Obviously, Sundholm's derivation exactly mirrors the one of the Knower Paradox, using 'knowability' as the interpretation of K. This becomes even more apparent if we paraphrase what is going on in the formal derivations by using natural language. For the Knower Paradox, there is the following formulation by Égré (2005), pp.18-19, adapted from Tymoczko (1984) (my emphases are bold):

---

[9] Whereas the Liar results from the sentence 'This sentence is false,' the Strengthened Liar originates from 'This sentence is not true.'
[10] Sundholm (2008), p.375.
[11] Cf. Kaplan and Montague (1960) and Montague (1963).
[12] Cf. Égré (2005), p.18.

"Suppose someone **knows** this statement to be true; then this statement is true, otherwise it couldn't be **known**; therefore, 'nobody **knows** this statement to be true' is true, that is nobody **knows** this statement to be true. So nobody **knows** this statement to be true. This is what the statement says, hence it is true. But hold on! I have just *proved* this statement to be true. Hence someone (at least me) **knows** this statement to be true! Now this contradicts what has just been established."

To capture the reasoning behind Sundholm's Paradox we only have to slightly modify this passage:

"Suppose someone **can know** this statement to be true; then this statement is true, otherwise it couldn't be **knowable**; therefore, 'nobody **can know** this statement to be true' is true, that is nobody **can know** this statement to be true. So nobody **can know** this statement to be true. This is what the statement says, hence it is true. But hold on! I have just *proved* this statement to be true. Hence someone (at least me) **knows** (**and** *a fortiori* **can know**) this statement to be true! Now this contradicts what has just been established."

So, Sundolm's Paradox is not really a novel, genuine paradox, it is rather a variant of the Knower Paradox, but it is an important variant as the notion of knowability plays such an important role in recent debates.

Next, we will have a look at some more variations of the Knower Paradox before we turn our attention again to the notion of knowability.

4 Variations of the Knower Paradox

The Knower Paradox is often discussed in connection with related results about provability. For example, assuming certain properties of the provability predicate P, Löb's theorem says:[13]

$$T \vdash P\phi \to \phi \text{ only if } T \vdash \phi.$$

Put slightly differently:[14] Given a theory T formulated within a sufficiently rich language, and P a provability predicate for T, then, if P is factive ($P\phi \to \phi$ for every sentence ϕ), T is inconsistent.

Given the following correspondences

ϕ has been proved $\approx \phi$ is known
ϕ is provable $\approx \phi$ is knowable

it might make more sense, instead of discussing the results about provability in connection with the proper Knower Paradox involving the notion of knowledge, rather to compare them to the knowability version of the Knower Paradox.

Another question is whether there are versions of the Knower Paradox that don't assume the relevant predicate to be factive. As one of the main differences between

---

[13]Cf. Löb (1955).
[14]Cf. Gödel (1933).

knowledge and belief is that knowledge is factive whereas belief is not, this leads us to the so-called Believer Paradox and variations of it. Indeed, there is a result by Thomason (1980) involving the notion of belief and corresponding assumptions about this notion which is, in a certain sense, a generalisation of the Knower Paradox.[15]

It is interesting to remark that we have two knowability paradoxes: the Church-Fitch Paradox and Sundholm's Paradox. Sundholm's Paradox is a version of the Knower Paradox which makes use of factivity of knowledge/knowability and factivity of knowledge is also used in the standard version of Church-Fitch. Now, as we have just seen there are variations of the Knower Paradox, Believer Paradoxes, not relying on factivity but on weaker principles. The question is whether there is also a variation of Church-Fitch involving the non-factive notion of belief instead of the factive notion of knowledge? And indeed, recently Michael Fara has published a variant of the Church-Fitch Paradox for belief.[16]

5 Knowability and Factivity

Let's come back to the notion of knowability. So far, we have always presupposed that knowability is factive, that knowability entails truth. But, is this justified? Is it clear that knowability is factive? I doubt it. Whereas it is pretty clear that knowledge is factive,[17] it might be different with knowability.

The most straightforward understanding of knowability analyses it as the composition of a standard possibility operator and a standard knowledge operator. But, neither if the possibility operator is supposed to express metaphysical possibility, nor if it is supposed to express epistemic possibility, the resulting conception of knowability turns out to be factive.

If we analyse knowability as $\Diamond K$ with "$\Diamond$" expressing metaphysical possibility, a proposition is knowable iff it could have been the case that it was known. Or, using possible worlds talk: if there is a possible world such that in that world somebody knows the proposition. Of course, given factivity of knowledge, then the proposition has to be true in the respective possible world, but as this required possible world need not be the actual world, it might well be that the proposition isn't true in the actual world (despite being known in another possible world).

Things do not look much different when we turn our attention to epistemic modalities. In order to consider the possible worlds as epistemic possibilities instead of metaphysical possibilities, we can use the following heuristic: an epistemically possible world is a possible state of the world that is compatible with all that

---

[15] For a proof of the result, see Égré (2005), pp.23-24. Other variants of the Believer Paradox are due to Turner (1990) and Koons (1992).

[16] Cf. Fara (2007).

[17] That knowledge is factive is doubtful according to some variants of contextualism and subjectivism. But potential counterexamples only concern formulas with nested knowledge operators. Cf. Rebuschi and Lihoreau (2008).

one knows (or believes). For a proposition to be knowable (from my perspective) then means that somebody knowing this proposition is compatible with all that I know. If I know a proposition to be false, it follows that this proposition is not knowable in the relevant sense, of course. But, what about cases in which I neither know the proposition to be true nor know it to be false? Then it is compatible with my knowledge that the proposition might be true and also known, thus it is knowable, even if it is in fact false.

So, we have examined two conceptions of knowability and we have seen that both are non-factive. The question remains whether there are other sensible conceptions of knowability that are factive. One might appeal to two main sources in order to motivate the belief in the existence of factive knowability:

a) ordinary language use;

b) anti-realists' use of knowability.

a) Knowability in Ordinary Discourse

Knowability expert Joe Salerno is convinced that there are uses of 'knowability' (and related expressions) in natural language that clearly involve a factive notion of knowability. Let's have a look at one of his examples:

> "Does a factive conception of knowability figure in ordinary use? There is some reason to think so. 'Knowable' and related terms such as 'discoverable,' 'observable,' and 'verifiable' all seem to operate factively in ordinary discourse. Consider the following example, a dialog between colleagues A and B:
>
> > A: We could be discovered.
> > B: Discovered doing what?
> > A: Someone might discover that we're having an affair.
> > B: But we are not having an affair!
> > A: I didn't say that we were.
>
> A's remarks sound contradictory. In this context the factivity of 'someone might discover that' explains this fact. So there is some reason to believe that knowability and related modalities are factive in ordinary use."[18]

I agree that A's remarks are definitely odd. Whether they are contradictory in a strict sense is a different question, though. If it is possible to explain the oddness of A's remarks without having to appeal to a factive notion of knowability (or discoverability), the example loses much of its force. Indeed, I think, that the oddness of A's remarks can be explained by bringing in pragmatic considerations. If A and B in fact don't have an affair, as B claims, then it is completely irrelevant to speculate about a possible discovery of this (non-existing) affair. So, at least A is harshly violating Grice's maxim of relevance.

---

[18] Brogaard and Salerno (2006), p.261.

But, even worse, if we assume that A and B know that they don't have an affair (usually people know whether they have an affair with someone else or not!), then it is simply not true that it is epistemically possible for them that someone might know that they have an affair. Because somebody knowing that they have an affair would be incompatible with their knowledge. So, what A says is plainly and rather obviously false, and A can also be accused to violate Grice's maxim of quality. Epistemic knowability of p entails that the relevant persons don't know p to be false. But this does not amount to knowability of p entailing p to be true, as p might be false and its falsity simply unknown to the relevant persons.

So, the example does not establish that there is a factive notion of knowability in ordinary discourse. The oddness of A's remarks can be explained differently. The situation can be compared to examples in connection with Moore's Paradox. Somebody says: "I believe that it rains, but it is not raining." Such an utterance is definitely odd. But nobody would conclude because of such examples that the notion of belief as used in ordinary discourse be factive. Again, the oddness can be explained differently.

On the other hand, it is not too difficult to construct ordinary language examples involving true knowability claims with a clearly non-factive notion of knowability:

> Steve is a candidate in a TV quiz show. He is asked questions and each time he can choose between four possible answers A, B, C and D. The last question was about dinosaurs and Steve has chosen answer B. In the audience there are Mary, Steve's wife, and Angela, Steve's sister. The following dialog between them makes perfect sense:
>
>> Mary: Is it possible that Steve knows that B is the correct answer?
>> Angela: Yes, that's possible. In his youth Steve has read lots of books about dinosaurs, so he might have chosen B because he knows that B is the correct answer. But, of course, it is also possible that he had to guess. I myself I don't know the correct answer, it might be B or one of the other three.
>> Mary: Me neither. Let's hope that Steve knew that B was the correct answer or that he at least made a correct guess
>> ...

It is clear that the notion of knowability involved in this example is non-factive. I don't think that the notion of knowability involved in the Brogaard/Salerno example is any different. The difference between the examples is, what Mary and Angela are saying is perfectly reasonable, whereas what A is saying is strange. But the strangeness and oddness of A's utterances is definitely compatible with his notion of knowability being non-factive, as the oddness can be explained differently.

b) Knowability and Anti-Realism

When developing their semantic and epistemological theories and views, many anti-realists or constructivists make ample use of the notion of knowability. But their notion is definitely different from the ones discussed above, combinations of a standard possibility operator (be it metaphysical or epistemic) and a standard knowledge operator. The notion of knowability used by anti-realists is factive.

Sundholm himself is quite explicit in equating truth with knowability and related notions:

> "The true statements are the evidenceable, **knowable**, warrantable, justifiable, …ones."[19]

At another place he distinguishes truth in the actual sense from truth in the potential sense:

> "When a judgement has become known, or evident, in virtue of its being demonstrated, it is clearly true. Truth is here taken in an actual sense of being known, or evident. However, before a mathematical judgement became known it could become known. In the potential sense a judgement is true when it is demonstrable (evidenceable, **knowable**, assertible, justifiable, warrantable)."[20]

In his framework, Sundholm uses the notion of knowability to develop his ideas, but he does not say much how to exactly analyse and explicate this notion itself. Another anti-realist, Neil Tennant, is more explicit concerning the notion of knowability he uses and its factivity. To demarcate it from the other notions of knowability we discussed above, he calls it feasible knowability:

> "[I]f φ is any contingent falsehood (such as, say, 'Grass is purple'), then it is not feasibly knowable that φ, in the sense of feasibility with which we are here concerned. Feasibility is not at all like the alethic modality of possibility. Another way of putting this last moral is to say that not only the epistemic operator K, but also its modalization $\diamond K$, is 'factive'."[21]

The idea behind feasible knowability maybe becomes a little bit clearer by invoking an analogy with dispositional predicates like 'water-soluble.' That a certain object is water-soluble does not mean that there is a possible world, and that in this possible world the object dissolves in water. After all, the object might have a different physical microstructure in another possible world. What is rather meant is that the object, its physical micro-structure unchanged, would dissolve if put into water. So, when considering other possibilities, certain features of the object have to remain fixed and unchanged, as for example the physical microstructure, in our case.

---

[19] Sundholm (2004), p.452; my emphasis.
[20] Sundholm (1997), p.202; my emphasis.
[21] Tennant (2000), p.829.

Correspondingly, an important idea behind feasible knowability seems to be that when considering other possibilities, the non-epistemic facts have to remain fixed and unchanged (so it would be excluded that the knowable proposition is false in the actual world, but true in a feasible possibility we consider), and only the epistemic facts, what we know or what we don't know, might change. Such a conception of knowability is certainly interesting,[22] but it still remains to be worked out in detail. Recently there has been quite some promising progress in such a direction.[23]

Until this is accomplished (not only for knowability, but also for other notions used by anti-realists), misunderstandings are almost unavoidable. The anti-realists develop their semantic-epistemological framework(s) by using certain modal expressions, but they disapprove of standard (non-anti-realistic) semantic theories for modalities like possible worlds semantics. So, when someone who was philosophically raised with possible worlds etc. reads the expositions of the framework(s) by anti-realists, he is inclined to understand the modal notions used by the anti-realists in a way that is unacceptable for them.

Sundholm suspects that Fichte might be right, that

> "the point is moot whether there is a neutral background position from which the issue between realists and idealists [or anti-realists; H.R.] can be adjudicated."[24]

Nevertheless, even if there is no neutral background, we all should at least try to clarify the concurring positions and views as far as possible.

6 Acknowledgments

I am very thankful to Paul Égré (Paris) who not only made me aware of the similarities between Sundholm's Paradox and the Knower Paradox in the first place, but who also provided me very helpful remarks and comments in personal communications. Finally, I have to thank the editors of this volume, Giuseppe Primiero and Shahid Rahman, for their almost endless patience.

Dear Göran, I am pretty sure that you will disagree with some (most?) things I said in this little paper. Nevertheless, I hope you will like it, if only (small) parts of it.

BIBLIOGRAPHY

Brogaard, B. and Salerno, J. (2002/2004). "Fitch's Paradox of Knowability", Stanford Encyclopedia of Philosophy (online entry).

---

[22] The main problems I see with such a conception concern the feasible knowability of epistemic facts themselves, and that the epistemic facts seem to supervene on the non-epistemic facts, such that changing the epistemic facts automatically also means changing at least some non-epistemic facts.

[23] See Fara (2009), Zardini (2009).

[24] Sundholm (2004), p.440.

Brogaard, B. and Salerno, J. (2006). "Knowability and a Modal Closure Principle", *American Philosophical Quarterly*, 43(3), pp.261-270.

Égré, P. (2005). "The Knower Paradox in the Light of Provability Interpretations of Modal Logic", *Journal of Logic, Language and Information*, 14, pp.13-48.

Fara, M. (2007). "The Paradox of Believability", *Review of Contemporary Philosophy*, 6, pp.13-17.

Fara, M. (2009). "Knowability and the Capacity to Know", forthcoming in Salerno, J. (ed.), *Knowability and Beyond* (special issue of Synthese).

Fitch, F. (1963). "A Logical Analysis of some Value Concepts", *The Journal of Symbolic Logic*, 28, pp.135-142.

Gödel, K. (1933). "Eine Interpretation des intuitionistischen AussagenkalkÃijls", *Ergebnisse eines mathematischen Kolloquiums*, 4, pp.39-40, reprinted in Gödel, K (1986): *Collected Works, I, Publications 1929-1936*, edited by Feferman, S., Kleene, S., Moore, G., Solovay, R. and van Heijenoort, R., pp.300-301, Oxford University Press, USA.

Kaplan, D. and Montague, R. (1960). "A Paradox Regained", *Notre Dame Journal of Formal Logic*, 1, pp.79-90.

Koons, R. (1992). *Paradoxes of Belief and Strategic Rationality*, Cambridge University Press, Cambridge.

Kvanvig, J. (2006). *The Knowability Paradox*, Oxford University Press, Oxford.

Löb, M. (1955). "Solution of a Problem of Leon Henkin", *Journal of Symbolic Logic*, 20, pp.115-118.

Montague, R. (1963). "Syntactical Treatments of Modality, with Corollaries on Reflexion Principles and Finite Axiomatizability", *Acta Philosophica Fennica*, 16, pp.153-167.

Rebuschi, M. and Lihoreau, F. (2008). "Contextual Epistemic Logic", in Dégrémont, C., Keiff L. and Rückert, H. (eds.): *Dialogues, Logics and other Strange Things. Essays in Honour of Shahid Rahman*, pp.305-335, College Publications, London.

Rückert, H. (2004). "A Solution to Fitch's Paradox of Knowability", in Rahman, S., Symons, J., Gabbay, D. and Van Bendegem, J. (eds.): *Logic, Epistemology and the Unity of Science, LEUS 1*, Kluwer, pp.351-380, Springer Verlag.

Salerno, J. (ed.) (2009a). *New Essays on the Knowability Paradox*, Oxford University Press, Oxford.

Salerno, J. (2009b). "Knowability Noir: 1945-1963", in Salerno (2009a), pp.29-48.

Sundholm, B.G. (1997). "Implicit Epistemic Aspects of Constructive Logic", *Journal of Logic, Language and Information*, 6, pp.191-212.

Sundholm, B.G. (2004). "Antirealism and the Roles of Truth", in Niiniluoto, I., Sintonen, M. and Wolenski, J. (eds.): *Handbook of Epistemology*, pp.437-466, Kluwer Academic Publisher, Dordrecht.

Sundholm, B.G. (2008). "A Novel Paradox?", in Dégrémont, C., Keiff L. and Rückert, H. (eds.): *Dialogues, Logics and other Strange Things. Essays in Honour of Shahid Rahman*, pp.375-377, College Publications, London.

Tennant, N. (2000). "Anti-Realist Aporias", *Mind*, 109, pp.825-854.

Thomason, (1980). "A Note on Syntactical Treatments of Modality", *Synthese*, 44, pp.391-395.

Turner, R. (1990). *Truth and Modality for Knowledge Representation*, MIT Press, Cambridge Mass.

Tymoczko, T. (1984). "Un Unsolved Puzzle about Knowledge", *Philosophical Quarterly*, 34, pp.437-458.

Zardini, E. (2009). "Two Diamonds are More than One. Transitivity and the Factivity of Feasible Knowability", forthcoming in Keiff, L., Marion, M. and Rahman, S. (eds.): *Anti-Realism in the Abstract Sciences*, Springer, Berlin.

On Dialogues and Natural Deduction

SHAHID RAHMAN, NICOLAS CLERBOUT, LAURENT KEIFF

ABSTRACT. The dialogical approach to logic has well known relations with other proof methods, like sequent calculi and semantic tableaux. In this paper we explore the connection between dialogues and natural deduction, an area that is much less understood. We define a system of dialogical games that captures the notion of validity of Johansson's minimal logic, and show how to extend the system for intuitionistic logic and classical logic. Then we describe an algorithm that transforms a dialogical proof into a Fitch-style natural deduction proof, and discuss the relation between the two approaches.

For you, dear Herr Urteil

1 Dialogue Games and Dialogical Logic

The dialogical tradition, as it stems from Lorenzen and Lorenz (1978) aimed at providing a new approach to the notion of *meaning* in logic that should build a conceptual link between languages games, argumentation and validity. The point was to understand logic as a special kind of linguistic interaction. Nowadays, a very dynamic and powerful stream of research explores this notion of interaction in the interface between mathematical game theory and logic. In our paper we place dialogical logic in the framework of what are called in mathematical game theory 'extensive games' in order to develop some points that were not systematically clarified in the dialogical tradition. It should be noted that such games are not a proof system proper. They are meant to give an account of how a rational argumentation about a given (logically complex) claim should be conducted, in the sole virtue of its logical form. The system assumes a notion of justification for logically elementary statements, then proceeds in showing how these elementary justifications can be used in the construction of a justification for complex statements. In this section we present the basic game system[1], then show how the usual notion of validity of a formula A (in minimal, intuitionistic and classical logic) can be expressed in terms on the existence of a certain kind of strategy in the game associated with A.

---

[1] For a textbook presentation (in French), see Fontaine and Redmond (2008)

1.1 Speech Acts

The fundamental idea behind the dialogical approach to logic is that a proof is a certain kind of very simple language game. Such games are built out of two fundamental types of speech act, namely *assertion* and *request*. The first can be thought, as for instance in Sellars (1997) and Brandom (2000), as a commitment to provide justifications of a certain kind, and the latter as an imperative to fulfill an assertive commitment. The dialogical tradition takes it that speech acts, and consequently proofs, are best understood when conceived as fundamental forms of *interaction*. This idea is the very core of the growing influence of game-theoretical ideas in logic, witnessed for instance by active research programs such as Hintikka-style Game-Theoretical Semantics, of e.g. Hintikka and Sandu (1997), the computation oriented Game Semantics of, e.g., Abramsky (1997), Blass (1992), Hyland and Ong (2000), Japaridze (2003), Girard's new research program called Ludics, in e.g. Girard (2003), and the dialogical tradition steming from Lorenzen and Lorenz (1978).

Let us first define a language the well-formed formulas (wff) of which are an adequate content for the assertions in our dialogical games.

DEFINITION 1.1.1 (THE ASSERTIVE LANGUAGE) Our language $\mathcal{L}$ is built upon a countable set of elementary formulas $\mathbb{P} = \{p_0, p_1, ...\}$ together with a set of connectives $\{\wedge, \vee, \rightarrow\}$, and a couple of brackets (and). Let $\bot \in \mathbb{P}$ be a distinguished elementary formula. Assume $p \in \mathbb{P}$. The set of wff is as usual freely generated over $\mathbb{P}$ by the grammar:

$$A := p \mid \bot \mid (A \vee B) \mid (A \wedge B) \mid (A \rightarrow B)$$

We sometimes write $\neg A$ as a shorthand for $A \rightarrow \bot$

Moves in a dialogical game are speech acts, and are referred to by expressions specifying an *agent* (i.e. the player making the move), a *force* of the move that can be either an assertion (for which we use the fregean notation $\vdash$) or a request (noted ?), and a *content*. Formally:

DEFINITION 1.1.2 (DIALOGICAL EXPRESSIONS) A *dialogical expression* is an instance of $\langle \mathbf{X} f e \rangle$ where $\mathbf{X} \in \{\mathbf{O}, \mathbf{P}\}$, $f \in \{\vdash, ?\}$ and $e \in \{L, R, \vee\} \cup \mathcal{L}$.

We will now define the rules for a system of dialogical games that we call *minimal dialogues*, which we will use as a basis for expressing the notions of validity in Johansson's minimal logic, and for the translation algorithm into Fitch-style natural deduction proofs.

1.2 Local Semantics

Argumentation forms (or particle rule) give the dialogical semantics of the connectives. Such forms give an abstract description of the way a formula, according to its outmost form, can be criticized, and how to answer the critique. The description is abstract or *local* in the sense that it can be carried out without reference to the context other than the presence of an assertion of a given formula in it. Informally, a *dialogical history* is the history of the dialogue, i.e. a sequence of dialogical expressions, together with an indication of the player who is to play. We give first two general definitions.

DEFINITION 1.2.1 (DIALOGICAL HISTORY) A dialogical history $\mathbb{H}$ is a tuple $\langle \Sigma, \mathbf{X} \rangle$ where Σ is a set of dialogical expressions and $\mathbf{X} \in \{\mathbf{O}, \mathbf{P}\}$.

DEFINITION 1.2.2 (ARGUMENTATION FORM) An *argumentation form* is an ordered triple (p, c, d) of dialogical expressions where p is the *precondition*, c the *challenge* and d the *defence*.

argumentation forms should be understood as follows. In any history $\mathbb{H} = \langle \Sigma, \mathbf{Y} \rangle$ in which player $\mathbf{X}$ asserted the precondition p, player $\mathbf{Y}$ [2] may challenge this assertion, yielding a new history $\mathbb{H}' = \langle \Sigma \frown c, \mathbf{X} \rangle$. In a history $\mathbb{H}$, where a $\mathbf{X}$-assertion has been challenged according to some argumentation form (p, c, d), $\mathbf{X}$ may answer to the challenge, yielding a new history $\mathbb{H}'' = \langle \Sigma \frown d, \mathbf{Y} \rangle$.

Precondition	$\langle \mathbf{X} \vdash A \wedge B \rangle$	$\langle \mathbf{X} \vdash A \wedge B \rangle$	$\langle \mathbf{X} \vdash A \rightarrow B \rangle$
Challenge	$\langle \mathbf{Y} ? L \rangle$	$\langle \mathbf{Y} ? R \rangle$	$\langle \mathbf{Y} \vdash A \rangle$
Defence	$\langle \mathbf{X} \vdash A \rangle$	$\langle \mathbf{X} \vdash B \rangle$	$\langle \mathbf{X} \vdash B \rangle$

Precondition	$\langle \mathbf{X} \vdash A \vee B \rangle$	$\langle \mathbf{X} \vdash A \vee B \rangle$
Challenge	$\langle \mathbf{Y} ?_\vee \rangle$	$\langle \mathbf{Y} ?_\vee \rangle$
Defence	$\langle \mathbf{X} \vdash A \rangle$	$\langle \mathbf{X} \vdash B \rangle$

Local semantics and choice: Let # be a propositional connective. The set of the argumentation forms the precondition of which is an assertion of a formula with # as the main connective is the dialogical local semantics for #. There are two rules for conjunction and for disjunction, and players may choose which one they will use. In a history where a conjunction has been asserted by $\mathbf{X}$, $\mathbf{Y}$ may request any of the conjuncts and in a history where his assertion of a conjunction has been challenged by such a request, $\mathbf{X}$ may assert the relevant conjunct. In the case of disjunction, both rules admit the same challenge, so $\mathbf{X}$ will have the choice of the

---

[2] Through the whole paper we assume $\mathbf{X} \in \{\mathbf{O}, \mathbf{P}\}$, $\mathbf{Y} \in \{\mathbf{O}, \mathbf{P}\}$ and $\mathbf{X} \neq \mathbf{Y}$. We will often refer to dialogical moves as $\mathbf{X}$-moves, or when suitable as $\mathbf{X}$-assertions.

rule he wants to follow in order to defend, which amounts to choose and assert one of the disjuncts. There is only one rule for the conditional, but notice that in a history where **Y** asserted a complex formula (say A) in order to challenge the **X**-assertion of a conditional (say $A \to B$), **X** has a choice between defending the conditional according to the rule, and challenging the antecedent according to the rule that corresponds to A's main connective.

1.3 Structural Rules

The dialogical games are meant to capture situations where a protagonist in a language-game commits himself to justify a claim (that we call the *thesis*) in a context where his opponent is committed to the justification of a set of *initial hypotheses*. The following structural rules define the two main aspects of the game:

- what a game is, i.e. the way argumentation forms may be used in order to produce a dialogue;

- the payoff function of the game, i.e. a winning criterion for game histories: A player who must move and can not has lost.

In order to deal with both aspects, we first need a couple of definitions. As most other games, dialogical games should prevent loops, i.e. the indefinite repetition of the same situation. We call a move initiating a loop a redundant move. Formally, redundancy is defined with respect to types of moves: Repetition of a challenge is not redundant if a new type of move has been made between the first and the second occurences.

DEFINITION 1.3.1 (REDUNDANT MOVES) We distinguish between redundancy of a challenge and redundancy of a defence:

Challenge Let $A, B \in \mathcal{L}$. Let $\mathbb{H} = \langle \Sigma, \mathbf{X} \rangle$ be a dialogical history such that $\langle \mathbf{Y} \vdash A \rangle \in \Sigma$. Let $\langle \mathbf{X} \, f \, e \rangle \in \Sigma$ be a challenge against $\langle \mathbf{Y} \vdash A \rangle$. Let $\mathbb{H}_0$ be the prefix of $\mathbb{H}$ with $\langle \mathbf{X} \, f \, e \rangle$ as last element. We say that challenge $\langle \mathbf{X} \, f \, e \rangle$ is *redundant in* $\mathbb{H}$ iff there is no assertion $\langle \mathbf{Y} \vdash B \rangle \in \mathbb{H} - \mathbb{H}_0$ such that $\langle \mathbf{Y} \vdash B \rangle \notin \mathbb{H}_0$.[3]

Defence Any repetition of a defense is redundant.

---

[3]That is, if there is no new assertion by **Y** after challenge $\langle \mathbf{X} \, f \, e \rangle$, where an assertion is new only if it did not occur in $\mathbb{H}_0$.

We also need the following terminology.

DEFINITION 1.3.2 Let $\mathbb{H} = \langle \Sigma, \mathbf{X} \rangle$ be a dialogical history. We say that $\mathbb{H}$ is **X**-*terminal* with game rules **D** iff there is no move available to **X** according to the rules in **D**.

[SR-0] (Initial History) Let $\Delta \subset L$ be a finite set of formulas and $A \in L$ a formula. The *initial position* of a dialogue for A under hypotheses Δ (notation: $\mathcal{D}(\Delta, A)$) is a history $\mathbb{H}_0 = \langle(\langle \mathbf{O} \vdash \Delta \rangle, \langle \mathbf{P} \vdash A \rangle), \mathbf{O}\rangle$ [4].

[SR-1] (Gameplay) Let $\mathbb{H} = \langle \Sigma, \mathbf{X} \rangle$ be a dialogical history. **X** is to play in $\mathbb{H}$. The set of available moves for player **X** in history $\mathbb{H}$ is the set of non-redundant challenges specified by argumentation forms applicable to the **Y**-assertions in $\mathbb{H}$, together with the set of non-redundant defences against **Y**'s *last* challenge in $\mathbb{H}$, as specified by the argumentation forms. No other move is allowed.

[SR-2] (Winning) Player **Y** wins in a terminal history $\mathbb{H}$ iff $\mathbb{H}$ is **X**-terminal. In a terminal history where **X** wins, **Y** looses.

Let $\mathbb{AF}$ denote the set of argumentation forms given in the previous section. The game system for minimal dialogues is the set of rules

$$\mathbf{D}_{min} = \mathbb{AF} \cup \{\mathbf{SR\text{-}0}, \mathbf{SR\text{-}1}, \mathbf{SR\text{-}2}\}.$$

DEFINITION 1.3.3 A dialogue $\mathcal{D}(\Delta, A)$ in a rule system **D** is the set of all terminal histories with the initial position of $\mathcal{D}(\Delta, A)$ as a prefix and such that any dialogical expression in it is legal in virtue of the rules of **D**.

Histories of the game correspond to what one understands as dialogues in the usual (non logical) sense of the term. Terminal histories are (usual) dialogues which are complete as far as the logical form of the thesis is concerned. Dialogues as defined here contain all possible complete (usual) dialogues. We discuss our motivations in the following section.

Notice that the argumentation forms as we defined them in the previous section ensure that all moves following the initial position will be played by **O** and **P** alternately. A terminal history of a dialogue $\mathcal{D}(\Delta, A)$ is thus a pair $\langle \Sigma, \mathbf{X} \rangle$ where Σ has $(\langle \mathbf{O} \vdash \Delta \rangle, \langle \mathbf{P} \vdash A \rangle)$ as a prefix, followed by a (possibly empty) sequence of alternating **O**- and **P**-moves. The following theorem shows that the payoff function is correctly defined, i.e. that any history in $\mathcal{D}(\Delta, A)$ will reach a terminal position:

THEOREM 1.3.1 *Let $\Delta \subset L$ be a finite set of formulas and $A \in L$ a formula. Any terminal history in the dialogue $\mathcal{D}(\Delta, A)$ is of finite length.*[5]

---

[4] For any finite set of formula $\Delta = \{A_0, ..., A_n\}$, we write $\langle \mathbf{X} \vdash \Delta \rangle$ as a shorthand for the sequence $(\langle \mathbf{X} \vdash A_0 \rangle, ..., \langle \mathbf{X} \vdash A \rangle)$.

[5] The *length* of a history $\mathbb{H} = \langle \Sigma, \mathbf{X} \rangle$ is the number of elements of Σ.

Proof : The proof relies on the following considerations. First of all, we remark that any terminal history has $\mathbb{H}_0$ as a prefix, which is of finite length. Now the local rules feature the subformula property in this sense: For any argumentation form (p,c,d) where the degree[6] of the precondition p is n, the degree of the defence d is at most $n-1$. So any sequence of moves where a player systematically challenges the defence against his previous challenge ends up with an elementary defence after finitely many moves. Now there is no local rules for elementary assertions, so the sequence of challenges must end there. Finally the rules taking redundancies in charge ensure that no infinite loop may occur in $\mathcal{D}(\Delta,A)$. ∎

Example We give here the dialogue $\mathcal{D}(\varnothing,(A \vee B) \to C)$:

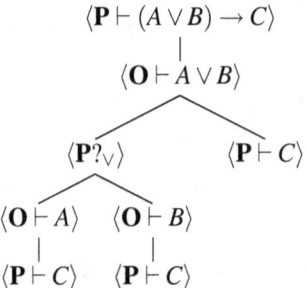

A notational remark Readers familiar with the dialogical literature (as e.g. Rahman and Keiff 2004) may be surprised not to find the usual notation for dialogues. Such a notation displays a dialogue in the form of a 2-column table, one for each player. Each column is divided in three sub-columns, the outer bearing a number for the move, the central bearing the formula or the attack marker, and the inner bearing the number of the challenged move when the move is a challenge. A challenge and the corresponding defence are written on the same line. Here's an example:

	O			P	
				$(A \wedge B) \to A$	0
1	$A \wedge B$	0		A	4
3	A		1	$?_L$	2

Actually such a notation seems to be the most natural way to represent a dialogue, as it shows how the argumentation forms structure the dialogue. The most important point is that it makes it clear when a given assertion is a defence, and when it is a challenge (against a conditional). Our notation does not show the difference in

---
[6]By *degree* we mean the number of connectives in it.

the syntax.[7] However, the table notation is not very convenient when the aim is to describe strategies, where splittings occur. Since dialogical validity is defined in terms of strategies and since we would like to show how to extract strategies from the extensive form of games, we have chosen to use the tree-like notation.

1.4 Dialogues and Validity

The basic game-theoretical tool to study the properties of games is their representation in extensive form. A simple version of the usual definition, retaining only what is useful for our purpose would have the following definition:

DEFINITION 1.4.1 (EXTENSIVE FORM) The *extensive form* $\mathcal{E}$ of a dialogical game $\mathcal{D}(\Delta, A)$ is the smallest rooted tree[8] such that:

i. The root is labelled with the initial history of $\mathcal{D}(\Delta, A)$, with $\langle \mathbf{P} \vdash A \rangle$ as its *active* expression.

ii. Let h be a branch in $\mathcal{E}$, n the last node of h, $\langle \mathbf{X}\, f\, e \rangle$ the active expression of n, and Σ the sequence of dialogical expressions labelling the nodes of h. For any dialogical expression e' such that e' denotes a legal move in a history $(\Sigma, \mathbf{Y})$ according to the rules of $\mathcal{D}(\Delta, A)$, there is a node in the tree which is a successor of n and labelled with e.

iii. Any leaf n of $\mathcal{E}$ bears as an extra label $(1,-1)$ if n belongs to a branch that is **P**-terminal (**O** wins) and $(-1,1)$ otherwise.

It is easy to see that a dialogue $\mathcal{D}(\Delta, A)$ as we defined it in the previous section *is* an extensive form. Each terminal history in $\mathcal{D}(\Delta, A)$ is a *play* of the game, that is a possible course of the actual argumentation. Let us now give a formal definition of a strategy:

DEFINITION 1.4.2 (STRATEGY) A *strategy* for **X** (or **X**-strategy) $S_\mathbf{X}$ in $\mathcal{D}(\Delta, A)$ is a subtree of $\mathcal{D}(\Delta, A)$ such that:

i. $S_\mathbf{X}$ contains only maximal branches of $\mathcal{D}(\Delta, A)$.

ii. Any node labelled with a **X**-move which has at least one successor in $\mathcal{D}(\Delta, A)$ has exactly one successor in $S_\mathbf{X}$.

iii. For any node n labelled with a **Y**-move, if m is a successor of n in $\mathcal{D}(\Delta, A)$ then m belongs to $S_\mathbf{X}$.

---

[7] See the example at the end of section 1.5.

[8] A *tree* is a set of nodes together with an irreflexive relation of successor, such that (i) there is a single node that is the successor of no other node; (ii) for any two distinct nodes in the tree, it is not the case that they share a successor. A *branch* of the tree is any sequence of nodes begining with the root and linearly ordered by the successor relation. A *leaf* of the tree is a node that has no successor.

A **X**-strategy is *winning* iff it contains no **X**-terminal history. In other terms, anytime in the course of the game where **X** is to choose a move, the strategy indicates a move such that if **X** plays it, he remains in a play of the game where he is sure to win. We can now give a precise formulation of the triviality of $\mathbf{D}_{min}$:

THEOREM 1.4.1 *In any dialogue $\mathcal{D}(\Delta,A)$ with the rules $\mathbf{D}_{min}$, any **P**-strategy is winning.*

The reason for this is simple. In a dialogue $\mathcal{D}(\Delta,A)$, **O** moves first, and the only moves available to him (if any) are challenges. The argumentation forms always allow for a defense against a challenge, and as long as **O** will have non-redundant challenges, **P** will have non-redundant defences. Since the argumentation forms ensure that the defence of an assertion is always logically simpler than the precondition, and since there is no challenge against elementary assertions, **O** will necessarily run out of challenges, and **P** wins asserting the last non-redundant defence. The meaning of a winning strategy $S_\mathbf{P}$ in $\mathcal{D}(\Delta,A)$ is that statement A is justifiable in the context of hypotheses Δ *provided* **P** *knows how to justify the set of elementary assertions he made in the course of the game*.[9] Clearly, winning strategies talk about (conditional) justifiability and not (actual) justification.

The notion of winning strategy is thus not enough to define the class of statements that one may consider as logically valid. In the conceptual framework of dialogical logic, validity is demonstrated by a property of the game associated with a formula. The idea is that some statements trigger an argumentation process that is, *for inherently interactive reasons*, sufficient to consider them *actually* justified. This property of the argumentation comes from the dynamics of the commitment to elementary justifications.

Assume that I enter a debate against someone with respect to some claim A. My strategy in the debate tells me what are the elementary claims I should be able to justify in order to justify my claim. Assume that one of them, say p, is such that in the history $\mathbb{H}$ where I assert p, my opponent has already asserted p. If we grant that argumentation games for elementary proposition are of perfect information and that if a player is showed how to make a move he can always reproduce it, then in $\mathbb{H}$ I have what we may call after Abramsky a *copycat* strategy. Such a strategy consists simply in replicating against his assertion of p any challenge from my opponent against my own, and answering his challenges by replicating his answers. In such game situations, either my opponent will fail to justify his claim, or he will succeed, but doing so he will show me how to justify my own claim.

From these considerations stems the dialogical appraoch to *formality*. We say that an elementary assertion $\langle \mathbf{X} \vdash p \rangle$ in a dialogical history $\mathbb{H} = \langle \Sigma, \mathbf{X} \rangle$ is *contentious* iff $\langle \mathbf{Y} \vdash p \rangle \notin \Sigma$. So an elementary assertion of **X** is not contentious iff **Y**

---

[9]More precisely, **P** must know how to justify every elementary assertion he must make in any maximal history of S_P.

is already committed to the justification of p by having asserted it. Thus a winning
P-strategy will give an actual justification to the thesis provided it does not contain
any contentious claim. Such a strategy would show that any critical party trying to
refute the thesis will in the course of the argumentation commit itself to accept the
justifications for the thesis. We say that a strategy is *formal* just when it contains
no contentious assertion.

The reader familiar with the dialogical tradition will notice that the approach
of this paper diverges here with the standard view on dialogical logic, in that we
consider formality as a property of *strategies* while it is usually seen as a property
of *games*. However this does not mean that we take it that the signification of
formal winning strategies *reduces* to validity. As we will discuss in the last section
of this paper, the usual Introduction/Elimination rules of natural deduction do not
reflect all that one finds in a FWS**P**.

Although the term 'copycat' denotes a certain way to build a strategy, it is also
a property of the games. It can be stated thus:

DEFINITION 1.4.3 (COPYCAT GAMES) Let $\Delta \subset \mathcal{L}$ be a finite set of formulas and
$A \in \mathcal{L}$. If $A \in \Delta$, then there is a FWS**P** in $\mathcal{D}(\Delta, A)$.

What the copycat really says is that in any game situation where **X** should defend an assertion of a formula A that the other player has also asserted, he can
provide a complete justification[10] of A without any contentious move. The reason
is simple: Any **Y**-challenge in the complete justification of $\langle \mathbf{X} \vdash A \rangle$ can be replicated by **X** against $\langle \mathbf{Y} \vdash A \rangle$, and any **Y**-defense can also be replicated, with the
difference that **Y** always moves first, so none of **X**'s replication of a **Y**-move is
contentious. In that respect, one may well extend the definition of redundant move
with any challenge against a non-contentious assertion.

Our interest in Johansson's *minimal logic* (**ML**) is determined by the fact that
it gives a basis from which one can build intuitionistic and classical logics by
extension of the set of inference rules or, equivalently, of the set of axioms. Let
us now state the connection between strategies and validity in the case of **ML**. We
will come in the next section to its Fitch-style natural deduction system. As an
axiomatic system, **ML** is defined as the set of axioms:

1. $p \to (q \to p)$

2. $((p \to (p \to q))) \to (p \to q)$

3. $(p \to q) \to ((q \to r) \to (p \to r))$

---
[10] A complete justification of A is the set of all the defences of A and all the defences of these defences
and so on.

4. $(p \wedge q) \to p$

5. $(p \wedge q) \to q$

6. $p \to (q \to (p \wedge q))$

7. $p \to (p \vee q)$

8. $q \to (p \vee q)$

9. $(p \to r) \to ((q \to r) \to ((p \vee q) \to r))$

10. $(p \to q) \to ((q \to \bot) \to (p \to \bot))$

11. $(p \wedge (p \to \bot)) \to \bot$

together with the inference rules Uniform Substitution and Modus Ponens.

We claim that the notion of winning formal **P**-strategies (FWS**P**) in $\mathbf{D}_{min}$ is correct for minimal logic in the following sense:

THEOREM 1.4.2 *Let $\Delta \subset \mathcal{L}$ be a finite set of formulas and $A \in \mathcal{L}$. There is a proof of $\Delta \Rightarrow A$*[11] *in minimal logic iff there is a FWS**P** in $\mathcal{D}(\Delta, A)$.*

Proof : The left to right part is fairly easy. It suffices to show that (i) there is a FWS**P** in the dialogue $\mathcal{D}(\varnothing, A)$ for any axiom A of minimal logic and (ii) inference rules of minimal logic preserve the existence of a FWS**P**.

The right to left part is an immediate corollary of the correction of our translation algorithm. Assume there is a FWS**P** in $\mathcal{D}(\Delta, A)$. Then we translate it into a Fitch-style proof of A from hypotheses Δ. If the Fitch-style natural deduction system for minimal logic is correct, then there is an axiomatic proof of $\Delta \Rightarrow A$. ∎

1.5 Intuitionistic logic and Classical Dialogues

When presented in Natural Deduction, the relation between Johansson minimal logic and intuitionistic logic is very simple: The latter is the result of the addition of *Ex Falso Sequitur Quodlibet* to the inference rules.[12] This rules stipulates that any formula can be infered from $\bot$. There is a way to extend our dialogical games that is equally simple.

---

[11] As usual, a proof of $\Delta \Rightarrow A$ in an axiomatic system **S** is a sequence of formulas such that each of them is either an axiom, or the result of applying an inference rule to previous formulas in the sequence.

[12] See section 2 for details.

[SR-3] (*Ex Falso Quodlibet*) Let $\mathbb{H} = \langle \Sigma, \mathbf{X} \rangle$ be a dialogical history such that $\langle \mathbf{Y} \vdash \bot \rangle \in \Sigma$. Player $\mathbf{X}$ may challenge $\langle \mathbf{Y} \vdash \bot \rangle$ with a move $\langle \mathbf{X} ? A \rangle$ for any $A \in \mathcal{L}$. In any dialogical context $\mathbb{H}' = \langle \Sigma', \mathbf{Y} \rangle$ such that $\langle \mathbf{X} ? A \rangle \in \Sigma'$, $\mathbf{Y}$ may play $\langle \mathbf{Y} \vdash A \rangle$.

Notice that SR-3 amount litterally to consider $\bot$ as a nullary connective with the following argumentation form:

Precondition	$\langle \mathbf{X} \vdash \bot \rangle$
Challenge	$\langle \mathbf{Y} ? A \rangle$
Defence	$\langle \mathbf{X} \vdash A \rangle$

The obvious drawback of rule SR-3 is that it introduces potentially redundant challengees against $\bot$ that should be ruled out. We extend definition 1.3.1 with the following clause:

DEFINITION 1.5.1 (REDUNDANT CHALLENGE AGAINST $\bot$) Let $\mathbb{H} = \langle \Sigma, \mathbf{X} \rangle$ be a dialogical history such that $\langle \mathbf{Y} \vdash \bot \rangle \in \Sigma$. $\langle \mathbf{X} ? B \rangle$ is redundant in $\mathbb{H}$ if there is a challenge $\langle \mathbf{X} ? A \rangle \in \Sigma$ against $\langle \mathbf{Y} \vdash \bot \rangle$ and $\langle \mathbf{X} \vdash A \rangle \notin \Sigma$.

That is, in order to challenge $\bot$ again, player $\mathbf{X}$ should find a way to use the first formula he requested. Given definitions 1.3.1 and 1.5.1, we define our system for intuitionistic dialogues as:

$$\mathbf{D}_{int} = \mathbf{D}_{min} \cup \{\mathbf{SR\text{-}3}\}$$

Clearly, in any dialogical context where $\mathbf{X}$ has asserted contentiously $\bot$ and $\mathbf{Y}$ has made no contentious assertion, $\mathbf{Y}$ has a formal strategy. So if in a dialogical game $\mathbf{P}$ has a strategy to force $\mathbf{O}$ to concede $\bot$ contentiously, he has a FWS. This explains in particular why $\bot \to A$ is valid for any A.

To transform the natural deduction presentation of intuitionistic logic into classical logic, it suffices to add the rule of excluded middle.[13] The usual interpretation of the difference between intuitionistic logic and classical logic is very similar for dialogue systems and sequent calculi: The latter vary in the number of formulae to the right, while the former vary in the number of challenges a player can recall.[14] But our goal here is to see to what extent dialogues and natural deduction can be understood in the same perspective. So we will follow the natural deduction route to classical logic:

---

[13] See section 2 for details.
[14] Recall that SR-1 says that one can only defend the last challenge. This makes $\mathbf{D}_{min}$ fundamentally intuitionistic.

[SR-4] (**Excluded Middle**) Let $p \in \mathbb{P}$ be an elementary formula, $A \in \mathcal{L}$, and $\Delta \subset \mathcal{L}$ a finite set of formulas. Let $\mathbb{H} = \langle \Sigma, \mathbf{X} \rangle$ be a dialogical history in $\mathcal{D}(\Delta, A)$. Then $\mathbf{X}$ may play $\langle \mathbf{X}\ ?\ p \vee (p \to \bot) \rangle$. In any dialogical history $\mathbb{H} = \langle \Sigma, \mathbf{X} \rangle$ such that $\langle \mathbf{Y}\ ?\ p \vee (p \to \bot) \rangle \in \Sigma$, $\mathbf{X}$ may play either $\langle \mathbf{X} \vdash p \rangle$ or $\langle \mathbf{X} \vdash (p \to \bot) \rangle$.

SR-4 is clearly equivalent to a presupposition of determinacy for the atoms of the language, in the sense that for any atom p, either p or $p \to \bot$ should be justifiable. Of course, since a rule such as SR-4 will introduce infinitely many possible redundant moves in the game, we shall update our definition of redundancy:

DEFINITION 1.5.2 (REDUNDANT EXCLUDED MIDDLE MOVES) A challenge $\langle \mathbf{X}\ ?\ p \vee (p \to \bot) \rangle$ is redundant in a dialogical context $\mathbb{H} = \langle \Sigma, \mathbf{X} \rangle$ iff one of the following holds:

i. (*relevance*) p is not a subformula of the formulas in the initial history;

ii. (*repetition*) $\langle \mathbf{X}\ ?\ p \vee (p \to \bot) \rangle \in \Sigma$.

Given definitions 1.3.1, 1.5.1 and 1.5.2, we define our system for classical dialogues as:

$$\mathbf{D}_{cl} = \mathbf{D}_{int} \cup \{\mathbf{SR\text{-}4}\}$$

The core of strategies In spite of the fact that strictly redundant moves are forbidden by the rules of a dialogue, a FWS**P** contains a lot of moves that are redundant *from the point of view of validity*, i.e. when what is at stake is to show that a given strategy is *formal*. While it is desirable to formulate dialogues rules with respect only to considerations of argumentation, when one aims at translating a dialogical proof (that is to say a FWS**P**) into some other kind of proof, it is much better to eliminate all those undesirable moves. We call the result of this elimination the *core* of a strategy.

The core of a FWS**P** should only retain such moves that are important to determine wether the inference from premisses to the thesis is valid or not, i.e. wether the **P**-strategy is formal or not. Clearly, any (non redundant) repetition of a **O**-challenge would make no difference, for if the first defence was formal, so will be the second. Any **O**-move that is legal in virtue of SR-3 or SR-4 will also be without incidence. For SR-3, consider a dialogical history containing $\langle \mathbf{P} \vdash \bot \rangle$. Then either this move was contentious, and the strategy is not formal whatever happens next, or it is not contentious, hence the history contains also $\langle \mathbf{O} \vdash \bot \rangle$. Then for any challenge against his own assertion of $\bot$, **P** can challenge **O** in the same way and win formally by copycat. For SR-4, the reasoning is the same: Any **O** challenge would be followed by the same **P** move, and **P** will have a copycat strategy to defend formally. We introduce a last simplification on FWS**P** to define its core: Everytime **O** has a choice between two moves, he will (when the rules allow) try the second move in the same history where he tried the first before the game reaches a terminal position. In order to show that a winning **P**-strategy is formal, it is sufficient to consider after each **O**-split only the moves relevant to one of **O**'s options. Indeed, the concatenation of a two sequences of moves where **P** plays formally is also a sequence of **P**-formal moves. Figure 1 below is an example of the core of a FWS**P** in $\mathbf{D}_{cl}$ for Peirce's Law. The extensive form of the game is much too rich to be conveniently displayed as an example, since it contains every possible strategy for both players. Here we are only concerned with a small fragment of it, namely one specific **P**-strategy, from which the only moves we retain are the one that are relevant with respect to validity.

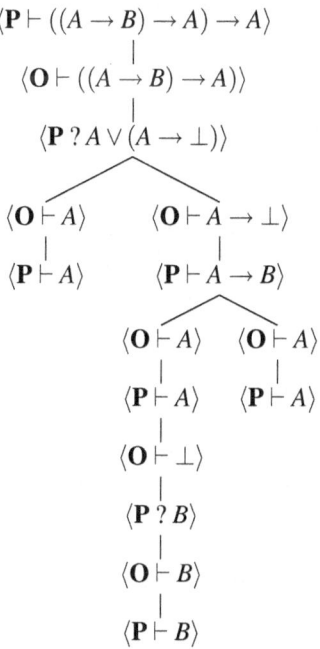

Figure 1.

An interesting feature of this strategy is the split after $\langle \mathbf{P} \vdash A \rightarrow B \rangle$. In both branches **O** moves with $\langle \mathbf{O} \vdash A \rangle$, but in the left branch this is a challenge against $\langle \mathbf{P} \vdash A \rightarrow B \rangle$, while in the right branch it is a defence of $\langle \mathbf{O} \vdash (A \rightarrow B) \rightarrow A \rangle$.[15]

---

[15]This difference does not show up here. This is the price we have to pay for the tree-like notation. The problem could easily be fixed, adding to the definition of a dialogue a function associating to each move a unique number, to each challenge the number of the precondition allowing it, and to each defence the number of the challenge it answers to. This would involve us in some kind of hybrid or labelled system. A move that, for the sake of simplicity, we would like to avoid.

In the right branch, the last **O**-challenge is the second move and **P** can defend the thesis, while on the left branch the last challenge from **O** is against $\langle \mathbf{P} \vdash A \to B \rangle$, leaving no other *formal* option than the use of **SR-3** to defend the conditional.

It is a well known fact that formal strategies are the dialogical equivalent for the usual notion of validity, in the sense that there exists a FWS$_\mathbf{P}$ in $\mathcal{D}(\Delta, A)$ in $\mathbf{D}_{int}$ (resp. $\mathbf{D}_{cl}$) iff $\Delta \Rightarrow A$ is valid in intuitionistic (resp. classical) logic. See Felscher (1985) and Rahman (1993) for the proofs.

1.6 Dialogues and Tableaux

Our purpose in the remaining of the section is to show the way one can extract from a FWS$_\mathbf{P}$ a tree-like structure of assertions that actually is a tableau proof.

All through our presentation of dialogical rules, we insisted in giving strictly *symmetrical* rules, in the sense that – except for the definition of the initial history – players in a dialogue have exactly the same rights. But the study of validity, i.e. of winning formal **P**-strategies, introduces an obvious asymmetry: The only splits in the strategy are **O**'s choices. But such choices can be understood in two different ways. On the one hand, as we did up to now, one can define the objects of the choices as moves. On the other hand, as we suggested when we explained the notion of choice with respect to the conditional, one can think that when choosing a move, the player is actually choosing a *rule*.

This change of perspective may be motivated by the following consideration. If one accepts that the business of logic is to provide a theory of *inference*, and that inference is a relation between assertions, then our argumentation forms may be considered as building blocks for such a theory. The relation of justification is defined in a dialogue as holding between an assertion and the set of its defences against all possible challenges. As a consequence, from this point of view, challenges are transitory indications about the course of the game that, while crucially important when one considers (dialogical) proof as a *process*, play no role in the *result* of the process. So one may want to represent a complete dialogue from the perspective of inference understood as a set of justificatory relations between assertions and their defences.

Actually, for disjunction and conjunction there is no difference between both perspectives. But there is an important one for the conditional. The extensive form of a local game for a conditional is:

$$\langle \mathbf{X} \vdash A \to B \rangle$$
$$|$$
$$\langle \mathbf{Y} \vdash A \rangle$$
$$\langle \mathbf{X} \, f \, e \rangle \quad \langle \mathbf{X} \vdash B \rangle$$

(where $\langle \mathbf{X}\, f\, e \rangle$ stands for the relevant challenge against $\langle \mathbf{Y} \vdash A \rangle$, according to A's outmost form.) But from the perspective of argumentation forms, $\langle \mathbf{Y} \vdash A \rangle$ is ambiguous as it is both a precondition in the left branch and a challenge in the right branch. Thus the correct topology of a tree representing the relation of inference as defined above (that is with respect to preconditions and defences only) should be:

$$
\begin{array}{c}
\langle \mathbf{X} \vdash A \to B \rangle \\
\diagup \qquad \diagdown \\
\langle \mathbf{Y} \vdash A \rangle \qquad \langle \mathbf{X} \vdash B \rangle \\
\diagup \diagdown \\
\cdots \quad \langle \mathbf{Y}\, f\, e \rangle
\end{array}
$$

(where $\langle \mathbf{Y}\, f\, e \rangle$ stands for one of the possible *defences* of A.)

Now when one is interested in proofs, the extensive forms are no longer considered *in abstracto* but are, borrowing the term from Girard's Ludics,[16] *incarnated*. Indeed, from the point of view of FWS$_\mathbf{P}$, the splits reflect $\mathbf{O}$'s choices. $\mathbf{P}$ may well play both possible rules in the same course of the game, and actually, *it is in his strategical interest to do so*.

From the preceding considerations, it is easy to give rules for building a tree that represent the inference relation for a dialogue $\mathcal{D}(\Sigma, A)$. The only complication comes from the SR-1 rule stipulating that only the last challenge may be answered. From the point of view of argumentation forms, this amounts to consider that one can apply a $\mathbf{X}$-rule only if its precondition is the last $\mathbf{X}$-assertion, and we consider all applicable $\mathbf{X}$-rules as applied at once. Here are the rules:

---

[16] See Girard (1999) for instance.

O-rules	Prules
$\langle \mathbf{O} \vdash A \vee B \rangle$ ⟋⟍ $\langle \mathbf{O} \vdash A \rangle \quad \langle \mathbf{O} \vdash B \rangle$	$\langle \mathbf{P} \vdash A \vee B \rangle$ | $\langle \mathbf{P} \vdash A \rangle$ | $\langle \mathbf{P} \vdash B \rangle$
$\langle \mathbf{O} \vdash A \wedge B \rangle$ | $\langle \mathbf{O} \vdash A \rangle$ | $\langle \mathbf{O} \vdash B \rangle$	$\langle \mathbf{P} \vdash A \wedge B \rangle$ ⟋⟍ $\langle \mathbf{P} \vdash A \rangle \quad \langle \mathbf{P} \vdash A \rangle$
$\langle \mathbf{O} \vdash A \rightarrow B \rangle$ ⟋⟍ $\langle \mathbf{P} \vdash A \rangle \quad \langle \mathbf{O} \vdash B \rangle$	$\langle \mathbf{P} \vdash A \rightarrow B \rangle$ | $\langle \mathbf{O} \vdash A \rangle$ | $\langle \mathbf{P} \vdash B \rangle$

One will obviously recognise here the tableaux rules for signed formulas as in Smullyan (1968) when one interprets **O** as T and **P** as F. See Rahman (1993) for a thorough presentation of the connection between dialogues and tableaux in the propositional and first-order cases of intuitionistic and classical logic.

2 Fitch-style Natural Deduction

DEFINITION 2.0.1 (FITCH-STYLE DEDUCTION) A Fitch-style deduction is a sequence Σ of tuples, each of the form (l, A, m), where l is taken from some ordered set of labels $\mathbb{L}$, A is a derived formula, m is the justification of the derivation, and such that:

- a justification m is of one of the following foms:

form of m	justification rule
–	No justification. The formula is an assumption
$\wedge$E, l	$\wedge$ elimination on the formula of line l
$\wedge$I, l_1, l_2	$\wedge$ introduction on the formulae of lines l_1 and l_2
$\rightarrow$E, l_1, l_2	$\rightarrow$ elimination on the formulae of lines l_1 and l_2
$\rightarrow$I, l_i-l_j	$\rightarrow$ introduction from the block of lines l_i-l_j
$\vee$E, l_1, l_i-l_j, l_k-l_n	$\vee$ elimination on the formula of line l_1, the block of lines l_i-l_j, and the block of lines l_k-l_n
$\vee$I, l	$\vee$ introduction on the formula of line l
R, l	reiteration rule on the formula of the line l

- Σ obeys the following rules[17]:

$$
\begin{array}{c|c} l \\ \vdots \end{array} \qquad \begin{array}{c|c} & A \\ \hline & \vdots \end{array}
$$

$$
\begin{array}{c|c} l_1 & A \\ \vdots & \vdots \\ l_2 & B \\ \vdots & \vdots \\ l & A \wedge B \end{array} \quad \wedge\text{I}, l_1, l_2 \qquad
\begin{array}{c|c} l & A \wedge B \\ \vdots & \vdots \\ l' & A \end{array} \quad \wedge\text{E}_1, l
$$

$$
\begin{array}{c|c} l & A \wedge B \\ \vdots & \vdots \\ l' & B \end{array} \quad \wedge\text{E}_2, l
$$

$$
\begin{array}{c|c} l_1 & \\ \vdots & \begin{array}{|c} A \\ \vdots \\ B \end{array} \\ l_2 & \\ l & A \to B \end{array} \quad \to\text{I}, l_1, l_2 \qquad
\begin{array}{c|c} l_1 & A \\ \vdots & \vdots \\ l_2 & A \to B \\ l & B \end{array} \quad \to\text{E}, l_1, l_2
$$

$$
\begin{array}{c|c} l & A \\ \vdots & \vdots \\ l' & A \vee B \end{array} \quad \vee\text{I}, l \qquad
\begin{array}{c|c} l & A \\ \vdots & \vdots \\ l' & A \end{array} \quad \text{R}, l
$$

---

[17] We present these rules in graphical notation, for the sake of clarity. For a more formal, non graphical, account of derivation rules in Fitch-style Natural Deduction, see Geuvers and Nederpelt (2004)

```
i    │ A ∨ B
⋮    │ ⋮
j₁   │   │ A
     │   │───
⋮    │   │ ⋮
j₂   │   │ C
k₁   │   │ B
     │   │───
⋮    │   │ ⋮
k₂   │   │ C
l    │ C              ∨E, i, j₁–j₂, k₁–k₂
```

There are different notational conventions for Fitch-style Natural Deduction. One can be found in Gamut (1992). We use one which is used in Garson (2006) or Barwise and Etchemendy (1993) for example. This notation is caracterized by two kinds of graphic features: vertical lines and horizontal lines, the latter being called 'Fitch bars.' A Fitch bar is placed right beneath each assumption, with one exception: The premises of the derivation we want to show are all put together above a sole Fitch bar. A vertical line starts with an assumption and is used to indicate how long the assumption is available.

The rules →I and ∨E make use of one of the most appealing features of Fitch-style Natural Deduction: subderivations. Each time an assumption not belonging to the set of premises is done, a new subderivation is created. A vertical line is drawn until the assumption is discharged, and the rules →I and ∨E show how such a discharge can occur. We call a subderivation *finished* when the assumption starting it is discharged. No individual step of a finished subderivation can be used to apply a rule outside of the subderivation, but an individual step of an unfinished subderivation can be used in a subderivation of 'lower level.' Rules →I and ∨E involve the use of finished subderivations as a whole.

We define the (Fitch-style) Natural Deduction system for minimal logic as the set of rules

$$\mathbf{F}_{min} = \{\wedge E, \wedge I, \to E, \to I, \vee E, \vee I, R\}.$$

The Natural Deduction system for intuitionistic logic is defined as the set

$$\mathbf{F}_{int} = \mathbf{F}_{min} \cup \{EFSQ\},$$

where EFSQ is the following rule:

$$
\begin{array}{c|c}
l & \bot \\
\vdots & \vdots \\
l' & A \qquad \text{EFSQ}, l
\end{array}
$$

Finally, $\mathbf{F}_{cl}$, the Natural Deduction system for classical logic, is defined as

$$\mathbf{F}_{cl} = \mathbf{F}_{int} \cup \{EM\},$$

where EM is the following rule:

$$
\begin{array}{c|c}
i & \begin{array}{|l} A \\ \vdots \end{array} \\
\vdots & \\
j & \begin{array}{|l} C \end{array} \\
k & \begin{array}{|l} A \to \bot \\ \vdots \end{array} \\
\vdots & \\
l & \begin{array}{|l} C \end{array} \\
m & C \qquad \text{Excluded Middle, } i\text{–}j, k\text{–}l
\end{array}
$$

DEFINITION 2.0.2 *(conclusion, premisses, Fitch-style proof)*
Let Σ be a Fitch-style deduction in $\mathbf{F}_{min}$, $\mathbf{F}_{int}$, or $\mathbf{F}_{cl}$

1. The last line of Σ is called the conclusion of Σ.

2. The premisses of Σ are any formulae A such that A is an assumption of the subderivation which ends with the conclusion of Σ.

3. Let Δ be a set of premisses and A a formula. A Fitch-style deduction with A as a conclusion and the members of Δ as premisses is called a (Fitch-style) proof of A from Δ.

3 From Dialogues to Fitch-style Proofs

In this section, we give a procedure to translate the core of a FWS$_\mathbf{P}$ in $\mathcal{D}(\Delta, A)$ into a Fitch-style Proof of A (from Δ).

3.1 The procedure

The algorithm takes the core of a formal **P**-winning strategy FWS$_\mathbf{P}$ in one of the game systems we defined ($\mathbf{D}_{min}$, $\mathbf{D}_{int}$, $\mathbf{D}_{cl}$) and translate it into a Fitch-style proof in the corresponding Natural Deduction system ($\mathbf{F}_{min}$, $\mathbf{F}_{int}$, $\mathbf{F}_{cl}$, respectively). The mechanism is rather simple and consists in arranging the content of the assertive moves of FWS$_\mathbf{P}$'s core in a linear order such that the sequence of formulae complies with the natural deduction rules. Each member of the sequence is then labelled with a suitable number and a justification. We describe a procedure that translates the core of a FWS$_\mathbf{P}$ in $\mathbf{D}_{cl}$ into a Fitch-style proof in $\mathbf{F}_{cl}$. The algorithm for intuitionnistic logic is obtained by removing the clauses relative to the rule for Excluded Middle. The algorithm for minimal logic is obtained by removing also the clauses relative to Ex Falso Quodlibet.

Let us begin with some terminology. We say that, in a FWS$_\mathbf{P}$'s core, an assertion $\langle \mathbf{X} \vdash C \rangle$ *depends on* a move $\langle \mathbf{Y} \vdash A \vee B \rangle$ if $\mathbf{X}$ can formally defend C only after $\mathbf{Y}$ defended $A \vee B$. In a similar way, we say that $\langle \mathbf{X} \vdash B \rangle$ depends on application of rule SR-4 if $\mathbf{X}$ can formally defend it only after he played $\langle \mathbf{X} \, ? \, A \vee (A \rightarrow \bot) \rangle$ and $\mathbf{Y}$ answered this move.

Initial stage

First we place the members $\phi_1, \ldots, \phi_n$ of Δ as the premises of the deduction and the thesis A as the conclusion. If A depends on a **O**-disjunction, then the justification of A is the application of a $\vee$E Rule. If it depends on application of SR-4, it is the result of rule EM. Otherwise it is the introduction rule corresponding to its main connective.

$$
\begin{array}{c|c}
p_1 & \phi_1 \\
\vdots & \vdots \\
p_n & \phi_n \\
\vdots & \vdots \\
n & A
\end{array}
$$

Incomplete Justification forms

Until the end of the procedure, no complete justification can be formed. Only the name of the rule applied can be given (see below). The lines of the steps to which the rule is applied will complete the justifications once every move is used by the procedure.

- Any move $\langle \mathbf{O} \vdash A \rangle$ challenging a move $\langle \mathbf{P} \vdash A \rightarrow B \rangle$ is introduced as a new assumption.

- Any pair of moves $(\langle \mathbf{O} \vdash A \rangle, \langle \mathbf{O} \vdash B \rangle)$ played in virtue of the local rule for disjunction is introduced as a pair of assumptions opening parallel sub-derivations.

- Any pair of moves $(\langle \mathbf{O} \vdash A \rangle, \langle \mathbf{O} \vdash A \rightarrow \bot \rangle)$ played in virtue of rule SR-4 is also introduced as a pair of assumptions opening parallel subderivations.

- Any move $\langle \mathbf{O} \vdash A \rangle$ used in virtue of rule SR-3 is introduced as the result of EFSQ Rule.

- Any other move $\langle \mathbf{O} \vdash A \rangle$ is introduced as the result of an Elimination Rule. The main connective of the formula defended by $\mathbf{O}$ with this move determines which Elimination Rule is used.

- Any move $\langle \mathbf{P} \vdash A \rangle$ occurring after $\mathbf{O}$ already performed the move $\langle \mathbf{O} \vdash A \rangle$ is the result of a Reiteration Rule.

- Any move $\langle \mathbf{P} \vdash A \rangle$ which depends on a move $\langle \mathbf{O} \vdash B \vee C \rangle$ is the result of a $\vee$E Rule.

- Any move $\langle \mathbf{P} \vdash A \rangle$ which depends on application of rule SR-4 is introduced as the result of EM Rule.

- Any other move $\langle \mathbf{P} \vdash A \rangle$ is the result of an Introduction Rule. The main connective of A determines which Introduction Rule is used.

Generalities

Once the initial stage is done, the procedure follows the order of the moves but ignores non assertoric expressions, which are specific to the dialogical approach. In general, moves of the form $\langle \mathbf{O} \vdash A \rangle$ will be placed upwards, that is such that its label immediately follows the one of the last assertion placed upwards; Moves of the form $\langle \mathbf{P} \vdash A \rangle$ will be placed downwards, that is such that its label (not necessarily immediately) precedes the last assertion placed downwards in the current subderivation.

A $\mathbf{O}$-challenge against a conditional assertion is placed upwards as a new assumption which is available up to the assertion lastly placed downwards, leaving it outside the new subderivation. When there is a split in the dialogical proof, these general conventions about placement must be adapted.

Split: case 0

Whenever a move $\langle \mathbf{P} \vdash A \rangle$ depends on the application of rule SR-4, it is placed as the result of EM Rule. Morever, the moves $\langle \mathbf{O} \vdash B \rangle$ and $\langle \mathbf{O} \vdash B \to \bot \rangle$ must be dealt with immediately after $\langle \mathbf{P} \vdash A \rangle$, even if other moves occur in FWS$_P$'s core between $\langle \mathbf{P} \vdash A \rangle$ and the split. These other moves must be placed in both subdeductions opened with B and $B \to \bot$.

Suppose C is the assertion lastly placed upwards and D the one lastly placed downwards (notice that D can possibly be A). The last formula of both subderivations is an occurrence of A (which incomplete justification form can consist in a reiteration or Introduction rule). The occurrence of A associated with the label l cannot be used again this way. We obtain:

h	C	
i		B
$\vdots$		$\vdots$
i_2		A
j_1		$B \to \bot$
$\vdots$		$\vdots$
j_2		A
k	D	
$\vdots$	$\vdots$	
l	A	Excluded Middle

Split: case 1

Suppose we have to place the move $\langle \mathbf{P} \vdash A \rangle$ that counts as a challenge against a move $\langle \mathbf{O} \vdash A \to B \rangle$. This **P** move is followed by a **O** split: The left branch starts with the proper challenging move $\langle \mathbf{O}\text{-}f\text{-}e \rangle$ against $\langle \mathbf{P} \vdash A \rangle$, the right branch starts with the defensive move $\langle \mathbf{O} \vdash B \rangle$. The procedure places A and B in the current subderivation as follows.

$$
\begin{array}{r|ll}
i & C & \ldots \\
\vdots & \vdots & \\
j & A & \ldots \\
k & B & \to E \\
\vdots & \vdots & \\
l & D & \ldots \\
\end{array}
$$

where C is the last assertion placed upwards and D is the last assertion placed downwards. The procedure will place the remaining assertions of the left branch between C and A, and the remaining assertions of the right branch between B and D, following the appropriate clauses.

Split: case 2

Consider the case where FWS$_P$'s core comes to a split in which the left branch starts with the move $\langle \mathbf{O}?_L \rangle$ and the right branch $\langle \mathbf{O}?_R \rangle$ challenging a preceding move $\langle \mathbf{P} \vdash A \wedge B \rangle$. The respective defensive moves are $\langle \mathbf{P} \vdash A \rangle$, $\langle \mathbf{P} \vdash B \rangle$ and are placed in the following way:

$$
\begin{array}{r|ll}
i & C & \ldots \\
\vdots & \vdots & \\
j & A & \ldots \\
\vdots & \vdots & \\
k & B & \ldots \\
k+1 & D & \ldots \\
\end{array}
$$

where C is the last assertion placed upwards and D the last assertion placed downwards by the procedure. The remaining assertions of the left branch will be placed between C and A, while the remaining assertions of the right branch will be placed between A and B.

Split: case 3

Suppose FWS$_P$'s core comes to a split where the left branch starts with the move $\langle \mathbf{O} \vdash A \rangle$ and the right branch starts with $\langle \mathbf{O} \vdash B \rangle$, such that these moves are the two possible defences of a preceding move $\langle \mathbf{O} \vdash A \vee B \rangle$. Those two assertions are placed as follows:

$$
\begin{array}{l|ll}
i & C & \ldots \\
\\
j_1 & \quad \Big| A \\
\vdots & \quad \Big| \vdots \\
\\
k_1 & \quad \Big| B \\
\vdots & \quad \Big| \vdots \\
\\
l & D & \ldots
\end{array}
$$

where C is the last assertion placed upwards and D the last assertion placed downwards. The remaining assertions of the left branch are placed within the subderivation opened by A, those of the right branch within the subderivation opened by B.

The last formula of both subderivations is an occurrence of the first formula occurring in the sequel of the proof which was given the justification form $\vee$E and which has not been used this way yet. Let E be such a formula, we obtain something of the form:

$$
\begin{array}{l|ll}
j_1 & \quad \Big| A \\
\vdots & \quad \Big| \vdots \\
\\
j_2 & \quad \Big| E & \ldots \\
k_1 & \quad \Big| B \\
\vdots & \quad \Big| \vdots \\
\\
k_2 & \quad \Big| E & \ldots \\
\vdots & \quad \Big| \vdots \\
\\
l & E & \vee\text{E}
\end{array}
$$

The occurrence of E associated with the label l must not be used again this way.

Completing the justifications

When every assertion in FWS$_\mathbf{P}$'s core has been dealt with by the procedure, the useless dots are removed, and the steps are numbered up-down, starting from 1. Then the justifications of each step can be completed. In what follows, the formula of which we complete the justification is called A. Recall that assumptions get no justification.

- *Ex falso*: The full justification of A is $EFSQ, l$ where l is the label given to the move $\langle \mathbf{O} \vdash \bot \rangle$ which is challenged in virtue of SR-3 to play the move $\langle \mathbf{P} \vdash A \rangle$.

- *Reiteration*: The full justification of A is R, l where l is the label given to the previous move $\langle \mathbf{O} \vdash A \rangle$.

- *Conditional Introduction*: The full justification is $\rightarrow$I, l_1-l_2 where l_1 is the label given to the move $\langle \mathbf{O} \vdash C \rangle$ challenging A, and l_2 is the label allocated to the move $\langle \mathbf{P} \vdash D \rangle$ which defends it.

- *Other Introduction Rules*: The full justification is obtained by adding the label(s) of the move(s) $\langle \mathbf{P} \vdash D \rangle$ defending A in FWS$_\mathbf{P}$'s core.

- *Disjunction Elimination*: The full justification of A is $\vee$E, i, j-k, l-m where i is the label given to the disjunctive move $\langle \mathbf{O} \vdash C \vee D \rangle$ A depends on, j-k denotes the labels of the subderivation initiated by the first disjunct and l-m denotes the labels of the subderivation initiated by the second disjunct.

- *Other Elimination Rules*: The full justification is obtained by adding the label given to the move $\langle \mathbf{O} \vdash B \rangle$ defended by the move $\langle \mathbf{O} \vdash A \rangle$.

3.2 Correction of the algorithm

THEOREM 3.2.1 *Let S be a formal winning strategy for $\mathbf{P}$ in $\mathcal{D}(\Delta, A)$. $\mathcal{F}$, the result of applying our algorithm to the core of S, is a correct Fitch-style proof of A from Δ.*

Proof: For the sake of brievety, we only sketch a proof that is fairly easy but fastidiously long. Here are the main ideas:

(i) The result of the procedure is a sequence of tuples, each of the form (l, A, m) with l a label, A a formula and m a justification, linearly ordered in a sequence by their labels.

(ii) Now consider this sequence as obtained from the empty sequence by successively adding the tuples, following the order on the labels. It can be shown by an inductive reasonning that this construction observes at each step the rules of Fitch-style Natural deduction as given in section 2^{18}. Once this is done, we have shown that $\mathcal{F}$ is a Fitch-style deduction.

(iii) After this, it is easy to show that $\mathcal{F}$ also observes the clauses of definition 2.0.2, which allows us to conclude that $\mathcal{F}$ is a correct Fitch-style proof of A from Δ. ∎

---

[18] In fact, this is very clear – but very long – when one works with formal definitions similar to the one used in Geuvers and Nederpelt (2004).

3.3 A simple example

Let us consider the formula $(a \vee b) \rightarrow (a \vee b)$ which is valid in minimal logic. Here is the core of a FWS$_P$ for it:

$$\langle \mathbf{P} \vdash (a \vee b) \rightarrow (a \vee b) \rangle$$
$$|$$
$$\langle \mathbf{O} \vdash a \vee b \rangle$$
$$|$$
$$\langle \mathbf{P} \vdash a \vee b \rangle$$
$$|$$
$$\langle \mathbf{O}?_\vee \rangle$$
$$|$$
$$\langle \mathbf{P}?_\vee \rangle$$

$$\langle \mathbf{O} \vdash a \rangle \qquad \langle \mathbf{O} \vdash b \rangle$$
$$| \qquad\qquad |$$
$$\langle \mathbf{P} \vdash a \rangle \qquad \langle \mathbf{P} \vdash b \rangle$$

The initial stage is:

$$\begin{array}{c|c} \vdots & \vdots \\ l & (a \vee b) \rightarrow (a \vee b) \qquad \rightarrow \text{I} \end{array}$$

The next assertion is a **O**-assertion that counts as a challenge against the thesis. We obtain:

$$\begin{array}{c|cc} l_1 & a \vee b & \\ \vdots & \vdots & \\ l & (a \vee b) \rightarrow (a \vee b) & \rightarrow \text{I} \end{array}$$

The next assertion is the corresponding **P**-defence which depends itself on a **O**-disjunction. So we obtain

$$\begin{array}{c|cc} l_1 & a \vee b & \\ \vdots & \vdots & \\ l_2 & a \vee b & \vee \text{E} \\ l & (a \vee b) \rightarrow (a \vee b) & \rightarrow \text{I} \end{array}$$

Next we have a **O** split between two possible defences of a disjunction.

$$
\begin{array}{ll}
l_1 & \quad a \vee b \\
l_3 & \qquad a \\
\vdots & \qquad \vdots \\
l_4 & \qquad a \vee b \qquad \vee \mathrm{I} \\
l_5 & \qquad b \\
\vdots & \qquad \vdots \\
l_6 & \qquad a \vee b \qquad \vee \mathrm{I} \\
l_2 & \quad a \vee b \qquad \vee \mathrm{E} \\
l & (a \vee b) \rightarrow (a \vee b) \quad \rightarrow \mathrm{I}
\end{array}
$$

In both branches, there remains a **P** assertion which is a repetition of a previous **O**-assertion and which counts as a defence of the move which depends on the **O**-disjunction. Thus:

$$
\begin{array}{ll}
l_1 & \quad a \vee b \\
l_3 & \qquad a \\
l_7 & \qquad a \qquad\qquad \mathrm{R} \\
l_4 & \qquad a \vee b \qquad \vee \mathrm{I} \\
l_5 & \qquad b \\
l_8 & \qquad b \qquad\qquad \mathrm{R} \\
l_6 & \qquad a \vee b \qquad \vee \mathrm{I} \\
l_2 & \quad a \vee b \qquad \vee \mathrm{E} \\
l & (a \vee b) \rightarrow (a \vee b) \quad \rightarrow \mathrm{I}
\end{array}
$$

Finally, we number the steps of this Fitch-style deduction and complete the justification forms:

```
1  │ a ∨ b
2  │  │ a
3  │  │ a           R, 2
4  │  │ a ∨ b       ∨I, 3
5  │  │ b
6  │  │ b           R, 5
7  │  │ a ∨ b       ∨I, 6
8  │ a ∨ b          ∨E, 1, 2–4, 5–7
9  │ (a ∨ b) → (a ∨ b)   →I, 1–8
```

3.4 Comments and discussion

Let us resume our take on the definition of meaning as it stems from the previous sections. The dialogical perspective offers means to distinguish between three aspects of meaning:

i. The argumentation forms associated with a connective # determine what may be called its *local meaning*, by giving rules for an interaction in a sellarsian game of assertions and requests. Such rules are abstract triples (p, c, d) which give constraints on games without defining what a game is. They just describe what one is committing oneself to when performing a #-assertion. One may say that the local meaning of # is the meaning of the connective *stricto sensu*. Take disjunction as an example. Its local meaning is given by the forms:

Precondition	$\langle \mathbf{X} \vdash A \vee B \rangle$	$\langle \mathbf{X} \vdash A \vee B \rangle$
Challenge	$\langle \mathbf{Y} \, ? \, \vee \rangle$	$\langle \mathbf{Y} \, ? \, \vee \rangle$
Defence	$\langle \mathbf{X} \vdash A \rangle$	$\langle \mathbf{X} \vdash B \rangle$

Local meaning spells out a set of ways to deal with committment in the assertoric language game, independently of any specific form such a game may take (i.e. retaining as the only relevant feature of the game the connexion between commitment and request instanciated by the structure Precondition–Challenge–Defense).

It is extremely important to build the theory of meaning at this abstract level because, as advocated by Wittgenstein[19], many games can be played without a complete definition of the rules. Hence not only the same argumentation forms are a common basis for many different language games (as exemplified in this paper), but you can even imagine a game where such forms are all you know about the rules.

ii. The local meaning of the connective # may then be contrasted with its *strategical meaning*, i.e. the way argumentation forms combine into extensive forms of a game associated with a #-assertion. This aspect of meaning is determined by the combination of argumentation forms *and* structural rules, and is given by the set of terminal histories in the dialogue associated with the assertion. Structural rules bear this name for they structure the use of the argumentation forms, in the sense that they give constraints for a set of move to be an actual play of the game. The strategical meaning of a disjunctive statement is thus given by the following schematic extensive form:

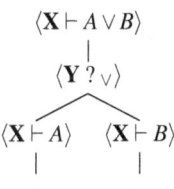

Strategical meaning recursively combines into an extensive form for an arbitrary statement, which spells out the *conditions on the justifiability* of the statement in terms of a disjunction of justifications for sets of elementary assertions, one such set for each terminal history.

iii. The notions of copycat strategy and of non-contentious assertion allow to single out a specific case of conditions on the justifiability of a thesis, namely when those conditions are null, which corresponds precisely to the cases when the thesis is a valid consequence of the initial concessions. With respect to disjunction and formal strategies, one might observe the following. Let C be a complex formula and $\langle \mathbf{Y} \, ? \, C \rangle$ denote the appropriate challenge against $\langle \mathbf{X} \vdash C \rangle$. Assume a game situation where **P** has no formal justification of his assertion of $\langle \mathbf{P} \vdash C \rangle$ before **O** defended $A \vee B$. Here's a fragment of a **P** strategy:

---

[19] See Wittgenstein's *Lectures* 110.

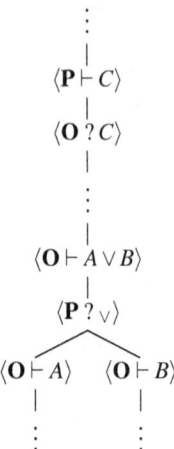

One may observe that, for **P**, to have a formal justification of ⟨**P** ⊢ *C*⟩ exactly amounts to have one given the **O**-assertion of *A* on the one hand, and to have one given the **O**-assertion of *B* on the other hand. And this is precisely what the Elimination rule for disjunction says.

Thus in general one may introduce a notion of *inferential behaviour* for the connectives:

DEFINITION 3.4.1 (INFERENTIAL BEHAVIOUR) Let # be a connective and *A* a #-formula. The *inferential behaviour* of # is compound of:

(a) the relation between the assertion ⟨**P** ⊢ *A*⟩ and the set of assertions of formulas *B* such that **P** needs a formal justification of ⟨**P** ⊢ *B*⟩ in order to formally justify the assertion of *A*;

(b) the relation between the assertion ⟨**O** ⊢ *A*⟩ and the set of assertions of formulas *B* such that **P** can formally justify ⟨**P** ⊢ *B*⟩ given the concession ⟨**O** ⊢ *A*⟩.

In other terms, the inferential behaviour of # is given by the role a #-formula will play, as either a **P**- or a **O**-assertion, *in a game where **P** wants to stay under nullary conditions on the justification he provides.*

Let us conclude with some remarks on the relation between dialogues and natural deductions. The local semantics in dialogical logic and the Eliminiation-Introduction rules in Natural Deduction may be considered as rival approaches to define the meaning of logical constants, when one understands, in the wittgensteinian tradition, meaning as use.

One may then observe that there is no dialogical equivalent to the introduction rules, and thus that the local semantics is incomplete in so far as it does not reflect completely the behaviour of a connective when it constitutes the outmost form of either a conclusion or a premiss. But this argument is somehow misleading, for dialogical *logic* and natural deduction are in agreement indeed on both aspects – as our algorithm and the previous remarks show clearly. There is a difference, though, and an important one. The argumentation forms are intended to give an account of the meaning of connectives in the context of sellarsian games of asking for and giving reasons (or justifications), which we take as the fundamental form of language game, or at least one of them. Such forms may then be studied from a strategical point of view, and a requisit of formality may be added in order to reach the level of their inferential behaviour.[20]

We take it as one of the main contribution of the dialogical approach to show how this behaviour is *explained* by argumentation forms (i.e. meaning proper) plus strategical considerations, whereas Introduction/Elimination rules rather *describe* it. Actually, while one can seek in the general patterns of interaction as described by argumentation forms on the one hand and structural rules on the other hand the reasons why inference rules are like they are, such a question is meaningless when one take them as a definition of the meaning of connectives.

In the core of a FWS$_\mathbf{P}$, we consider *incarnated* local rules, in the sense that the tree structure generated by application of argumentation forms to a precondition may be defined differently when the precondition is a **P**- or a **O**-assertion. This is the whole point of our remarks about dialogues and tableaux. In such a strategy, a **O**-argumentative form[21] decomposes a complex **O**-assertion in simpler one(s), which is exactly what does an elimination rule in a Natural Deduction proof. But when we restrict our attention to FWS$_\mathbf{P}$'s cores, there is no more symmetry in the signification of game phases than there is in the rules. **O**-forms eliminate complex assertions in order to reach elementary ones because, in the context of a formal strategy, such **O**-elementary assertions (and the associated justifications) should be understood as *resources* for **P**. Now **P**-forms have a different signification: They are the proper justifications, since through the whole game the burden of the justification of the thesis lies on **P**'s side. Therefore a **P**-argumentative form gives the justification of the assertion of a complex formula by means of (the assertion of one of) its immediate subformulas, which just amounts to the same as applications of Introduction rules. In other terms, to see how they relate to natural deduction proofs, **P**-forms should be read bottom-up (i.e. from elementary justification to the justification of the thesis) while **O**-forms should be read top-down (i.e. from the concession of complex ressources to the concession of elementary ones). Most of

---

[20]Notice that such a division between the meaning and the inferential behaviour of a connective relies crucially on the decision to introduce formality as a property of strategies and not of games.

[21]That is, a form the precondition of which is a **O**-assertion.

the clauses of the algorithm give an account of these relations between dialogical local rules and Elimination-Introduction rules in Natural Deduction.

There is an exception however to these considerations, of which we have already said something: the elimination rule for disjunction, which is associated with a **P** case in the procedure. Contrary to the other elimination rules, the formula derived may not be a subformula of the disjunction. But such a rule is clearly of a meta-logical nature. Since it consists in drawing a formula if it is derived in two parallel subderivations (each starting with one of the disjuncts), it clearly relies on considerations about the combination of proofs and not only on the meaning of disjunction, as witnessed by the three premises in the (non Fitch) elimination rule.

Clearly, in an *argumentation* (i.e. a terminal history in the dialogue, as represented in iii. above) only one of the disjuncts will be asserted, which is as it should be: Whenever I assert a disjunction, I just claim I can provide a justification to one of the disjuncts. It is only when it comes to define conditions on *formal strategies* in the game that the *behaviour* of disjunction coincides with its elimination rule.

We see the difference between explaining and (only) describing an inferential behaviour as a strong argument in favor of the claim that Natural Deduction rules are not, as they stand, meaning-constituting. In fact, other arguments for this claim were already given in discussions about the *tonk* connective introduced in the famous Prior (1961): the usual introduction of harmony constraints[22] indicates that Introduction/Elimination rules are not sufficient to constitue the meaning of the connectives. The dialogical perspective offers another way to understand what goes wrong with *tonk*, namely an endogenous way.

For the sake a brevity, we restrict ourselves to a summary[23]. First of all, let us recall the Natural Deduction rules for *tonk*:

$$
\begin{array}{l|l}
l & A \\
\vdots & \vdots \\
l' & A \text{ tonk } B
\end{array}
\qquad \text{tonk I}, l
$$

$$
\begin{array}{l|l}
l & A \text{ tonk } B \\
\vdots & \vdots \\
l' & A
\end{array}
\quad \text{tonk E}_1, l
\qquad\qquad
\begin{array}{l|l}
l & A \text{ tonk } B \\
\vdots & \vdots \\
l' & B
\end{array}
\quad \text{tonk E}_2, l
$$

---

[22] As pointed out by Belnap (1962) and others.

[23] A thorough discussion on this subject is to be presented in a forthcoming paper

It is impossible to give local rules which will match the Introduction/Elimination ones through our algorithm. To see this, assume that there were such rules. Then they should specify whose player (challenger or defender) may choose between a continuation of the game with A or with B. So let us suppose that the choice is the challenger's. The local rules for *tonk* would then be of the form:

Precondition	$\langle \mathbf{X} \vdash A \text{ tonk } B \rangle$	$\langle \mathbf{X} \vdash A \text{ tonk } B \rangle$
Challenge	$\langle \mathbf{Y} \, ? \, L \rangle$	$\langle \mathbf{Y} \, ?R \rangle$
Defence	$\langle \mathbf{X} \vdash A \rangle$	$\langle \mathbf{X} \vdash B \rangle$

But consider what happens when this form is incarnated as a **P**-case in a FWS**P** and translated through our algorithm: we obtain the following rule:

$$
\begin{array}{c|l}
l_1 & A \\
\vdots & \vdots \\
l_2 & B \\
\vdots & \vdots \\
l' & A \text{ tonk } B \qquad \text{tonk I}, l_1, l_2
\end{array}
$$

However, this is not the Introduction rule for *tonk*, which is of disjunctive type while the one we obtain is of conjunctive type. Well, then how about letting the choice be the defender's? Unfortunately, a similar problem appears.

Precondition	$\langle \mathbf{X} \vdash A \text{ tonk } B \rangle$	$\langle \mathbf{X} \vdash A \text{ tonk } B \rangle$
Challenge	$\langle \mathbf{Y} \, ? \, \text{tonk} \rangle$	$\langle \mathbf{Y} \, ?\text{tonk} \rangle$
Defence	$\langle \mathbf{X} \vdash A \rangle$	$\langle \mathbf{X} \vdash B \rangle$

If we take the incarnated form of a **O**-case in a FWS**P** and translate it by means of our procedure, we obtain a rule which is not the Elimination rule for *tonk*: The rule obtained will be much like the Elimination rule for disjunction while we need an Elimination rule of a conjunctive type.

To summarize, from the dialogical point of view, the reason why the inferential behaviour expressed by Natural Deduction rules for *tonk* makes no sense is that these rules do not let us decide who has the choice. Indeed this decision must be taken at the local level, but such a decision will not yield the intended Natural Deduction rules. In other words there are no player-independant rules for *tonk*: This behaviour does not reflect any form of interaction between players in a language game. Player-independance is a necessary feature of local rules since there is no mean to specify who the players are at this level[24]. Since local rules cannot be

---
[24] Players are defined by the structural rules.

given for *tonk*, this connective has no meaning *stricto sensu*, and this is straightforward in the dialogical approach: The very notion of local meaning has a form of in-built harmony.

Acknowledgments

The authors are grateful to an anonymous referee for helpful corrections and insightful remarks.

BIBLIOGRAPHY

Abramsky S. (1997). "Semantics of Interaction: an introduction to Game Semantics". *Proceedings of the 1996 CLiCS Summer School, Isaac Newton Institute*, P. Dybjer and A. Pitts (Eds.), pp.1–31, Cambridge University Press, Cambridge.

Barwise, J., and J. Etchemendy (1993). *The language of First-Order Logic*. CSLI Lectures Notes.

Belnap, N. (1962). "Tonk, Plonk and, Plink". *Analysis 22*, pp.130–134.

Blass, A. (1992). "A game semantics for linear logic". *Annals of Pure and Applied Logic 56*, pp.183–220.

Brandom, R. B. (2000). *Articulating Reasons*. Harvard University Press, Harvard.

Felscher, W. (1985). "Dialogues, strategies and intuitionistic logic provability". *Annals of pure and applied logic 28.*, pp.217–254.

Fontaine, M. and Redmond J. (2008). *Logique Dialogique: une introduction. Volume 1 Méthode de Dialogique: Règles et Exercices,* in *Cahiers de logique et d'Epistémologie 3,* College Publications, London.

Gamut, L.T.F. (1992). *Logic, Language, and Meaning*. Volume 1 Introduction to Logic. University of Chicago Press, Chicago.

Garson, J. W. (2006). *Modal Logic for Philosophers*. Cambridge University Press, Cambridge.

Geuvers, H., and R. Nederpelt (2004). "Rewriting for Fitch style natural deductions". Proceedings of RTA 2004. *Lecture Notes in Computer Science 3091*, pp.134–154.

Girard, J.-Y. (1999). "On the meaning of logical rules I : syntax vs. semantics". *Computational Logic*, Berger and Schwichtenberg (Eds.), Springer-Verlag, Heidelberg, pp.215–272.

Girard, J.-Y. (2003). "From foundations to ludics". *Bulletin of Symbolic Logic 9.2*, pp.131–168.

Hintikka, J. and Sandu, G. (1997). "Game-theoretical semantics". *Handbook of Logic and Language*, J. van Benthem and A. ter Meulen (Eds.), Elsevier, Amsterdam, pp.361–410.

Hyland, J. M. E. and C.-H. L. Ong (2000). "On Full Abstraction for PCF". *Information and Computation 163*, pp.285–408.

Japaridze, G. (2003). "Introduction to computability logic", *Annals of Pure and Applied Logic 123*, pp.1–99.

Lorenzen, P., and K. Lorenz (1978). *Dialogische Logik*. Wissenschaftliche Buchgesellschaft, Darmstadt.

Prior, A. (1961) "The runabout inference ticket". *Analysis 21*, pp.38–39.

Rahman, S. (1993). *Uber Dialoge, Protologische Kategorien und andere Seltenheiten*. Peter Lang, Frankfurt am Main.

Rahman S. and L. Keiff (2004). "On how to be a dialogician" in *Logic, Thought and Action*, D. Vanderveken (Ed.) Logic, Epistemology and the Unity of Science series, Springer, Dordrecht.

Sellars, W. (1997). *Empiricism and the Philosophy of Mind*. Harvard University Press, Cambridge.

Smullyan, R. (1968). *First-Order Logic*. Dover Publications, New York.

'π' in the Sky

BJØRN JESPERSEN, MARIE DUŽÍ, PAVEL MATERNA

ABSTRACT. We argue for a procedural semantics for mathematical constants, using 'π' as a hard test case. The semantics of 'π' consists in 'π' expressing as its sense a procedure producing the number π. What is semantically salient about 'π' is only what it means and not also what it denotes. In the final analysis, 'π' is shorthand for, and therefore synonymous with, a definite description expressing a definition of π and denoting the number so defined. But for each $definition_n$ of π there is going to be a pair $\langle$'π', $definition_n(\pi)\rangle$. We avoid homonymy by assigning each such pair to a particular conceptual system.

1 'π' denotes π - but how?

Consider numerical constants like '1' and 'π'. What is their semantics? We are going to argue in favour of a realist procedural semantics, according to which sense and denotation are correlated as procedure and product. So it is obvious that our procedural semantics bears similarities to Moschovakis' as based on algorithm and value. At the same time we are in stark opposition to Kripke's unrealistic realist contention that the semantics of 'π' consists in nothing other than 'π' rigidly denoting π. Yes, 'π' does denote π – indeed, 'π' qualifies as a strongly rigid designator of π –[1] but there is substantially more to the semantics of 'π' than merely the denotation relation. In this paper we focus on 'π', since our general top-down strategy is to develop a semantics for the hardest (or a very hard) case and then generalise downwards to increasingly less hard cases from there.

In outline, our procedural semantics says that 'π' expresses as its sense a procedure whose product is π. The procedure is, as a matter of mathematical convention, a *definition* of π and the product is, as a matter of mathematical fact, the (transcendental) *number* so defined. For comparison, '1' expresses as its sense the procedure consisting in applying the successor function to 0 once and denotes whatever (natural) number emerges as the product of this procedure.

The upside of a procedural semantics for 'π' is that to *understand*, as a reader or hearer, and exercise *linguistic competence*, as a writer or speaker, one must merely understand a particular numerical definition and need not know which number it

---

[1]Cf. Kripke (1980), p.48.

defines. Procedural semantics, whether realist or idealist, construes sense as an *itinerario mentis* abstracting from the itinerary's destination. Making the denotation of a numerical constant irrelevant to understanding and linguistic competence is not pressing in the case of '1', but is so in the case of 'π'. The downside, however, is that at least two equivalent, but obviously distinct, definitions of π are vying for *the* role as the sense of 'π'. One is *the ratio of a circle's area and its radius squared*; The other is *the ratio of a circle's circumference to its diameter*. They are equivalent, because the same number is harpooned by both definitions. But the procedures are conceptually different, so they should not both be assigned to 'π' as its sense on pain of installing homonymy. This kind of predicament has become historically famous. Says Frege,

> "Solange nur die Bedeutung dieselbe bleibt, lassen sich diese Schwankungen des Sinnes ertragen, wiewohl auch sie in dem Lehrgebäude einer beweisenden Wissenschaft zu vermeiden sind und in einer vollkommenen Sprache nicht vorkommen dürften."[2]

We shall suggest a solution to this predicament. The crux of the solution is to relegate each definition of π to individual *conceptual systems*. Since an interpreted sign such as 'π' is a pair whose elements are a character (in this case the Greek letter 'π') and a sense, there will be as many such pairs as there are conceptual systems defining π. Disambiguation of 'π'-involving discourse will consist in making explicit which particular π-defining system should supply the sense of a token of the character 'π'.

A related predicament, which we shall also address, is whether 'π' is best construed as a *name* for π or as a shorthand for a *definite description*. If a name, the sense of 'π' will, in our semantics, be a primitive procedure consisting in the instruction to obtain, or access, π in one step. The procedure will not tell us *how* to obtain π, but only *that* π is to be obtained. This does not sit well with π being something as complicated as a transcendental number. But it does sit well with 'π' being itself a primitive, or simple, character not disclosing any information about its denotation. So at least on a literal analysis, according to which syntactic and semantic structures are by and large isomorphic, 'π' should be paired off with a non-complex sense. If 'π' is a definite description (in disguise), the sense of 'π' will, in our semantics, be a complex procedure consisting in the instruction to manipulate various mathematical operations and concepts in order to define a number. Only the problem, as we just pointed out, is, *which* procedure? Is it the instruction to calculate the ratio of a circle's area and its radius squared, or is it the instruction to calculate the ratio of a circle's circumference and its diameter, or is it some yet other instruction? Whichever it may be, though, the grammatical constant 'π' will be *synonymous* with the definite description 'the ratio...' chosen. The problem of homonymy does not rear its head in case the sense of 'π' is a

---

[2]Frege (1892), n. 2, p.42.

primitive procedure, for then 'π' is only *equivalent* (co-denoting) with a particular definition. In fact, since all the variants of definitions co-define the same number, 'π' will be equivalent with all such descriptions.

Whether 'π' be a name or a disguised definite description, it holds that its denotation needs to be *defined* and that an *algorithm* is required to bridge between definition and number. By showing how to calculate π, the algorithm shows, *ipso facto*, what the denotation of 'π' is. Our underlying semantic schema comes in two variants, one pure, the other impure. The pure one is

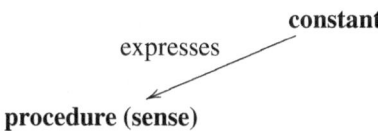

The relation *a priori* of expressing as obtaining between constant and sense exhausts the pure semantics of the constant. Only its sense is semantically salient, so a semantic analysis of 'π' must make its sense explicit. However, as soon as a procedure is explicitly given, its product (if any) is implicitly given, for the relation from procedure to product is an internal one: A procedure can have at most one product, and that product is invariant. The pure schema depicts a constant expressing its sense and not also what the constant denotes. An impure schema includes not only constant and sense, but also denotation:

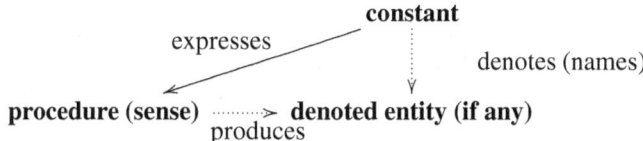

The procedure will produce its product independently of any algorithm; This is because the relation between procedure and product is an internal one. But for epistemological reasons we will need some way or other of calculating its product to learn what it is, so we need a π-calculating algorithm to show us what number satisfies whatever π-defining condition. Such an algorithm will, *ipso facto*, reveal to us what the denotation of 'π' is. The number 3.14159... which is π is itself no player in the pure semantics of 'π'. π is just whatever number rolls out as the value of the given procedure. The number 3.14159... is itself of little mathematical interest and of no semantic import. The *properties* of π, by contrast, are of great interest; e.g., it is interesting to know whether π is normal in some base; and establishing that π is transcendental (and not just rational) was a major mathematical achievement.

An algorithm may appear in one of two capacities. Either it is an intermediary between the definition and the number so defined: then the algorithm (whichever it is) is no player in the pure semantics of 'π'. Or an algorithm is the very sense of 'π': then the algorithm is a player in the pure semantics of 'π'. Our procedural semantics allows that a π-calculating algorithm may itself be elevated to playing the role of sense of 'π'. In such a case 'π' will have as its sense one particular way of calculating π. An algorithm is a particular kind of procedure and can as such figure as a linguistic sense relative to a procedural semantics.

In the former case, if the definition is a *condition* then the algorithm will calculate the *satisfier* of the condition. Full competence with respect to the definition *the ratio...* will yield knowledge of a condition to be satisfied by a real number, but will not yield knowledge of which number satisfies it. So the definition is, strictly speaking, a definition of something for a number to be; namely, the ratio of two geometric proportions. Hence, three players need to be kept separate in the impure semantics of 'π': constant, sense and number. A sense is a logical procedure defining π, while an algorithm is a mathematical procedure calculating π. Thus, if the sense defines π as the ratio between the area of a circle and its radius squared, a matching algorithm must calculate this ratio. Full linguistic competence with respect to 'π' neither presupposes, nor need involve, knowledge of how to calculate π. What competence consists in depends on whether the sense of 'π' is a primitive or complex procedure. If primitive, competence requires knowing which transcendental real 3.14159... is π. If compound, competence requires understanding the concept *the ratio of*, as well as either the concepts *the area of*, *the radius of*, *the square of*, or the concepts *the circumference of* and *the diameter of*, together with knowledge of how to mathematically manipulate them. A school child will understand such a complex procedure; It takes a professional mathematician to develop and comprehend a π-calculating algorithm. The task facing the mathematician is to come up with an algorithm equivalent with the definition defining the given ratio.

In the latter case, where an algorithm is the sense of 'π', full linguistic competence with respect to 'π' is to understand a definition of π and, again, not of the number so defined. But since the algorithm is now not an intermediary between definition and number, linguistic competence will be harder to come by, since the sense of 'π' is now likely to involve much more complicated mathematical notions than just, say, those of ratio, area, and circumference, such as the limit of an infinite series.

2 Beyond Benacerraf

Assume that the truth-condition of '...π...' requires π to exist as an independent, abstract entity. Assume, further, that we can have no epistemic access to entities that we can have no causal interaction with. Then next stop is Benacerraf's dilemma as formulated for π: We do not know what number is π; Yet we want to dub π 'π' in order to talk about π in '...π...'. So how is 'π' to be introduced into mathematese? Moreover, now that 'π' has actually been introduced into standard mathematical vocabulary and been in use for three hundred years, what would a realist (as opposed to constructivist or otherwise idealist) construal of its semantics look like?

We propose placing our procedural semantics within the general Fregean programme of explicating sense (*Sinn*) as the *mode of presentation* (*Art des Gegebenseins*) of the entity (*Bedeutung*) that a sense determines.[3] Muskens correctly points out that "[t]he idea was provided with extensive philosophical justification in Tichý [Tichý (1988)]" and that "[Tichý's] notion of senses as *constructions* essentially captures the same idea."[4] Going with this Fregean programme, however, raises a batch of questions deserving and demanding to be answered. Just how finely are senses sliced? What is the ontological status of a sense? What does a sense 'look like'; In particular, what is its structure? And how does a sense *determine* something?

We agree with Moschovakis' conception of sense ("referential intension," in his vernacular) as "an (abstract, idealized, not necessarily implementable) algorithm which computes the denotation of [a term]."[5] Moschovakis outlines his conception thus:

"The starting point [is] the insight that a correct understanding of programming languages should explain the relation between a program and the algorithm it expresses, so that the basic interpretation scheme for a programming language is of the form

(50)   program $P \to$ algorithm $(P) \to$ den(P).

It is not hard to work out the mathematical theory of a suitably abstract notion of algorithm which makes this work; and once this is done, then it is hard to miss the similarity of (50) with the basic Fregean scheme for the interpretation of a natural language,

---

[3]We know we are cutting corners here by paraphrasing 'Bedeutung' as 'entity'. We are doing so in order not to get bogged down in the ongoing discussion of how best to render 'Bedeutung'. The standard translation has been 'reference', but this fails to do justice to Frege's idiosyncratic distinction between 'Sinn' and 'Bedeutung', which are pretty much synonymous nouns in ordinary German, barring idiomatic usage; e.g., 'sinnlos' and 'bedeutungslos' are certainly not synonymous adjectives. The best verbatim translation would have been 'meaning', to be contrasted with 'sense'. But the idea of Frege being the *meaning* of 'Frege' sits very poorly indeed on the ears. Besides, 'Bedeutung' comes with a suggestion of pointing at an entity – 'deuten auf' – that 'meaning' lacks. Fortunately, we can afford to be offhand about 'Bedeutung', since we are so strongly biased toward *Sinn*.

[4]Muskens (2005), p.474.

[5]Moschovakis (2006), p.27; See also Moschovakis (1994). Moschovakis' notion of algorithm borders on being too permissive, since algorithms are normally understood to be effective. See Cleland (2002) for discussion. Tichý separates algorithms sharply from constructions: "The notion of construction is [...] correlative not with the notion of algorithm itself but with what is known as a particular algorithmic *computation*, the sequence of steps prescribed by the algorithm when it is applied to a particular input. But not every construction is an algorithmic computation. An algorithmic computation is a sequence of *effective* steps, steps which consist in subjecting a manageable object [...] to a feasible operation. A construction, on the other hand, may involve steps which are not of this sort." Tichý (1986), p.526; Tichý (2004), p.613.

(51)   term $A \rightarrow$ meaning $(A) \rightarrow$ den(A).

> This suggested at least a formal analogy between algorithms and meanings which seemed worth investigating, and proved after some work to be more than formal: When we view natural language with a programmer's eye, it seems almost obvious that we can represent the meaning of a term A by the algorithm which is expressed by A and which computes its denotation."[6]

In modern jargon, TIL belongs to the paradigm of *structured meaning*. However, Tichý does not reduce structure to set-theoretic sequences, as do Kaplan and Cresswell.[7] Nor does Tichý fail to explain how the sense of a molecular term is determined by the senses of its atoms and their syntactic arrangement (as Moschovakis objects to 'structural' approaches[8]).

In general, a procedure is a structure encompassing one or more steps that individually detail how to determine a product and jointly detail how to determine the product of the procedure that they are sub-procedures of. (This holds even for one-step procedures.) Structures are needed as molecular units in which to organise atomic sub-procedures in a particular order. A compound structure constitutes a hierarchy of sub-procedures. The philosophical idea informing our procedural semantics is that since senses are procedures, any two senses are indistinguishable just when they are, roughly speaking, procedurally indistinguishable. (We shall individuate senses in terms of *procedural isomorphism*; See below.) Intuitively, any two procedures are identical just when they are instructions to do the same to the same things in the same order.

TIL constructions are procedures. Constructions divide into atomic and compound, according as they encompass one or more steps. The atomic ones are *Variable* and *Trivialization*; the compound ones, *Composition* and *Closure*.[9] A variable x constructs an object relative to a valuation function pairing variables and entities off, such that x constructs the value assigned to it. The Trivialization ^0X constructs the entity X (which may be whatever sort of entity found in the ontology of TIL). A Composition is the procedure of applying a function at one of its arguments to obtain the value (if any) at that argument; The functional value is the product of that procedure. A Closure is the procedure of arranging objects $x_1,..,x_n$ and y as functional arguments and values, respectively; The resulting function is

---

[6] Moschovakis (2006), p.42.

[7] Kaplan may well have been the one to reintroduce the notion of structured meaning into mainstream analytic philosophy of language. See his Kaplan (1978), written in 1970; but see also Lewis Lewis (1972). Cresswell, in Cresswell (1985), has become the standard point of reference. All three agree that structure, especially a structured proposition, is (or can be modelled as) an ordered n-tuple. This won't do, however, since sequences underdetermine structure and so cannot solve Russell's old problem of propositional unity.

[8] See Moschovakis (2006), p.27.

[9] And four others – *Execution, Double Execution, Tuple, Projection* – that we do not need here.

the product of that procedure. If the sense of 'π' is simple, its sense is the Trivialization of π: $^0π$. If complex, it is a Composition. In either case the product of the respective procedure is the same transcendental number.

3 Logical foundations

Here follows an outline of the logical backbone of our procedural semantics for 'π'. TIL constructions, as well as the entities they construct, all receive a logical (as opposed to linguistic) *type*.

DEFINITION 1 (TYPE OF ORDER 1)
Let B be a *base*, where a base is a collection of pair-wise disjoint, non-empty sets. Then:

i) Every member of B is an elementary *type of order* 1 *over B*;

ii) Let $α, β_1, \ldots, β_m (m > 0)$ be types of order 1 over B. Then the collection $(αβ_1 \ldots β_m)$ of all m-ary partial mappings from $β_1 \times \ldots \times β_m$ into $α$ is a functional *type of order* 1 *over B*;

iii) Nothing is a *type of order* 1 *over B* unless it so follows from (i) and (ii). □

DEFINITION 2 (CONSTRUCTION)

i) The *Variable x* is a *construction* that constructs an object O of the respective type dependently on a valuation v; It v-constructs O;

ii) *Trivialization*: Where X is an object whatsoever (an extension, an intension or a *construction*), ^0X is the *construction Trivialization*. It constructs X without any change;

iii) The *Composition* $[XY_1 \ldots Y_m]$ is the following *construction*. If X v-constructs a function f of a type $(αβ_1 \ldots β_m)$, and $Y_1, \ldots, Y_m$ v-construct entities $B_1, \ldots, B_m$ of types $β_1, \ldots, β_m$, respectively, then the *Composition* $[XY_1 \ldots Y_m]$ v-constructs the value (an entity, if any, of type $α$) of f on the tuple-argument $\langle B_1, \ldots, B_m \rangle$. Otherwise, the *Composition* $[XY_1 \ldots Y_m]$ does not v-construct anything and so is v-*improper*;

iv) The *Closure* $[λx_1 \ldots x_m Y]$ is the following *construction*. Let $x_1, x_2, \ldots, x_m$ be pairwise distinct variables v-constructing entities of types $β_1, \ldots, β_m$ and Y a construction v-constructing an entity of type $α$. Then $[λx_1 \ldots x_m Y]$ is the construction $λ$-*Closure* (or *Closure*). It v-constructs the following function f of type $(αβ_1 \ldots β_m)$. Let $v(B_1/x_1, \ldots, B_m/x_m)$ be a valuation identical with v at least up to assigning objects $B_1, \ldots, B_m$ of types $β_1, \ldots, β_m$, respectively, to variables $x_1, \ldots, x_m$. If Y is $v(B_1/x_1, \ldots, B_m/x_m)$-improper (see iii), then f is undefined on $\langle B_1, \ldots, B_m \rangle$. Otherwise, the value of f on $\langle B_1, \ldots, B_m \rangle$ is the entity of type α $v(B_1/x_1, \ldots, B_m/x_m)$-constructed by Y;

v) Nothing is a *construction*, unless it so follows from (i) through (iv). □

DEFINITION 3 (RAMIFIED HIERARCHY OF TYPES)
Let B be a base. Then:

- T_1 (*types of order* 1): defined by Def.1;

- C_n (*constructions of order n*):

    i) Let x be a variable ranging over a type of order n. Then x is a *construction of order n over B*;

    ii) Let X be a member of a type of order n. Then $^0X, ^1X, ^2X$ are *constructions of order n over B*;

    iii) Let $X, X_1, \ldots, X_m (m > 0)$ be constructions of order n over B. Then $[XX_1 \ldots X_m]$ is a *construction of order n over B*;

    iv) Let $x_1, \ldots, x_m, X (m > 0)$ be constructions of order n over B. Then $[\lambda x_1 \ldots x_m X]$ is a *construction of order n over B*;

    v) Nothing is a construction of order n over B unless it so follows from C_n (i)-(iv).

- T_{n+1} (*types of order n* + 1)

    Let $\star_n$ be the collection of all constructions of order n over B. Then

    i) $\star_n$ and every type of order n are *types of order n* + 1;

    ii) If $0 < m$ and $\alpha, \beta_1, \ldots, \beta_m$ are types of order $n + 1$ over B, then $(\alpha \beta_1 \ldots \beta_m)$ (see T_1 ii)) is a *type of order n* + 1 *over B*;

    iii) Nothing is a *type of order n* + 1 *over B* unless it so follows from (i) and (ii). □

The Trivialisation ^0X of an object X constructs X without changing anything about X.[10] One may wonder whether this construction is dispensable. For can X not construct itself? No, it cannot. Trivialization is indispensable, for the following reasons. TIL observes a strict demarcation between procedures and their products; i.e., constructions and what they construct. In particular, improper constructions,

---

[10] A comparison with programming languages might be helpful. The Trivialisation ^0X and its use might be compared to the mechanisms of a *(fixed) pointer* to an entity X and its *dereference*, respectively.

which construct nothing, are still something. Thus an object and a construction of it are different entities. We opt for homogeneity within constructions: The only admissible constituents of a construction are constructions. We would abhor allowing Mont Blanc into a construction, just as much as Frege would abhor accepting Mont Blanc into a *Sinn*. Moreover, just as the constituents of a program are its subprograms and not the particular abstract/concrete objects on which the program operates, so the constituents of a construction are not concrete objects or abstract objects different from constructions. The constituents are sub-constructions and the objects they operate on, including lower-order constructions, must must be supplied by a primitive construction like Trivialisation. The Closure $[\lambda x_1 \ldots x_m Y]$ corresponds to declaring a procedure with the formal parameters $x_1, \ldots, x_m$ and a body Y in a computer program. The Composition $[X Y_1 \ldots Y_m]$ corresponds to a computer program calling the procedure X with the actual values $b_1, \ldots, b_m$ produced by sub-procedures $Y_1, \ldots, Y_m$ and assigned to its parameters.[11]

The ontological status of a construction is as an objective, abstract, structured procedure residing in a Platonic realm. Constructions are not inherently linguistic senses, for they exist prior to and independently of language. But they may be made, via linguistic conventions, to serve as linguistic senses. That is, in true realist fashion, TIL considers language a *code*.[12] Programmatically stated, our semantics for 'π' complements the ontology for π put forward in Brown (1990).

A construction *determines* what it constructs by *constructing* it. So the logic of determination consists in the constructional descent from a procedure to its product, as specified for each particular kind of construction in Def. 2. Constructions are too finely individuated to figure as linguistic senses, since some of the procedural differences they embody are logically insignificant and are not encoded linguistically. Most obviously, two α-equivalent constructions like $\lambda x[^0+x\,^0 1]$ and $\lambda y[^0+y\,^0 1]$ are just that – two constructions of the successor function and not one; Yet the difference between the λ-bound variables x and y is procedurally irrelevant. The solution to the granularity problem consists in forming equivalence classes of *procedurally isomorphic* constructions and privileging a member of each such class as *the* procedural sense of a given unambiguous term or expression. Technically speaking, the quest is for a suitable degree of extensionality in the λ-calculus. Needless to say, it remains an open research question exactly what the desirable calibration of linguistic senses should be, but our current thesis is that procedures, and hence senses, should be identified up to α- and η-equivalence.[13]

The slightly coarser calibration of senses in terms of procedural isomorphism is obtained by identifying senses with *concepts*. A concept is a *normalized closed*

---

[11] We thank Giuseppe Primiero for urging the need for further explanation of Trivialization in particular.

[12] See Tichý (1988), pp.228ff.

[13] See Duží et al. (forthcoming) §. 2.2 or Jespersen (forthcoming). For discussion of Frege's quest for the right calibration of *Sinn*, see Sundholm (1994) and Penco (2003).

construction. Let a *closed* construction be one containing no free variables. Let the normalization procedure consist in generating the normal form of construction c – $NF(c)$ – by privileging a particular member of the set of constructions procedurally isomorphic to c. Let the privileged member be the alphabetically first one without an η-*contractum*; formally, $NF(c) = NF^{\alpha}((NF^{\eta}(c))$. Then $NF(c)$ is the concept induced by c.

Two examples to fix ideas. Of $\lambda x[^0Xx]$ and ^0X the latter is the normal form of the constructions constructing X (whatever X may be), for the former is an η-redex: Its contractum is ^0X. The Closures $\lambda x[^0+\ x\ ^01]$ and $\lambda z[\lambda y[^0+\ y\ ^01]z]$, x,y,z ranging over natural numbers, qualify as procedurally isomorphic constructions of the successor function, but the former is privileged as the normal form of the constructions of that function. The latter can be first η-reduced to $\lambda y[^0+\ y\ ^01]$, which is α-equivalent to $\lambda x[^0+x\ ^01]$.[14]

4 Kripke's 'π' and ours

Central to Kripke's denotational semantics is the distinction between *fixing the reference* and *giving the meaning/a synonym*.[15] One of Kripke's illustrations is this:

> ['π'] is not being used as *short* for the phrase 'the ratio of the circumference of a circle to its diameter' [...] It is used as a *name* for a real number, which in this case is necessarily the ratio of the circumference of a circle to its diameter.[16]

Kripke's semantics for 'π' is simple (simplistic, as it turns out):

$$\text{`}\pi\text{'} \longrightarrow \pi$$
$$\text{rigidly designates}$$

The description 'the ratio...' serves to single out the unique ratio shared by all circles, after which that number is baptised 'π'. The description is subsequently kicked off and so does not form part of the semantics proper of Kripke's 'π'. This is problematic. Nobody knows of some one particular real that it is π. So nobody knows of some one particular real that it is the reference of 'π'. So it is obscure what linguistic competence with respect to 'π' would consist in. Note that it is not an option to say that 'π' designates whatever real is the ratio of a circle's circumference to its diameter, for this uniqueness condition forms no part of Kripke's semantics for 'π'.[17] Kripke's introduction of 'π' is impeccable, and his 'π' does

---

[14] For further details, see Duží et al. (forthcoming) §. 2.2 and Horák (2002).

[15] A distinction anticipated at least by Geach in his (1969).

[16] Kripke (1980), p.60.

[17] The Kripkean can have recourse to some causal theory of reference in the case of words for empirical entities like tigers, lemons and gold. But Benacerraf's second horn blocks this avenue. We hypothesise that Kripkean rigid designation cannot meaningfully be extended to numerical constants and other terms denoting abstract entities.

denote π. But we cannot use his 'π' to denote π, nor can we understand anyone else's use of 'π', since we cannot know which particular transcendental number is π. In short, Kripke's 'π' has been severed radically from any humanly possible linguistic practice, so it is inoperative.

In the idiom of procedural semantics, Kripke focuses entirely on the *product* at the expense of the *procedure*. As a matter of mathematical fact, 3.14159... is π, but why introduce a non-descriptive name when that name severs the link between condition/procedure and satisfier/product? It seems that on Kripke's semantics it will be a discovery, and not a convention, that π is the ratio of a circle's circumference to its diameter. If so, it also seems that Kripke's 'π' misconstrues mathematical practice.[18]

Some π-producing procedure must figure in the semantics of 'π'; Only how? TIL faces a dilemma of its own, as we saw above. On the one hand, a literal analysis of 'π' would dictate that the sense of 'π' be $^0π$, yielding the schema

$$\text{'π'} \xrightarrow{\text{expresses}} {}^0π \xrightarrow{\text{constructs}} π$$

The advantage of this construal is that what looks like a constant *is* a constant (and not a definite description masquerading as one). However, this is too close to Kripke's 'π' for comfort. We would be reinstating the problem that the semantics of 'π' pairs no mathematical condition off with 'π'. To master 'π', $^0π$ would suffice. The Trivialization merely instructs us to construct π and not also how to construct it.

On the other hand, not least epistemic concerns dictate that the sense of 'π' ought to be an *ontological definition* of π, yielding the schema

$$\text{'π'} \xrightarrow{\text{expresses}} [\iota x[\forall y[x = [^0Ratio[\ldots y\ldots][\ldots y\ldots]]]]] \xrightarrow{\text{constructs}} π$$

(By 'ontological definition' we mean a compound construction (here, a Composition) that, in this case, constructs the number π, thereby laying down *what* π is.

---

[18] One novelty launched in Kripke (2008) is his notion of (immediately) revelatory sense (introduced in connection with his discussion of Frege and Church's hierarchy of indirect senses): "[A] sense is revelatory of its reference if one can figure out from the sense alone what the referent is. [...] [B]oth 'nine' and perhaps even 'the square of three' do have revelatory senses. Given that one can understand them, one can tell what the referent is," Kripke (2008), p.187. "[...] [O]ne might say that a sense is *immediately revelatory* if no calculation is required to figure out its referent. [...] 'Nine' ...is immediately revelatory," Kripke (2008), p.188. A procedural semantics will deny, however, that any term may have an immediately revelatory sense, since a procedure (calculation, broadly construed) is required to figure out its referent. A constant like 'nine' will have a revelatory sense, provided its sense is an algorithm rather than a definition. But 'π' cannot have a revelatory, let alone immediately revelatory, sense. So his (2008) sheds no light on what a non-dysfunctional Kripkean semantics for 'π' should look like.

An ontological definition contrasts with a linguistic definition, which introduces a new term synonymous with an existing term.) This makes 'π' a shorthand term synonymous with 'the ratio...', and its sense is an ontological definition of π. The advantage of this construal is that it pairs a mathematical condition off with 'π'; But again, which? There is no criterion to help decide which of the possible ontological definitions should be *the* sense of 'π'. It would be arbitrary to select one and assign it as sense; but assigning them all introduces homonymy.

It would seem evident that a language-user needs to know at least one definition of π in order to use and understand 'π'. If we go with the Trivialization-based analysis of 'π', the first step toward enhancing it is to make the logico-semantic fact that $^0π$ is *equivalent* with $[λx[∀y[x = [^0Ratio[...y...][...y...]]]]]$ part of the semantics of 'π'. $^0π$ is indifferent to how π is constructed by this or that compound construction, so as far as the equivalence relation goes, any compound π-construction is as good as any other. '$^0π$' may be *introduced* as equivalent with $[λx[∀y[x = [^0Ratio[^0Area\ y][^0Square[^0Radius\ y]]]]]]$, or $[λx[∀y[x = [^0Ratio[^0Circumference\ y][^0Diameter\ y]]]]]$, or any other compound π-constructing construction. *Understanding* is another matter. One thing is to understand $[λx[∀y[x = [^0Ratio[^0Area\ y][^0Square[^0Radius\ y]]]]]]$; Another thing is to understand $[λx[∀y[x = [^0Ratio[^0Circumference\ y][^0Diameter\ y]]]]]$. One may well know that '$^0π$' is equivalent to this Composition without knowing, *ipso facto*, that it is equivalent to that Composition.

5 Realistic realism?

Both causal theory of reference and denotational semantics are neither here nor there as a theory of terms for abstract entities such as numbers. We are putting forward a procedural semantics as a rival theory in order not to get gored by Benacerraf's horns or turning linguistic competence with mathematical constants into an enigma.

We suggest, in the final analysis, that the semantics of 'π' ought to be that it is shorthand for, and therefore synonymous with, a definite description expressing a definition of π and denoting the number so defined. But for each definition$_n$ of π there is going to be a pair $\langle$'π', $definition_n(π)\rangle$. So how do we handle the resulting homonymy? *Schwankungen des Sinnes* are neither here nor there in a regimented language such as mathematese.

Our solution revolves around *conceptual systems*. By 'conceptual system' we mean a set of constructions that is fully determined by the chosen set of *simple concepts*. Simple concepts are Trivialisations of non-constructional entities of order 1. The compound concepts of a conceptual system are then all the compound constructions that are formed according to the rules of Def. 2 (plus perhaps involving additional constructions) using simple concepts and variables. The exact definition of *conceptual system* is as follows.

DEFINITION 4 (CONCEPTUAL SYSTEM) Let a finite set Pr of simple concepts $C_1, \ldots, C_k$ be given. Let *Type* be an infinite set of types induced by a finite base of ground types. Let *Var* be an infinite set of variables, countably infinitely many for each member of *Type*. Finally, let C be an inductive definition of constructions. By means of *Pr, Type, Var* and C, an infinite class *Der* is defined as the transitive closure of all the closed compound constructions derivable from *Pr* and *Var* using the rules of C, such that:

i) every member of *Der* is a compound concept;

ii) if $C \in Der$, then every sub-construction of C that is a simple concept is a member of *Pr*.

The set of concepts $Pr \cup Der$ is a *conceptual system* derived from *Pr*. The members of *Pr* are *primitive concepts*, while the members of *Der* are *derived concepts* of the given conceptual system. □

As is seen, *Pr* unambiguously determines *Der*. The *expressive power* of a given (stage of a) language L is then determined by the set *Pr* of the conceptual system underlying L.

Relative to a particular conceptual system, a pair $\langle$ 'π', $definition_n(\pi) \rangle$ is an unambiguous assignment of exactly one definition of π to 'π', provided the conceptual system is independent; i.e., its set of primitive concepts is minimal. Consequently, 'π' is not ambiguous, for this character must always be given together with a particular definition of π culled from a particular conceptual system. The appearance of ambiguity arises only when two or more conceptual systems are invoked in the course of a discourse in which tokens of 'π' occur.

The upshot of our solution is that there are several π-denoting constants sharing the same first element, 'π'. So when two mathematicians are both deploying tokens of 'π', there is a risk of them talking at cross purposes, until and unless they compare notes and, in case of invoking different conceptual systems, come to agree on the same definition of π in the interest of synonymy. Yet the mathematical results they may have individually obtained with respect to π are bound to be equivalent, for any two definitions of π are bound to converge in the same number. After all, the problem was always to do with *Schwankungen des Sinnes* and never *Schwankungen der Bedeutung*.

Acknowledgements

Versions of this paper were read by Marie at the Joint Paris/Arché Workshop *Philosophy of Mathematics and Abstract Entities*, ENS, Paris, February 28- March 1, 2008, and by Bjørn at LOGICA '08, Hejnice, June 16-20, 2008; at Department of Fuzzy Modelling, TU Ostrava, 25 June 2003; at Department of Philosophy, University of Milan, 22 April 2008; and at Department of Philosophy, University of Padua, 5 May 2008. Marie and Pavel were supported by Grant No. 401/07/0451, *Semantisation of Pragmatics*, of the Grant Agency of the Czech Republic. Portions of the present manuscript have been lifted from Duží et al. (forthcoming) §. 3.2.1; but this paper was written especially for this collection. We are grateful to Giuseppe Primiero and Shahid Rahman for the kind opportunity to contribute to Göran's Festschrift. All three of us have fond memories of lively debates with Göran about the pros and cons of realist (nay, Platonist!) and idealist/constructivist logic and semantics both at various instalments of LOGICA at various locations in the Czech Republic and at the University of Leiden in the Netherlands. Göran is, after all, the world's only constructivist who is also versed in TIL, see his (2000) for discussion. It has been fascinating and educational to notice a number of features shared both by our background theory and Göran's (Martin-Löf's type theory), not least concerning the idea of sense as procedure, the advocacy of interpreted logical syntax, and the proper treatment of inference. Yet as a keen critic of Platonism, we have little doubt that Göran is going to point out that our 'π' is no less of a pie in the sky than Kripke's.

BIBLIOGRAPHY

Benacerraf, P. (1973). "Mathematical truth", *Journal of Philosophy*, 70, pp.661–679.

Brown, J.R. (1990). "π in the sky", in A.D. Irvine, editor, *Physicalism in Mathematics*, pp.95–120, Kluwer Academic Publisher, Dordrecht.

Cleland, C.E. (2002). "On effective procedures", *Minds and Machines*, 12, pp.159–179.

Cresswell, M.J. (1975). "Hyperintensional logic", *Studia Logica*, 34, pp.25–38.

Cresswell, M.J. (1985). *Structured Meanings*, MIT Press, Cambridge.

Duží, M., Jespersen, B. and Materna, P. (forthcoming). *Procedural Semantics for Hyperintensional Logic*, forthcoming in the series *Logic, Epistemology and the Unity of Science*, Springer, Dordrecht.

Frege, G. (1892). "Über Sinn und Bedeutung", in G. Patzig, editor, *Funktion, Begriff, Bedeutung: Fünf logische Studien*, Vandenhoeck und Ruprecht, Göttingen.

Geach, P.T. (1969). "The perils of Pauline", *Review of Metaphysics*, 23, pp.287–300.

Horák, A. (2002). *The Normal Translation Algorithm in Transparent Intensional Logic for Czech*, PhD Thesis, Masaryk University, Brno. Retrievable at http://www.fi.muni.cz/~hales/

Jespersen, B. (forthcoming). "Hyperintensions and procedural isomorphism: Alternative (1/2)", in K. Kijania-Placek (ed.), *Proceedings of ECAP 6*, College Publications, London.

Kaplan, D. (1978). "Dthat" in P. Cole, editor, *Syntax and Semantics*, Academic Press.

Kripke, S. (1980). *Naming and Necessity*, Blackwell, Oxford.

Kripke, S. (2008). "Frege's theory of sense and reference: some exegetical notes", *Theoria*, 74, pp.181-218.

Lewis, D. (1972). "General semantics", in D. Davidson and G. Harman, editors, *Semantics of Natural Language*, pp.169–218, Reidel, Dordrecht.

Materna, P. (2004). *Conceptual Systems*, Logos-Verlag, Berlin.

Moschovakis, Y. (1994). "Sense and denotation as algorithm and value", in J. Väänänen and J. Oikkonen, editors, *Lecture Notes in Logic*, vol. 2, Springer, Dordrecht.

Moschovakis, Y. (2006). "A logical calculus of meaning and synonymy", *Linguistics and Philosophy*, 29, pp.27–89.

Muskens, R. (2005). "Sense and the computation of reference", *Linguistics and Philosophy*, 28, pp.473–504.

Penco, C. (2003). "Frege: two theses, two senses", *History and Philosophy of Logic*, 24, pp.87–109.

Sundholm, B.G. (1994). "Proof-theoretical semantics and Fregean identity criteria for propositions", *The Monist*, 77, pp.294–314.

Sundholm, B.G. (2000). "Virtues and vices of interpreted *classical* formalisms", in T. Childers and J. Palomäki, editors, *Between Words and Worlds: A Festschrift for Pavel Materna*, pp.3-12, Filosofia Publisher, Czech Academy of Sciences, Prague.

Tichý, P. (1986). "Constructions" *Philosophy of Science*, 53, pp.514–534. Republished in Tichý (2004).

Tichý, P. (1988). *The Foundations of Frege's Logic*, De Gruyter, Berlin.

Tichý, P. (2004). *Collected Papers in Logic and Philosophy*, V. Svoboda, B. Jespersen and C. Cheyne editors, Filosofia, Czech Academy of Sciences, Prague; University of Otago Press, Dunedin.

On Ramsey's 'Silly Delusion' Regarding Tractatus 5.53

KAI F. WEHMEIER

ABSTRACT. I supply a semantics and a tableaux calculus for a first-order logic based on Hintikka's strongly exclusive interpretation of the variables, and prove that the calculus is sound and complete with respect to the semantics.

1 Introduction.

In Tractatus 5.53, Wittgenstein, with characteristic brevity, proposes a notational convention that is intended to make the identity symbol superfluous in logical languages:

"Identity of the object I express by identity of the sign and not by means of a sign of identity. Difference of the objects by difference of the signs."

As it stands, the convention is ambiguous. Hintikka (1956) suggests two ways in which it might be interpreted; he refers to these as the 'weakly' and 'strongly exclusive interpretations of the variables,' respectively:

- **weakly exclusive:** From the range of a bound variable all objects are excluded that are values of variables occurring free within the scope of the associated quantifier.

- **strongly exclusive:** From the range of a bound variable all objects are excluded that are values of variables in whose scope the subformula governed by the associated quantifier occurs (where the scope of a bound variable is the scope of the quantifier that binds it, and the scope of a free variable is the entire formula in which it occurs free).

In 5.531 through 5.5321, Wittgenstein proceeds to show how the contents of various formulas in Russellian notation can be rendered in a language employing the convention of 5.53. In particular, in 5.5321 Wittgenstein explains that Russell's $\forall x(Fx \to x = a)$ may be rewritten as $(\exists xFx \to Fa) \wedge \neg\exists x\exists y(Fx \wedge Fy)$. As

Hintikka (1956), p.230, observes, this example shows that Wittgenstein cannot have had the strongly exclusive interpretation in mind. For, under the strongly exclusive reading, the Wittgensteinian formula says that, if anything other than a is F, then a is also F, and that there are no two objects x and y, both distinct from a, and both satisfying F. This would be true in a two-element domain in which both elements satisfy the predicate F.

Under the moniker 'Wittgensteinian predicate logic', I have discussed the weakly exclusive interpretation in Wehmeier (2004) and in Wehmeier (2008); it shall not detain us further. While not Wittgenstein's intended convention, Hintikka's strongly exclusive reading is interesting in its own right, but also because Frank Ramsey appears to have misinterpreted Tractatus 5.53 as proposing such a reading.

On November 11, 1923, Ramsey writes to Wittgenstein:

"Have you noticed the difficulty in expressing without = what Russell expresses by $\exists x(Fx \wedge x \neq a)$?"[1]

After Wittgenstein supplies the formula

$$(Fa \rightarrow \exists x \exists y(Fx \wedge Fy)) \wedge (\neg Fa \rightarrow \exists x Fx),$$

Ramsey sheepishly replies on December 27, 1923:

"I didn't think there was a real difficulty about $\exists x(Fx \wedge x \neq a)$, i.e., that it was an objection to your theory of identity, but I didn't see how to express it, because I was under the silly delusion that if an x and an a occurred in the same proposition the x could not take the value a."[2]

Two points about this exchange seem remarkable.

First, no matter whether one adopts the strongly or the weakly exclusive interpretation, Ramsey's troublesome formula can be rewritten simply as $\exists x((Fa \vee \neg Fa) \rightarrow Fx)$, so Ramsey should have been able to find an appropriate rendition even while under his 'silly delusion.'

Second, the 'silly delusion' appears to be exactly Hintikka's strongly exclusive interpretation: No matter where in a formula a free variable a occurs, its value is excluded from the range of all bound variables occurring in the same formula. In the following, I will thus refer to a first-order logical system using the strongly exclusive interpretation of the variables as *Ramsey predicate logic* or simply *R-logic*.

In what follows, I develop a model-theoretic semantics for R-logic, show its mutual interpretability with ordinary first-order logic with identity (FOL$^=$), and provide a tableaux calculus, for which proofs of soundness and completeness are sketched.

---

[1] Wittgenstein (1995), p.191.
[2] Wittgenstein (1995), p.194.

2 The Syntax and Semantics of R-Logic.

The language $\mathcal{L}$ of first-order logic (FOL), as understood here, has the following primitive symbols.

1. the propositional connectives $\neg, \wedge, \vee, \rightarrow$

2. the quantifier symbols $\forall$ and $\exists$

3. countably many bound individual variables $x, y, x_1, x_2, \ldots$

4. countably many free individual variables $a, b, a_1, a_2, \ldots$

5. for each positive integer n, countably many n-ary predicate symbols $P^n, Q^n, R^n, P_1^n, P_2^n, \ldots$

Atomic $\mathcal{L}$-formulas are strings of the form $Ra_1 \ldots a_n$, where R is an n-ary predicate symbol and the a_i are free variables. If A and B are $\mathcal{L}$-formulas, then so are $\neg A$, $A \wedge B$, $A \vee B$, and $A \rightarrow B$. If $F[a]$ is an $\mathcal{L}$-formula in which x does not occur, then $\forall x F[x]$ and $\exists x F[x]$ are also $\mathcal{L}$-formulas. The language of R-logic is $\mathcal{L}$.[3]

The language $\mathcal{L}^=$ of first-order logic with identity (FOL$^=$) arises from $\mathcal{L}$ by the addition of a designated binary predicate symbol $=$. Its formulas are constructed just like those of R-logic, except that there is the additional binary predicate symbol $=$ available.

A sentence is a formula without occurrences of free variables.

A structure $\mathcal{U}$ is a non-empty domain U together with an n-ary relation $R_{\mathcal{U}}$ over U for each n-ary predicate symbol R of $\mathcal{L}$. When considering $\mathcal{U}$ as a structure for FOL$^=$, the additional predicate symbol $=$ is to be interpreted as true identity on U. A $\mathcal{U}$-assignment is a mapping ν from the individual variables into U.

Before I give the official definition of the semantics for R-logic, let me illustrate one of its peculiarities. From the range of a bound variable we are to omit the values of all variables in whose scope the associated quantifier occurs, where the scope of a free variable is the entire formula in which it occurs free. By way of example, consider the formula $Pa \wedge \forall x Px$. The bound variable x occurs within the scope of the free variable a, and hence its range must not include the value assigned to a. If, on the other hand, we consider $\forall x Px$ not as a subformula of a larger context, but as a formula in its own right, the bound variable x does not occur within the scope of any other variable, and should hence range over the entire domain. This means that when we break down $Pa \wedge \forall x Px$ into components for the purpose of semantic evaluation, we must memorize the fact that the right conjunct originally occurred within the scope of a, or else we will interpret its bound variable incorrectly. In R-logic, the range of a bound variable thus depends

---
[3] In matters of syntax, we are largely following Schütte (1977), especially in the use of typographically distinguished free variables and of the theory of nominal forms.

on the context of the associated quantifier's occurrence within the formula, and it is this feature of R-logic that necessitates a somewhat more cumbersome semantic apparatus than those of either FOL$^=$ or W-logic.

Now let $\mathcal{U}$ be a structure with non-empty domain U. We regard every element u of U as a constant symbol with denotation u (i.e., we use elements from the domain as autonymous constants, again following Schütte 1977). The $\mathcal{L}(\mathcal{U})$-sentences arise from the $\mathcal{L}$-formulas by replacing every occurrence of a free variable a with a constant symbol $u \in U$ (and analogously for $\mathcal{L}^=(\mathcal{U})$-sentences). Alternatively, we may describe the $\mathcal{L}(\mathcal{U})$-sentences as the closed formulas of the first-order language obtained from $\mathcal{L}$ by the addition of an individual constant u, as an additional primitive symbol, for each element $u \in U$ (and similarly for the $\mathcal{L}^=(\mathcal{U})$-sentences). If v is a $\mathcal{U}$-assignment, and A is an $\mathcal{L}$-formula (or an $\mathcal{L}^=$-formula), then A^v is the result of replacing every individual variable a in A by the individual constant $v(a)$.

The semantics to be defined below is new in the sense that Hintikka (1956) only provides rules by means of which the quantifiers of R-logic can be successively eliminated in favor of standard quantifiers and the identity symbol, but does not specify an independent semantics for R-logic.

We first define the relation $\Vdash$ of R-truth of an $\mathcal{L}(\mathcal{U})$-sentence in a structure $\mathcal{U}$, relative to a finite set V of elements of U (these will represent the objects excluded from the range of bound variables).

1. $\mathcal{U} \Vdash \langle Ru_1 \ldots u_n, V \rangle$ iff $\langle u_1, \ldots, u_n \rangle \in R_\mathcal{U}$

2. $\mathcal{U} \Vdash \langle \neg A, V \rangle$ iff $\mathcal{U} \not\Vdash \langle A, V \rangle$

3. $\mathcal{U} \Vdash \langle A \wedge B, V \rangle$ iff $\mathcal{U} \Vdash \langle A, V \rangle$ and $\mathcal{U} \Vdash \langle B, V \rangle$, and similarly for the other Boolean connectives

4. $\mathcal{U} \Vdash \langle \forall x F[x], V \rangle$ iff for every $u \in U \setminus V$, $\mathcal{U} \Vdash \langle F[u], V \cup \{u\} \rangle$

5. $\mathcal{U} \Vdash \langle \exists x F[x], V \rangle$ iff for some $u \in U \setminus V$, $\mathcal{U} \Vdash \langle F[u], V \cup \{u\} \rangle$

Given an $\mathcal{L}(\mathcal{U})$-sentence A, we define $\mathcal{U} \Vdash A$ to mean $\mathcal{U} \Vdash \langle A, V \rangle$, where V is the set of elements of U that occur as constants in A. Since every $\mathcal{L}$-sentence is an $\mathcal{L}(\mathcal{U})$-sentence, we have thus defined R-truth of $\mathcal{L}$-sentences in structures $\mathcal{U}$ generally: An $\mathcal{L}$-sentence A is R-true in $\mathcal{U}$ iff $\mathcal{U} \Vdash \langle A, \emptyset \rangle$; It is R-valid if it is R-true in every structure $\mathcal{U}$. If S is a set of $\mathcal{L}$-sentences, and F an $\mathcal{L}$-sentence, we say that F is an R-consequence of S if every structure $\mathcal{U}$ in which every element of S is R-true also makes F R-true. Given an $\mathcal{L}$-formula A and a variable assignment v, we say that A is R-true in $\mathcal{U}$ under v if $\mathcal{U} \Vdash A^v$. Standard Tarskian truth $\models$ of $\mathcal{L}^=(\mathcal{U})$-sentences in $\mathcal{U}$ is defined as usual.

3   Interpreting R-Logic in FOL$^=$

Our first aim is to show that to every L-sentence A there corresponds an $L^=$-sentence $\psi(A)$ such that, for all structures $\mathcal{U}$, $\mathcal{U} \Vdash A$ iff $\mathcal{U} \models \psi(A)$, i.e., that R-logic is interpretable in FOL$^=$.

Hintikka (1956), pp.229-231, achieves this by providing rules for rewriting R-logic quantifiers by means of standard quantifiers and identity; A given L-formula is successively transformed, by application of these rules, into formulas containing both Ramseyan and standard quantifiers, as well as identity, until all Ramseyan quantifiers have disappeared. Our recursive translation avoids passing through such mixed formulas and must thus, for reasons explained in the preceding section, make an additional book-keeping effort with respect to free variables.

We first define a binary translation function ψ, taking as arguments an L-formula A and a finite set V of free variables.

1. For atomic formulas A, $\psi(A,V) := A$.

2. $\psi(\neg A, V) := \neg \psi(A,V)$

3. $\psi(A \wedge B, V) := \psi(A,V) \wedge \psi(B,V)$, and similarly for the other Boolean connectives

4. $\psi(\forall x F[x], V) := \forall x (\bigwedge_{b \in V} x \neq b \to \psi(F[x], V \cup \{x\}))$, where $\psi(F[x], V \cup \{x\})$ is short for $\psi(F[a], V \cup \{a\})^x_a$, i.e., the result of replacing a with x in $\psi(F[a], V \cup \{a\})$, where a is a fresh (neither in $\forall x F[x]$ nor in V) free variable.

5. $\psi(\exists x F[x], V) := \exists x (\bigwedge_{b \in V} x \neq b \wedge \psi(F[x], V \cup \{x\}))$

By abuse of language, we also define a unary translation function ψ as follows: For any L-formula A, $\psi(A) := \psi(A, \mathsf{FV}(A))$, where $\mathsf{FV}(A)$ is the set of free variables occurring in A. We write $\mathsf{v}[V]$ for the image of V under the mapping v, i.e., $\mathsf{v}[V] = \{\mathsf{v}(a) : a \in V\}$.

LEMMA 1 Let $\mathcal{U}$ be a structure, let V be a finite set of free variables, let v be a $\mathcal{U}$-assignment, and let A be an L-formula. Then

$$\mathcal{U} \Vdash \langle A^\mathsf{v}, \mathsf{v}[V] \rangle \Leftrightarrow \mathcal{U} \models \psi(A,V)^\mathsf{v}.$$

The proof of the lemma proceeds by induction on the formula A. The atomic case is trivial, as are the inductive cases involving the Boolean connectives. It remains to consider the quantifier cases; We discuss only the universal quantifier. Suppose that A is of the form $\forall x F[x]$. Then $\mathcal{U} \Vdash \langle \forall x F[x]^\mathsf{v}, \mathsf{v}[V] \rangle$ iff for every $u \in U \setminus \mathsf{v}[V]$, $\mathcal{U} \Vdash \langle F[u]^\mathsf{v}, \mathsf{v}[V] \cup \{u\} \rangle$, i.e., $\mathcal{U} \Vdash \langle F[a]^{\mathsf{v}^a_u}, \mathsf{v}^a_u[V \cup \{a\}] \rangle$, where v^a_u is just like v except that it maps a to u, a being a fresh free variable. By the induction hypothesis, this is the case iff for every $u \in U \setminus v[V]$, $\mathcal{U} \models \psi(F[a], V \cup \{a\})^{v_u^a}$. Since a is not in V, this holds iff for all $u \in U$, $\mathcal{U} \models (\bigwedge_{b \in V} a \neq b \to \psi(F[a], V \cup \{a\}))^{v_u^a}$. This, however, is equivalent to $\mathcal{U} \models \forall x (\bigwedge_{b \in V} x \neq b \to \psi(F[a], V \cup \{a\})^x_a)^v$, q.e.d.

COROLLARY 1 (INTERPRETABILITY OF R-LOGIC IN FOL$^=$)

1. For all $\mathcal{L}$-formulas A, all structures $\mathcal{U}$, and all $\mathcal{U}$-assignments v,
   $\mathcal{U} \Vdash A^v \Leftrightarrow \mathcal{U} \models \psi(A)^v$.

2. For all $\mathcal{L}$-sentences A and all structures $\mathcal{U}$, $\mathcal{U} \Vdash A \Leftrightarrow \mathcal{U} \models \psi(A)$.

Part 1 is the special case of lemma 1 where V is $FV(A)$, and part 2 follows from part 1 by noting that for $\mathcal{L}$-sentences A, A^v is the same expression as A.

4 Interpreting FOL$^=$ in R-Logic

In this section, we will establish a converse to corollary 1, that is, we will show that FOL$^=$ can be interpreted in R-logic.

There are two main differences between our translation and the one proposed by Hintikka (1956). First, in the quantifier case, Hintikka's translation procedure requires a prior rewriting of the formula to be translated into a particular normal form (schema (6) in Hintikka 1956, pp.231-232), which our translation does without. Second, the need for book-keeping of free variables involved in our translation manifests itself in Hintikka's procedure through the requirement that, for the purpose of translating a formula F, all rewritings of quantifiers are to be imagined as taking place inside the context of F (so that variables free in F but absent in quantified subformulas are still recognized as constraining the range of the subformulas' bound variables).

We first define a binary translation function ϕ, taking as arguments $\mathcal{L}^=$-sentences A and finite sets V of free variables, as follows:

1. Where a and b are distinct free variables, $\phi(a = b, V) := \bot(a, b)$, where $\bot(a, b)$ is some propositional contradiction in precisely the two free variables a and b (say $Rab \wedge \neg Rab$, for some binary relation symbol R).

2. $\phi(a = a, V) := \top(a)$, where $\top(a)$ is some propositional tautology in precisely the one free variable a (say $Pa \vee \neg Pa$, for some unary predicate symbol P).

3. $\phi(A, V) := A$ for all other atomic formulas A.

4. $\phi(\neg A, V) := \neg \phi(A, V)$

5. $\phi(A \wedge B, V) := \phi(A, V) \wedge \phi(B, V)$, and similarly for the other Boolean connectives.

6. $\phi(\forall x F[x], V) := \forall x \phi(F[x], V \cup \{x\}) \wedge \bigwedge_{a \in V} \phi(F[b], V)$, where $\phi(F[x], V \cup \{x\})$ is short for $\phi(F[a], V \cup \{a\})_a^x$, a being a fresh free variable.

7. $\phi(\exists x F[x], V) := \exists x \phi(F[x], V \cup \{x\}) \vee \bigvee_{b \in V} \phi(F[b], V)$

By abuse of language, we also let ϕ stand for the unary translation function defined as follows: $\phi(A) := \phi(A, \mathrm{FV}(A))$.

LEMMA 2 Let A be an $\mathcal{L}^=$-formula, let V be a finite set of free variables including $\mathrm{FV}(A)$, let $\mathcal{U}$ be a structure, and let v be a $\mathcal{U}$-assignment that is 1-1 on V. Then

$$\mathcal{U} \models A^v \Leftrightarrow \mathcal{U} \Vdash \phi(A, V)^v.$$

The proof proceeds by induction on the formula A. Only two cases are of interest.

First, suppose A is of the form $a = b$, with distinct free variables a and b. $\mathcal{U} \not\models (a = b)^v$, because v is 1-1 on $\{a, b\}$. Also $\mathcal{U} \not\Vdash \bot(a, b)^v$.

Second, suppose A is of the form $\forall x F[x]$ (the existential case is analogous). We have $\mathcal{U} \models \forall x F[x]^v$ iff for every $u \in U$, $\mathcal{U} \models F[u]^v$, i.e., for all $u \in U$, $\mathcal{U} \models F[a]^{v_u^a}$, where a is a fresh free variable. This in turn is equivalent to:

(*) For every $u \in U \setminus v[V]$, $\mathcal{U} \models F[a]^{v_u^a}$; and for every $u \in v[V]$, $\mathcal{U} \models F[a]^{v_u^a}$.

By the induction hypothesis, the first conjunct in (*) holds iff for every $u \in U \setminus v[V]$, $\mathcal{U} \Vdash \phi(F[a], V \cup \{a\})^{v_u^a}$, i.e., iff $\mathcal{U} \Vdash \phi(\forall x F[x], V)^v$. The second conjunct of (*) holds iff for every $b \in V$, $\mathcal{U} \models F[b]^v$, and hence by induction hypothesis iff $\mathcal{U} \Vdash \bigwedge_{b \in V} \phi(F[b], V)^v$. Putting things together, we see that (*) is equivalent to $\mathcal{U} \Vdash \phi(\forall x F[x], V)^v$, q.e.d.

COROLLARY 2 (INTERPRETABILITY OF FOL$^=$ IN R-LOGIC)

1. For all $\mathcal{L}^=$-formulas A, all structures $\mathcal{U}$, and all $\mathcal{U}$-assignments v that are 1-1 on $FV(A)$, $\mathcal{U} \models A^v$ iff $\mathcal{U} \Vdash \phi(A)^v$.

2. For all $\mathcal{L}^=$-sentences A and all structures $\mathcal{U}$, $\mathcal{U} \models A$ iff $\mathcal{U} \Vdash \phi(A)$.

Part 1 is an immediate consequence of the lemma, and part 2 follows from part 1 by noting that every variable assignment is 1-1 on the empty set of free variables, and that for sentences A, A^v is the same expression as A.

5 A Tableaux Calculus for R-Logic

Before we embark on the presentation of our tableaux calculus, it is perhaps appropriate to discuss briefly the calculus suggested by in Hintikka (1956) pp.235-236, for the system sketched there for the strongly exclusive interpretation (i.e., R-logic) appears to be unsound.

In order to see this, it is helpful to review some of Hintikka's apparatus, since his calculus is rather unlike any of those in use today. The deductive rules are framed in terms of a metalogical relation among $\mathcal{L}$-formulas called equivalence, which we shall write $\Leftrightarrow$. This relation is by stipulation transitive, and $\mathcal{L}$-formulas equivalent in the sense of $\Leftrightarrow$ are considered interchangeable in all contexts. A consequence relation $\Rightarrow$ is defined as follows: $A \Rightarrow B$ iff $A \Leftrightarrow A \wedge B$. An $\mathcal{L}$-formula A is said to be provable if $\neg A \Rightarrow A$ and refutable if $A \Rightarrow \neg A$.

Hintikka's rule (11)(c) (p.235) tells us that $\forall x Px \Rightarrow Py$, i.e., that $\forall x Px \Leftrightarrow \forall x Px \wedge Py$. So the $\mathcal{L}$-formulas on the left and on the right are interchangeable irrespective of context; in particular, then,

$$(*) \quad \forall x Px \wedge \neg Py \Leftrightarrow (\forall x Px \wedge Py) \wedge \neg Py.$$

The right hand side of $(*)$ is tautologically equivalent to the $\mathcal{L}$-formula $(\forall x Px \wedge \neg Py) \wedge \neg(\forall x Px \wedge \neg Py)$ and also contains the same free variables as the latter; hence by Hintikka's rule (9), we obtain

$$(**) \quad (\forall x Px \wedge Py) \wedge \neg Py \Leftrightarrow (\forall x Px \wedge \neg Py) \wedge \neg(\forall x Px \wedge \neg Py).$$

From $(*)$ and $(**)$ it follows by the transitivity of $\Leftrightarrow$ that

$$\forall x Px \wedge \neg Py \Leftrightarrow (\forall x Px \wedge \neg Py) \wedge \neg(\forall x Px \wedge \neg Py)$$

from which, by definition of $\Rightarrow$, we obtain

$$\forall x Px \wedge \neg Py \Rightarrow \neg(\forall x Px \wedge \neg Py),$$

i.e., that $\forall x Px \wedge \neg Py$ is refutable. This $\mathcal{L}$-formula, however, is satisfiable (it says that everything other than y is P, and y itself isn't P). An analogous result holds for provability instead of refutability. In Hintikka's calculus, the $\mathcal{L}$-formula $\forall xPx \to Py$ is provable[4], yet it is not R-valid on any reasonable definition of validity for R-logic (it fails to hold, for instance, in any one-element domain where P is interpreted by the empty set, and y is assigned the sole member of the domain as its value).

If the preceding argument is correct and Hintikka's calculus is indeed unsound for R-logic, then our tableaux system constitutes the first sound and complete proof procedure for R-logic.

While the tableaux calculus introduced below is in many ways similar to the Boolos-Burgess calculus of Boolos (1984) for FOL and to the W-procedure of Wehmeier (2008) for Wittgensteinian predicate logic, there is one obvious difference. The need for keeping tabs on free variables, which manifested itself already in the semantics for R-logic, also necessitates the labeling of tableaux with pairs $\langle A, V \rangle$ of $\mathcal{L}$-formulas A and finite sets V of free variables, rather than just $\mathcal{L}$-formulas. Note that in the context of tableaux, it is always understood that $\langle A, V \rangle$ is a pair consisting of an $\mathcal{L}$-formula A and a set V of free variables, whereas in the context of the semantics, A is taken to be an $\mathcal{L}(\mathcal{U})$-formula and V a subset of the domain of $\mathcal{U}$. Given a $\mathcal{U}$-assignment v, there is of course an obvious correspondence between the two notions.

The tableaux rules can be given schematically as follows (add as many of the usual propositional rules as you like).

1. Closure:

$$\frac{\langle \neg P, V \rangle \quad \langle P, W \rangle}{\times}$$

   where P is any atomic $\mathcal{L}$-formula.

2. Double Negation:

$$\frac{\langle \neg\neg A, V \rangle}{\langle A, V \rangle}$$

3. Conditional:

$$\frac{\langle A \to B, V \rangle}{\langle \neg A, V \rangle | \langle B, V \rangle}$$

---

[4] One can show generally that, if A is refutable, $\neg A$ is provable. We've just seen that $\forall xPx \wedge \neg Py$ is refutable, hence its negation, which is tautologically equivalent to $\forall xPx \to Py$, is provable.

4. Negated Conditional:

$$\frac{\langle \neg(A \to B), V \rangle}{\langle A, V \rangle}$$
$$\langle \neg B, V \rangle$$

5. Negated Universal:

$$\frac{\langle \neg \forall x F[x], V \rangle}{\langle \exists x \neg F[x], V \rangle}$$

6. Negated Existential:

$$\frac{\langle \neg \exists x F[x], V \rangle}{\langle \forall x \neg F[x], V \rangle}$$

7. Universal Instantiation:

$$\frac{\langle \forall x F[x], V \rangle}{\langle F[a], V \cup \{a\} \rangle}$$

where a is a free variable that does not occur in F, $a \notin V$, and either

(a) there is a pair $\langle B, W \rangle$ on the branch under consideration such that $a \in W$, or

(b) for all pairs $\langle B, W \rangle$ on the branch under consideration, $W = \emptyset$.[5]

8. Existential Instantiation:

$$\frac{\langle \exists x F[x], V \rangle}{\langle F[b_1], V \cup \{b_1\}\rangle | \ldots | \langle F[b_k], V \cup \{b_k\}\rangle | \langle F[a], V \cup \{a\}\rangle}$$

with the following proviso: If S is the union of all sets W such that some pair $\langle B, W \rangle$ occurs on the branch under consideration, then $\{b_1, \ldots, b_k\} = S \setminus V$, and the free variable a is new to the branch under consideration (i.e., for any $\langle B, W \rangle$ on the branch, a does not occur in B and $a \notin W$).

---

[5] It is possible that there is no free variable a not occurring in F and V even if case (2) doesn't obtain (viz., if all the elements of free-variable sets on the branch are also in V). In such a situation, universal instantiation is not applicable.

We say that a *branch* in a tableau is *closed* if it contains pairs $\langle P,V \rangle$ and $\langle \neg P, W \rangle$, where P is atomic and V and W are arbitrary. A *tableau* is *closed* if all its branches are closed. A *tableau for A* is a tableau beginning with $\langle \neg A, \mathsf{FV}(A) \rangle$. A closed tableau for A is also called a *proof of A*.

6 Soundness

Let $\mathcal{U}$ be a structure, let A be an $\mathcal{L}$-formula, let V be a finite set of free variables, let S be a set of pairs $\langle B, W \rangle$ where B is an $\mathcal{L}$-formula and W a finite set of free variables. We make the following definitions.

1. The pair $\langle A, V \rangle$ is R-satisfiable in $\mathcal{U}$ if for some $\mathcal{U}$-assignment v that is 1-1 on V, $\mathcal{U} \Vdash \langle A^{\mathsf{v}}, \mathsf{v}[V] \rangle$.

2. The $\mathcal{L}$-formula A is R-satisfiable in $\mathcal{U}$ if $\langle A, \mathsf{FV}(A) \rangle$ is.

3. The set S is simultaneously R-satisfiable if there is a $\mathcal{U}$-assignment v that is 1-1 on $\bigcup \{W : \text{for some } \mathcal{L}\text{-formula } B, \langle B, W \rangle \in S\}$ and such that $\mathcal{U} \Vdash \langle B^{\mathsf{v}}, \mathsf{v}[W] \rangle$ for every $\langle B, W \rangle \in S$.

4. Let T be a set of $\mathcal{L}$-formulas. Then T is simultaneously R-satisfiable if the set $\{\langle B, \mathsf{FV}(B) \rangle : B \in T\}$ is.

5. A branch in a tableau is simultaneously R-satisfiable if the set of all pairs $\langle B, W \rangle$ occurring on it is.

Our aim in this section is to prove the following

LEMMA 3 (SOUNDNESS LEMMA) *Let $\mathcal{U}$ be a structure, let V be a finite set of free variables, and let A be an $\mathcal{L}$-formula such that $\langle A, V \rangle$ is R-satisfiable in $\mathcal{U}$. Then every tableau starting with $\langle A, V \rangle$ has at least one branch that is simultaneously R-satisfiable in $\mathcal{U}$.*

The soundness theorem follows readily from the lemma: If A is a sentence and some tableau begun with $\langle \neg A, \emptyset \rangle$ closes, then $\langle \neg A, \emptyset \rangle$ is not R-satisfiable, hence $\mathcal{U} \Vdash A$ for all structures $\mathcal{U}$.

The proof of the soundness lemma proceeds by way of induction along the generation of tableaux. Since $\langle A, V \rangle$ is R-satisfiable in $\mathcal{U}$ by assumption, the induction basis is trivial. Applications of the propositional rules and the negated quantifier rules are easily handled by means of the induction hypothesis. Let us thus focus on the quantifier instantiation rules.

First, the universal case. Suppose given a simultaneously R-satisfiable branch $\mathcal{B}$ in a tableau T, and let T^* arise from T by extending $\mathcal{B}$ to $\mathcal{B}^*$ according to the universal instantiation rule. That is, there is a pair $\langle \forall x F[x], V \rangle$ on $\mathcal{B}$, and $\mathcal{B}^*$ results from $\mathcal{B}$ by writing the pair $\langle F[a], V \cup \{a\} \rangle$ below its last node. Here, a does not occur in F, $a \notin V$, and one of the following two cases obtains:

1. For some pair $\langle B, W \rangle$ on $\mathcal{B}$, $a \in W$. In this case, assuming that ν is the $\mathcal{U}$-assignment that simultaneously R-satisfies $\mathcal{B}$, ν will also simultaneously R-satisfy $\mathcal{B}^*$: Since $\langle \forall x F[x], V \rangle$ occurs on $\mathcal{B}$, we know that $\mathcal{U} \Vdash \langle F[u]^\nu, \nu[V] \cup \{u\} \rangle$ for every $u \in U \setminus \nu[V]$; But because $a \notin V$, we know that $\nu(a)$ is such a u. Also, because ν is 1-1 on the union of all free-variable sets occurring on $\mathcal{B}$, and no additional free variables occur on $\mathcal{B}^*$, the requisite injectivity condition is still fulfilled.

2. All free-variable sets occurring on $\mathcal{B}$ are empty. In this case, too, ν itself simultaneously R-satisfies $\mathcal{B}^*$. We already know that it satisfies every pair on $\mathcal{B}$. But since $\langle \forall x F[x], \emptyset \rangle$ is on $\mathcal{B}$, $\mathcal{U} \Vdash \langle F[u]^\nu, \{u\} \rangle$ for every $u \in U$, in particular, for $\nu(a)$. The injectivity condition is trivial.

Second, the existential case. Say the branch $\mathcal{B}$, simultaneously R-satisfied by the assignment ν, contains the entry $\langle \exists x F[x], V \rangle$ and has been extended into $k+1$ branches $\mathcal{B}_1, \ldots, \mathcal{B}_{k+1}$ according to the existential instantiation rule. We know the fresh free variable a occurring for the first time in the last entry on $\mathcal{B}_{k+1}$, i.e., in $\langle F[a], V \cup \{a\} \rangle$, is not an element of V. We also know that since ν R-satisfies $\langle \exists x F[x], V \rangle$, there is an element $u \in U \setminus \nu[V]$ such that $\mathcal{U} \Vdash \langle F[u]^\nu, \nu[V] \cup \{u\} \rangle$. Fix such an element u. We distinguish two cases:

1. For some free variable b_i that occurs in one of the free-variable sets on $\mathcal{B}$, but not in V, $\nu(b_i) = u$. I claim that ν then simultaneously R-satisfies $\mathcal{B}_i$: Clearly ν R-satisfies $\langle F[b_i], V \cup \{b_i\} \rangle$ in $\mathcal{U}$; It simultaneously R-satisfies $\mathcal{B}$ by induction hypothesis, and the injectivity condition also follows from the induction hypothesis, as no new variables occur on $\mathcal{B}_i$.

2. Otherwise, ν_u^a simultaneously R-satisfies the new branch B_{k+1} in $\mathcal{U}$: It clearly R-satisfies $\langle F[a], V \cup \{a\} \rangle$, and it equally clearly simultaneously R-satisfies $\mathcal{B}$, because a is new to $\mathcal{B}$. Further, ν_u^a is 1-1 on the union of all free-variable sets on $\mathcal{B}$ and $\{a\}$ because a is new to $\mathcal{B}$ and u is not the ν-image of any free variable occurring in any of the free-variable sets on $\mathcal{B}$.

7 Completeness

The completeness proof follows the usual strategy of constructing a model $\mathcal{U}$ for an open branch $\mathcal{B}$ in a 'systematic' tableau, where the domain of $\mathcal{U}$ is the set of free variables occurring on $\mathcal{B}$. In analogy with the Hintikka sets of Smullyan (1968), we define Ramsey sets, or R-sets for short, as follows.

Let S be a set of pairs $\langle A, V \rangle$, where A is an $\mathcal{L}$-formula and V a finite set of free variables. Let U be a non-empty set of free variables such that $\bigcup \{FV(A) \cup V : \langle A, V \rangle \in S\} \subseteq U$. Then S is an R-set for U if, whenever

1. $\langle A, V \rangle, \langle B, W \rangle \in S$ and A is atomic, then B is not $\neg A$;

2. $\langle\neg\neg A, V\rangle \in S$, then $\langle A, V\rangle \in S$;

3. $\langle A \wedge B, V\rangle \in S$, then $\langle A, V\rangle, \langle B, V\rangle \in S$;

4. $\langle\neg(A \wedge B), V\rangle \in S$, then $\langle\neg A, V\rangle \in S$ or $\langle\neg B, V\rangle \in S$; and similarly for the other Boolean connectives;

5. $\langle\forall xF[x], V\rangle \in S$, then for every $b \in U \setminus V$, $\langle F[b], V \cup \{b\}\rangle \in S$;

6. $\langle\neg\forall xF[x], V\rangle \in S$, then $\langle\exists x\neg F[x], V\rangle \in S$;

7. $\langle\exists xF[x], V\rangle \in S$, then for some $b \in U \setminus V$, $\langle F[b], V \cup \{b\}\rangle \in S$;

8. $\langle\neg\exists xF[x], V\rangle \in S$, then $\langle\forall x\neg F[x], V\rangle \in S$.

LEMMA 4 (HINTIKKA'S LEMMA FOR R-LOGIC) *If S is an R-set for U, then S is simultaneously R-satisfiable in some structure $\mathcal{U}$ with domain U.*

Toward a proof, let S be an R-set for U. We turn U into a structure $\mathcal{U}$ by defining, for any given n-ary relation symbol P, the relation $P_{\mathcal{U}}$ as follows: $\langle a_1, \ldots, a_n\rangle \in P_{\mathcal{U}} :\Leftrightarrow$ for some V, $\langle Pa_1 \ldots a_n, V\rangle \in S$.

Let v be the $\mathcal{U}$-assignment that maps every variable a in U to itself, and maps every other variable to an arbitrarily selected element of U. Note that, since v is the identity function on all relevant free variables, in what follows we needn't distinguish between $\mathcal{L}$-formulas A and $\mathcal{L}(\mathcal{U})$-sentences A^v, nor between finite sets V of free variables and their images $v[V]$ under v.

We claim that

(a) whenever $\langle A, V\rangle \in S$, then $\mathcal{U} \Vdash \langle A, V\rangle$, and

(b) whenever $\langle\neg A, V\rangle \in S$, then $\mathcal{U} \nVdash \langle A, V\rangle$.

The proof proceeds by induction on the $\mathcal{L}$-formula A. For atomic A, part (a) follows from the definition of $\mathcal{U}$. For part (b), suppose $\mathcal{U} \Vdash \langle Pa_1 \ldots a_n, V\rangle$. Then, by definition of $\mathcal{U}$, for some W, $\langle Pa_1 \ldots a_n, W\rangle \in S$. Since S is an R-set, it follows that $\langle\neg Pa_1 \ldots a_n, V\rangle \notin S$.

If A is of the form $\neg B$, then part (a) follows from the induction hypothesis, part (b). For part (b), if $\langle\neg\neg B, V\rangle \in S$, then since S is an R-set, also $\langle B, V\rangle \in S$, and so by the induction hypothesis, part (a), $\mathcal{U} \Vdash \langle B, V\rangle$, i.e., $\mathcal{U} \nVdash \langle\neg B, V\rangle$. The other Boolean connectives can be handled analogously, using the requisite induction hypotheses.

Now suppose A is $\forall xF[x]$. For part (a), assume $\langle\forall xF[x], V\rangle \in S$. Since S is an R-set for U, we know that for all $b \in U \setminus V$, $\langle F[b], V \cup \{b\}\rangle \in S$. Thus by part (a) of the induction hypothesis, for all $b \in U \setminus V$, $\mathcal{U} \Vdash \langle F[b], V \cup \{b\}\rangle$, i.e., $\mathcal{U} \Vdash \langle\forall xF[x], V\rangle$. For part (b), suppose $\langle\neg\forall xF[x], V\rangle \in S$. Since S is an R-set for

$U, \langle \exists x \neg F[x], V \rangle \in S$, and thus for some $b \in U \setminus V$, $\langle \neg F[b], V \cup \{b\} \rangle \in S$. Part (b) of the induction hypothesis then implies that for some $b \in U \setminus V$, $\mathcal{U} \not\Vdash \langle F[b], V \cup \{b\} \rangle$; hence $\mathcal{U} \not\Vdash \langle \forall x F[x], V \rangle$. The existential case is entirely analogous. Q.e.d.

Next we describe a systematic procedure for generating tableaux. For the purpose, we will assume that the free variables come ordered in an ω-sequence.

To systematically generate tableaux, iterate the following steps as long as possible.

1. Pick, if possible, the highest and leftmost pair $\langle A, V \rangle$ that has no checkmark next to it and such that A is non-atomic.

2. If the previous step is inapplicable, pick the highest and leftmost pair $\langle A, V \rangle$, with A of the form $\forall x F[x]$, that has the least number of checkmarks, and to which universal instantiation can be applied with respect to at least one branch passing through it.

3. If both previous steps are inapplicable, the systematic procedure terminates.

4. Apply the applicable rule to the chosen pair $\langle A, V \rangle$. If A is of the form $\forall x F[x]$, pick the instantiating free variable so that it is the first eligible free variable a such that the resulting instance $F[a]$ does not yet occur on the branch at hand. If no free variables occur on the branch, pick the first free variable.

5. Put a check mark next to the pair $\langle A, V \rangle$.

We now define a *systematic tableau* to be either

1. the result of the systematic procedure, if it terminates after finitely many iterations, or

2. if the systematic procedure does not terminate after any finite number of iterations, the limit (set-theoretic union) of the results of all finite stages of the systematic procedure.

LEMMA 5 *Every non-closed branch $\mathcal{B}$ of a systematic tableau is an R-set for $U_{\mathcal{B}} := \bigcup \{ W : \text{for some } \mathcal{L}\text{-formula B}, \langle B, W \rangle \text{ occurs on } \mathcal{B} \}$.*[6]

---

[6]Note that $U_{\mathcal{B}}$ cannot be empty: Let the initial node be $\langle A, \mathsf{FV}(A) \rangle$. If A isn't an $\mathcal{L}$-sentence, $\mathsf{FV}(A)$ and hence $U_{\mathcal{B}}$ isn't empty. Otherwise, since $\mathcal{L}$ contains no individual constants, A must contain a quantifier. Thus, after finitely many steps, the systematic procedure will attempt to apply either universal or existential instantiation in each open branch. Now suppose all free-variable sets were empty. In that case, (i) existential instantiation does not result in branching and introduces a fresh free variable into the branch, and (ii) universal instantiation is applicable (condition (b) of the rule being fulfilled) and similarly introduces a fresh free variable into the branch.

For a proof, note that condition (1) on R-sets is satisfied because $\mathcal{B}$ is an *open* branch. The other conditions, with the possible exception of (5), are obviously satisfied by construction. Now consider condition (5). Suppose that $\langle \forall x F[x], V \rangle$ occurs on $\mathcal{B}$ and that $b \in U_{\mathcal{B}} \setminus V$. The free variable b entered $U_{\mathcal{B}}$ through some application of either universal or existential instantiation. The systematic procedure ensures that, at some point after the first occurrence of b on $\mathcal{B}$, universal instantiation will be applied to $\langle \forall x F[x], V \rangle$ with respect to the free variable b. Hence $\langle F[b], V \cup \{b\} \rangle$ will also occur on $\mathcal{B}$, which concludes the proof of the lemma.

The completeness theorem now follows as usual:

THEOREM 1 *If a sentence A is R-valid, then there exists a proof of A (i.e., a closed tableau beginning with* $\langle \neg A, \emptyset \rangle$).

For if there are no closed tableaux for A, then there are no closed systematic tableaux for A. If the systematic tableau for A is finite, it follows immediately that it must contain a non-closed branch. If the systematic tableau for A is infinite, then by König's lemma, it contains an infinite branch, which is clearly non-closed. In either case, by the previous lemma, the systematic tableau for A contains a branch $\mathcal{B}$ that is an R-set for $U_{\mathcal{B}}$. By Hintikka's lemma for R-logic, there is thus a structure $\mathcal{U}$ in which all pairs on $\mathcal{B}$ are R-true. Hence $\langle \neg A, \emptyset \rangle$ is satisfiable, and A is not R-valid.

8 Acknowledgments

Thanks to Aldo Antonelli, who read and commented on the penultimate draft, and to Sam Hillier for several discussions regarding Hintikka's strongly exclusive interpretation. I am further grateful to an anonymous referee whose comments and questions led to substantial improvements in the paper.

BIBLIOGRAPHY

Boolos, G. (1984). "Trees and Finite Satisfiability: Proof of a Conjecture of Burgess", *Notre Dame Journal of Formal Logic*, vol. 25, no. 3, pp.193-197.

Hintikka, J. (1956). "Identity, Variables, and Impredicative Definitions", *Journal of Symbolic Logic*, vol. 21, pp.225-245.

Schütte, K. (1977). *Proof Theory*, Springer-Verlag, New York.

Smullyan, R.M. (1968). *First-Order Logic*, Springer-Verlag, New York.

Wehmeier, K.F. (2004). "Wittgensteinian Predicate Logic", *Notre Dame Journal of Formal Logic*, vol. 45, no. 1, pp.1-11.

Wehmeier, K.F. (2008). "Wittgensteinian Tableaux, Identity, and Co-Denotation", *Erkenntnis*, vol. 69, no. 3, pp.363-376.

Wittgenstein, L. (1922). *Tractatus Logico-Philosophicus*, transl. C. K. Ogden, Routledge and Kegan Paul, London.

Wittgenstein, L. (1995). *Cambridge Letters: correspondence with Russell, Keynes, Moore, Ramsey, and Sraffa*, ed. B. McGuinness and G. H. von Wright, Blackwell, Oxford.

INDEX

Åqvist, 131
Łukasiewicz, 82

Abelard, 184, 198–201, 204
  dictum, 198–200
Abstraction, 18, 60, 77, 79, 85, 86, 125, 201, 272, 305, 317, 330, 331, 336
Acceptance, 4, 90, 269, 271, 277, 283–287
Accessibility relation
  canonical, 234, 243
Ackermann, 10
Act
  illocutionary, 85, 95, 128, 134, 136
  stratified, 91
Adverbialization, 77
Aggregate, 4–6, 18, 141
Analogy, 3, 8, 10, 13–18, 20–23, 25–27, 100, 101, 133, 236, 244, 279, 299, 344, 366
Analytic philosophy, 74, 76, 80, 269, 270, 278, 344
Anti-realism, 73, 207, 208, 270, 291, 297, 299, 300
Argumentation, 22, 78, 121, 303, 305–311, 313, 314, 316, 317, 330, 331, 333, 334
Aristotle, 28, 29, 35, 36, 38, 41, 43–45, 47, 48, 54, 56, 57, 59, 62–68, 93, 121, 122, 124, 127, 128, 136, 287
Artifact, 87, 88

Assertion, 42, 43, 45, 56, 62, 75, 88, 90, 91, 132, 134, 136, 153, 184, 189–191, 193, 194, 196, 198–200, 202, 203, 208, 226, 269–271, 275, 279, 283, 285, 304, 305, 308, 310, 311, 313, 314, 316, 322–326, 328–333

Bain, 78
Bayart, 233, 234, 237, 238, 244, 246–248
Belief, 4, 5, 8, 10, 35, 36, 39–41, 46, 53, 55, 56, 76–78, 105, 106, 108, 124, 152, 153, 187, 188, 190, 218, 269, 270, 274–279, 281–287, 296–298
Benacerraf, 3, 20, 30, 342, 348, 350
Bergmann, G., 154
Bergmann, H., 100
Bergmann, J., 82, 83
Bernays, 159, 160, 176, 177
Beyer, 116, 121
Bolzano, 80, 89, 95–114, 116–132, 134–136, 185, 186, 276, 285
  analytic, 110, 118, 119
Boolos, 363
Brandl, 74
Bratman, 286
Brentano, 74–80, 89–93, 96–100, 113, 120, 150
Brouwer, 27, 207–210, 242, 246, 270, 284

Brouwerische system B, 242, 246
Buczyńska-Garewicz, 84, 93
Bulldozing, 234, 257
Burgess, 363

Canonical form, 131, 184, 194, 196, 198, 200, 201, 203, 272, 273
Canonical model, 249
Cantor, 4, 6, 8–10, 12, 95
Cardinality, 3, 26
    inaccessible cardinal, 9, 10
Carnap, 237, 247
Characteristic formula, 241, 246
Characteristica universalis, 4
Church, 291, 292, 296, 349
Closure, 172, 175, 176, 226, 241, 243, 246, 249, 301, 344, 345, 347, 348, 351, 363
    condition, 255
Cognitive act, 86, 209, 273, 274, 279
Cohen, 134, 269, 270, 283–287
Complete assignment, 239, 242
Completeness, 17, 233–235, 238, 240, 241, 243–248, 250, 251, 253, 257–259, 264, 265, 356, 366, 369
Consequence, 47, 53, 64, 68, 161, 164, 183, 185–189, 195, 196, 202, 212, 256, 331, 362
    R-consequence, 358
Constructive type theory, 202, 209, 211, 269, 271, 284, 285, 287, 352
Constructivism, 179, 207, 215
Content, 43, 62, 73–75, 77, 79–83, 85, 86, 88–90, 95, 97, 102, 106, 109, 115, 136, 145, 148, 154, 183–185, 189–203, 208, 209, 211, 216, 218–222, 224–226, 228, 229, 270, 304, 322
    conceptual, 76
    force-content distinction, 75, 77, 85, 87, 89, 91, 191, 197, 198, 201
    propositional, 84–86, 90, 91, 106, 196, 197, 208, 211, 214, 215, 218, 219, 221
    representation of, 198–201
Continuum hypothesis, 4
Contradiction, 15, 24, 44, 50, 51, 55, 185, 194, 237, 240, 244, 294, 360
    Principle of, 14, 17, 24, 26, 38, 41–43, 57, 93
    Proof by, 208, 244, 294
Copycat strategy, 314, 331
Countermodel, 234, 239, 240, 243–245, 251, 257, 260, 263, 264
    connected, 243

Decidability, 169, 246
Deduction theorem, 228, 236, 241
Deductively closed, 250
Demonstration, 22, 217, 276
Denial, 5, 75, 84–86, 88–91, 126
Descartes, 7, 15, 98, 283, 286
Destutt de Tracy, 131
Dialogues, 303, 304, 307–309, 313, 314, 316, 318, 322, 332, 333, 336
    Plato's dialogues, 36, 37, 39–41, 43–45, 53, 54, 57, 60, 66–68
Dispositionalism, 12, 78, 274, 277, 299
Dummett, 37, 73, 74, 92, 106, 184, 194, 208, 230, 288

Error, 21, 64, 99, 278, 282, 285, 286

Essence, 7, 8, 14, 15, 10, 19, 22, 24, 27–29, 80, 142, 145, 154
essential property, 15, 16, 19, 22, 23
Evidence, 10, 53, 59, 81, 86, 88, 90, 91, 123, 125, 126, 231, 277, 281, 284, 299
Existence (intentional), 77
Extensive form, 309, 316, 317, 331

Factivity, 291, 293, 294, 296–299
Feys, 242
Field, 76
Fitch, 291, 292, 296, 303, 304, 311, 312, 318–322, 327, 330, 334
Frege, 62, 73, 80, 89, 90, 96, 99, 100, 103, 131, 136, 145, 191–203, 208, 270, 271, 284, 304, 347, 353
Frege Point, 184, 190, 197

Gödel, 8–18, 20–23, 25–27, 29, 96, 159–161, 176, 214, 215, 236, 242, 301
Gentzen, 159–162, 165, 167, 169, 175–177, 185, 214, 244
God, 4, 6–8, 15, 22–24, 26–29, 42, 141, 146, 148, 155, 276, 282
Grice, 297, 298

Höfler, 78
Haldane, 76
Harmony
Principle of, 13, 14, 21–23, 26
Henkin, 233, 234, 243, 247, 248, 251, 257, 264, 265, 301
Heyting, 195, 207, 210
Heyting Arithmetic, 159, 161, 163

Hilbert, 17, 185, 238, 241, 246, 256, 270, 271, 284, 287
Hintikka, 37, 45, 66, 131, 218, 237, 247, 248, 355, 356, 358, 360, 362, 366
lemma, 367, 369
Hume, 78, 143, 144, 275, 277, 283
Husserl, 19, 26, 29, 61, 80, 84, 95–102, 104–106, 108, 112–128, 131–136, 143–146, 150, 154, 155
analytic, 110, 118–120

Idealism, 26, 73, 100, 154, 300
Identity, 76, 78, 81, 108–110, 121, 147, 193, 208, 209, 216, 219, 221, 228, 240, 355–359, 367
Principle of, 20, 24
Incompleteness, 160, 177
theorem(s), 4, 10, 159, 176
Independence results, 246
Indexicals, 81, 120, 135, 293
Inference, 15, 52, 59, 73, 163, 176, 183–189, 192, 195–198, 201–203, 209–211, 216, 270, 272, 273, 278, 279, 314, 316, 317, 352
'lucky', 187, 188
Infinite, Infinity, 4, 5, 9, 20, 22, 24, 25, 27, 56, 126, 133, 148, 164, 244–246, 254, 260, 263, 264, 342, 351, 369
Intentionalist (perspective on judgement), 75–79, 87–90
Interpretation, 235, 258, 259, 343
strongly exlcusive, 355, 356, 362
weakly exlcusive, 355, 356
Intimation (*Kundgabe*), 105
Intimation (*Kundgebung*), 84

Johansson, 303, 304, 311, 312
Judgement
   act of, 58, 75, 80, 84, 89, 124, 269, 271–276, 278, 279, 281, 285–287
   assertion conditions, 208–214, 216–219, 222–226
   candidate, 88, 271–273
   capacity to judge, 272, 275, 279, 287
   categorical, 75, 192, 193, 211–214, 218, 222, 223, 226, 228
   correct, 73, 75, 85, 90, 91, 184, 198, 218, 273, 274, 285
   evident, 90, 91, 273, 275
   hypothetical or dependent, 192, 193, 195, 207, 209–214, 218, 221–223, 225, 226, 228
   made, 78, 272, 273, 285, 299

König's lemma, 245, 369
Kanger, 237, 247
Kant, 17, 100, 112, 118, 120, 125, 208, 276
Kaplan, 233, 234, 238, 247, 248, 301, 344
Kerry, 96–98, 100
Kleene, 165
Knowability, 10, 212, 213, 224, 225, 228, 274, 291–300
Knowledge, 293–299
   'lucky', 187, 188
   demonstrative, 278
   logic as a tool to produce new knowledge, 184–189, 195, 203
   product, 273
Kreisel, 11, 30, 160, 176, 210
Kripke, 233–235, 237–239, 241, 243, 244, 246–248, 251, 257–259, 264, 265, 339, 348, 349, 352
   countermodel, 234, 251, 257, 260
   frame, 237, 238
   model, 237, 246, 250, 251, 263, 264
   semantics, 218, 228, 233–235, 237, 238, 348, 349

Löb, 236, 295, 301
Löwenheim-Skolem result, 241
Labelled formulas (system with), 233, 234, 251, 256, 258, 259, 261, 263, 265
Lehrer, 285
Leibniz, 3–8, 13, 15, 17–29, 98, 100, 248
Lemmon, 246, 247
Levy, 10
Lewis, 45, 131–135, 242, 344
Lindenbaum's Lemma, 249
Linguistic turn, 74, 76, 78, 79, 85
Locke, 269, 270, 274, 277–279, 281, 282, 287
Logic
   classical, 160, 161, 175, 234, 235, 246, 249, 303, 311, 313, 318, 321
   constructive, 207, 210
   deontic, 255
   dialogical, 37, 43, 44, 49, 50, 56, 60, 62, 65, 303, 310, 311, 332
   dynamic turn in, 58, 68, 89, 92, 229, 303
   epistemological conception of, 184
   extensions of basic modal logic, 209, 210, 213–215, 225, 228, 229, 234–238, 242, 246–248, 253, 265

intuitionistic, 11, 170, 207, 209–211, 228, 235, 237, 303, 311–313, 316, 318, 321, 336
  minimal, 303, 304, 311, 312, 321, 322, 328
  modal, 209, 215, 223, 228, 235, 242, 247–252, 257, 260, 264, 336
  notation (judgement-based vs. content-based), 183, 184, 189, 192–199, 201–203
  Ramsey predicate logic (R-logic), 356–363, 367, 369
  tense, 236
  Transparent Intensional (TIL), 344–347, 349, 352
  Wittgensteinian predicate logic (W-logic), 356, 358, 363
Lotze, 99, 120, 121

Martin-Löf, 184, 202, 211, 213
Mathematical constants, 339, 350
Matter, 7, 97, 98, 102–108, 113–118, 123–127, 136, 143–146, 275
Meaning, 18, 60, 62, 73–76, 79–82, 84–87, 107, 110, 128, 131, 133, 135, 141, 142, 145, 146, 154, 195, 209, 218, 269, 272, 277, 278, 287, 343, 344, 348
  explanation, 85, 210, 212, 216, 218, 219, 222, 251, 252
  meaningfulness, 86, 87, 212
  structured, 344
Meaning (as use), 332
Meinong, 98, 100, 102, 103, 144, 145
Mental
  act, 75–80, 82–87, 90, 91, 95, 103, 105–107, 121, 123–125, 128, 130, 132, 282
  state, 76, 95, 103, 105–107, 115, 121, 124, 274–276, 285
Metaphysics, 3–5, 13, 14, 24, 25, 27, 121
Mill, 63, 78, 276, 277
Mind, 3, 4, 6–8, 19–21, 23, 27, 29, 73, 77, 78, 87, 105, 125, 275, 277, 281, 285
  -independence, 73, 74, 89
Model, 8, 13, 52, 185, 195, 239, 240, 242–244, 248, 250, 263, 264, 356, 366
  canonical, 250
  connected, 243, 245
Monad, Monadology, 3–5, 8, 13–27, 29
Montague, 10, 106, 237, 247, 294, 301
Moore, 115, 134, 298
Multitude, 4, 5, 17, 28, 135

Natural Deduction, 160–163, 165–167, 170, 177, 178, 209, 228, 231, 235, 236, 303, 304, 311–313, 318, 322, 332, 333, 336
Naturalism, 76, 93, 274, 277

Opinion, 39, 40, 44, 46, 51, 63, 64, 269, 275–278, 281, 282, 287

Paradox
  Believer, 296
  Church-Fitch, 291, 292, 296
  Knowability, 291–293, 296
  Knower, 291, 294–296, 300
  Liar, 294
  Moore, 134, 298
  Russell, 119, 196, 199

Perception, 19–23, 76, 122, 133, 151, 278, 279, 287
Perry, 106
Phenomenology, 29, 125, 194
Popper, 284
Possibility
    epistemic, 80, 88, 228, 296, 299
    metaphysical, 296, 299
Prawitz, 37, 162, 184
Presentation (*Vorstellung*), 74–76, 78, 83, 87, 88, 91, 97, 98, 104, 105, 140, 152, 153
Prior, 237
Product, 186, 190, 272, 273, 278, 281, 339, 341, 344–347, 349
    Enduring, 83
    Mental or Psychic, 81, 83–86, 88, 92
    Non-enduring, 83
    Physical, 82, 88, 91
    Potential, 83
    Psychophysical, 82, 84–88
Proof
    object, 210, 211, 214, 215, 219, 221, 271–273
    provability, 160, 195, 208, 213–216, 220, 221, 228, 255, 265, 295, 362
    search, 234, 246, 251, 257, 258
Propositionalist theory of judgment, 87, 91
Psychologism (and anti-), 73, 74, 78, 91, 98, 99, 118, 270, 284

Quality, 91, 113, 114, 136, 146
Question, 43–49, 52, 54, 56, 57, 61, 63, 66, 68, 127–131, 134, 286
Quine, 77, 106, 111, 139

Ramsey, 356

Ramsey set (R-set), 366, 367, 369
Realism, 33, 148, 152, 154, 207, 270, 291, 300, 350
Reflection Priniciple, 3, 8–11, 13–15, 17, 18, 22, 23, 25, 27, 29, 30, 215
Reflexivity, 164, 238, 254
Relational atoms, 251, 256, 258, 261–264
Rosser, 238
Rule
    $\Box$-Rule, 220, 224, 227
    $\Diamond$-Rule, 222, 225, 227
    admissible, 161, 165, 171, 175, 251, 254, 256, 257
    box introduction, 235
    context dependent, 256
    geometric rule scheme, 255
    Hypothesis-Rule, 227, 228
    induction hypothesis, 359, 361, 365–367
    inference rule, 52, 238, 270–272, 311, 312, 333
    invertible, 248, 257
    mathematical, 251, 259, 260, 262–264
    Modus Ponens, 104, 192, 193, 241, 249, 256, 312
    Modus Ponens (dilemma), 194, 201, 203
    Modus Tollens, 56, 62, 68
    necessitation, 220, 222, 224, 225, 228, 235, 236, 241, 249, 256, 293, 294
    negation introduction, 52, 53, 90
    Premise-Rule, 227
    regular rule scheme, 255
    structural, 43, 56, 60, 218, 251

structural (contraction) 244,
251, 254, 257, 262
structural (cut), 251, 256, 257
structural (weakening), 251,
256
Russell, 5, 119, 120, 145, 146, 195,
196, 211, 344, 355, 356
Principia Mathematica, 120,
185, 195, 196

Sacks, 3
Salerno, 291, 297, 298, 301
Satisfiability, 239, 241–243
(simultaneously) R-satisfiable,
365, 367
Satz an sich, 89, 97, 99, 102, 108,
116, 121, 125, 128, 129
Schütte-style construction, 234
Schaar, vd., 91, 209
Scholastics, 29
Scott, 247
Searle, 106, 114
Self-reference, 293
Semantics
algebraic, 233, 246
possible worlds, 233, 247, 248,
251, 300
irreflexive accessibility relation, 234, 264
reflexive frame, 238, 263
procedural, 339, 340, 342–345,
349, 350
relational, 233, 235–237, 247,
252, 265
Sequent calculus, 160, 163, 165–
167, 170, 177, 185, 209,
214, 233, 234, 248, 251–
255, 257, 265
deep sequents, 265
derivable sequents, 166, 169,
171, 174–176, 256
end sequents, 166, 167

initial sequents, 166, 169, 171,
172, 174–177, 253, 257
irreducible sequents, 169
nested sequents, 264
reducible sequents, 170, 172–
174, 177
tree-hypersequents, 265
tree-sequents, 265
Set, Set Theory, 9–18, 20–23, 25, 27,
248, 249, 251, 344, 345
Maximal consistent set, 248–
251
Powerset, 9, 10
ZF, ZFC, 9, 10, 13
Smith, 84, 93, 150, 155, 157
Sound, Soundness, 5, 189, 221, 228,
249, 258, 356, 362, 365
Strawson, 123
Subterm property, 262
Sufficient reason
Principle of, 14, 20, 24, 25, 91
Sundholm, 30, 73, 93, 183–186,
188, 195–197, 202, 203,
205, 209, 211, 215, 219,
222, 229, 269, 291–296,
299, 300
Symmetry, 164, 243, 246, 254
Asymmetry between proponent
and opponent, 46, 49

Tableau(x)
(set of) closed, 240, 244
alternative, 234, 239–241, 243
auxiliary, 240, 241, 243, 244
Beth, 234, 239, 248
branch of, 364, 365
closed, 240, 244, 246, 364, 369
contiguous, 246
main, 234, 239–241, 243, 244
rules, 240, 243, 245, 246, 363
semantic, 243, 248, 303, 356,
362–365

subsidiary, 239
systematic, 366, 368, 369
Tennant, 291, 299, 302
Transcendent(al), 26–29, 80, 155, 339–342, 345, 349
Transitivity, 164, 243, 245, 254, 259, 302, 362
Tree, 163, 167, 195, 225, 234, 243–245, 257, 258, 260, 262, 263, 265
    reduction, 260, 262–264
    search, 234
Truth
    lemma, 250
    conditions, 37, 62, 98, 133, 209, 216, 219, 342
    contingent, 22, 80, 81, 143
    eternal, 7, 8
    factual, 143, 213
    logical, 27, 110–112, 118–120, 145
    mathematical, 7, 15, 25, 26, 29
    necessary, 22, 27, 143, 216, 217, 228, 250
    predicate, 294, 296, 299
    R-true, 358, 369
    table, 237, 241
    Tarskian, 358
    value, 38, 41, 53–55, 59, 75, 80, 90, 91, 104, 108, 110, 128, 190, 192, 196, 210, 242
    Verification Principle, 90, 208, 218, 291, 294, 299
Twardowski, 73, 76, 80–89, 91, 92, 96–98, 100–102, 104, 105, 108
Type
    Type Theory, 15, 202, 209, 211, 212, 218, 226, 228, 229, 269, 271, 284, 285, 287, 345, 346, 351, 352

type-token distinction, 30, 76, 78, 79, 83, 85, 123, 135

Universe, 8–13, 15–23, 29, 116
V, 8–12, 14–17, 23, 27
Utterance, 76, 80, 82, 84, 85, 87, 88, 104, 105, 110, 111, 119, 120, 131–136, 283, 298

Valid, Validity, 5, 13, 49–54, 65, 68, 80, 110–112, 119–121, 146, 184–189, 192, 193, 217, 220, 222, 224, 226, 234, 238–244, 246–251, 258–260, 264, 270, 272
    R-valid, 358, 362, 369
Valuation, 237, 238, 240, 243, 244, 250, 258, 260, 344, 345
von Neumann, 17, 159, 176
    Axiom, 16, 17, 23
von Wright, 242
Vorstellung an sich, 95, 97, 99, 107, 108

Wang, 3, 8, 10, 13, 14, 29
Wittgenstein, 134, 141, 142, 144–153, 155, 191, 195, 276, 332, 355, 356
    Tractatus Logico-Philosophicus, 141, 142, 146, 150, 153, 155, 195, 355, 356

Zermelo, 9, 11, 12

Authors
CONTACT INFORMATION

B. Göran Sundholm
Faculteit der Geesteswetenschappen, Instituut voor Wijsbegeerte
  Universiteit Leiden, the Netherlands
  http://www.hum.leiden.edu/philosophy/organisation/staff/sundholm.html

The Contributors

Mark VAN ATTEN
  Institut d' histoire et de philosophie des sciences et des techniques
  CNRS / Université Paris I, France
  mark.vanatten@univ-paris1.fr
  http://www-ihpst.univ-paris1.fr/6,mark_van_atten.html

Benoît CASTELNÉRAC
  Faculté de théologie, d'étique, et de philosophie
  Université de Sherbrooke, Quebec, Canada
  benoit.castelnerac@usherbrooke.ca
  http://www.usherbrooke.ca/fatep/dp/profs/bcastelnerac.html

Nicolas CLERBOUT
  Sciences Humaines, Lettres et Arts,
  Université de Lille 3, France
  nicolas.clerbout@etu.univ-lille3.fr
  http://stl.recherche.univ-lille3.fr/sitespersonnels/rahman/rahmanequipeclerbout.html

Jacques P. DUBUCS
Institut d' histoire et de philosophie des sciences et des techniques
CNRS / Université Paris I, France
jacques.dubucs@univ-paris1.fr
http://www-ihpst.univ-paris1.fr/3,jacques_dubucs.html

Catarina DUTILH NOVAES
Institut for Logic, Language and Computation
Universiteit van Amsterdam, the Netherlands
C.Dutilhnovaes@uva.nl
http://staff.science.uva.nl/~dutilh/

Marie DUŽÍ
Department of Computer Science
VSB - Technical University of Ostrava, Czech Republic
marie.duzi@vsb.cz
http://www.cs.vsb.cz/duzi/

Bjørn JESPERSEN
Faculty Technology, Policy and Management
Delft University of Technology, the Netherlands
b.t.f.jespersen@tudelft.nl
http://www.tbm.tudelft.nl/live/pagina.jsp?id=b6e1a21f-b522-445f-81a7-e124d460
en

Laurent KEIFF
Sciences Humaines, Lettres et Arts,
Université de Lille 3, France
laurent.keiff@gmail.com
http://stl.recherche.univ-lille3.fr/sitespersonnels/rahman/rahmanequipeKeiff.html

Wolfgang KÜNNE
Fachberich Philosophie
Universität Hamburg, Germany
wolfgang.kuenne@uni-hamburg.de
http://www.philosophie.uni-hamburg.de/Team/kuenne.html

Mathieu MARION
Département de Philosophie
Université du Québec à Montréal, Quebec, Canada
marion.mathieu@uqam.ca
http://www.er.uqam.ca/nobel/philuqam/dept/page_perso.php?id=32

Pavel MATERNA
Institute of Philosophy
Academy of the Sciences, Czech Republic
materna@phil.muni.cz
http://www.phil.muni.cz/~materna/index.html

Wioletta MISKIEWICZ
Institut d' histoire et de philosophie des sciences et des techniques
CNRS / Université Paris I, France
miskiewicz@ens.fr
http://www-ihpst.univ-paris1.fr/wmiskiewicz

Kevin MULLIGAN
Département de Philosophie
Université de Genève, Switzerland
kevin.mulligan@lettres.unige.ch
http://www.unige.ch/lettres/philo/enseignants/km/

Sara NEGRI
Department of Philosophy
University of Helsinki, Finland
sara.negri@helsinki.fi
http://www.helsinki.fi/~negri/

Jan VON PLATO
Department of Philosophy
University of Helsinki, Finland
jan.vonplato@helsinki.fi
http://www.helsinki.fi/~vonplato/

Giuseppe PRIMIERO
Fonds Wetenschappelijk Onderzoek - Vlaanderen
Centre for Logic and Philosophy of Science
University of Ghent, Belgium
Giuseppe.Primiero@UGent.be
http://logica.ugent.be/giuseppe/

Shahid RAHMAN
Sciences Humaines, Lettres et Arts,
Université de Lille 3, France
shahid.rahman@univ-lille3.fr
http://stl.recherche.univ-lille3.fr/sitespersonnels/rahman/accueilrahman.html

Helge RÜCKERT
Department of Philosophy
University of Mannheim, Germany
rueckert@rumms.uni-mannheim.de
http://www.phil.uni-mannheim.de/fakul/phil2/rueckert/index.html

Maria VAN DER SCHAAR
Faculteit der Geesteswetenschappen, Instituut voor Wijsbegeerte
Universiteit Leiden, the Netherlands
m.v.d.schaar@hum.leidenuniv.nl
http://www.hum.leiden.edu/philosophy/organisation/staff/schaar.html

Kai WEHMEIER
Philosophy Department, Faculty of Logic and Philosophy of Science
University of California, Irvine
wehmeier@uci.edu
http://www.lps.uci.edu/home/fac-staff/faculty/wehmeier/

www.ingramcontent.com/pod-product-compliance
Lightning Source LLC
Chambersburg PA
CBHW070933230426
43666CB00011B/2420